Lecture Notes in Physics

Editorial Board

R. Beig, Wien, Austria
B.-G. Englert, Ismaning, Germany
U. Frisch, Nice, France
P. Hänggi, Augsburg, Germany
K. Hepp, Zürich, Switzerland
W. Hillebrandt, Garching, Germany
D. Imboden, Zürich, Switzerland
R. L. Jaffe, Cambridge, MA, USA
R. Lipowsky, Golm, Germany
H. v. Löhneysen, Karlsruhe, Germany
I. Ojima, Kyoto, Japan
D. Sornette, Nice, France, and Los Angeles, CA, USA
S. Theisen, Golm, Germany
W. Weise, Trento, Italy, and Garching, Germany
J. Wess, München, Germany
J. Zittartz, Köln, Germany

Springer
Berlin
Heidelberg
New York
Barcelona
Hong Kong
London
Milan
Paris
Tokyo

Physics and Astronomy　　**ONLINE LIBRARY**

http://www.springer.de/phys/

Editorial Policy

The series *Lecture Notes in Physics* (LNP), founded in 1969, reports new developments in physics research and teaching -- quickly, informally but with a high quality. Manuscripts to be considered for publication are topical volumes consisting of a limited number of contributions, carefully edited and closely related to each other. Each contribution should contain at least partly original and previously unpublished material, be written in a clear, pedagogical style and aimed at a broader readership, especially graduate students and nonspecialist researchers wishing to familiarize themselves with the topic concerned. For this reason, traditional proceedings cannot be considered for this series, though volumes to appear in this series are often based on material presented at conferences, workshops and schools (in exceptional cases the original papers and/or those not included in the printed book may be added on an accompanying CD-ROM, together with the abstracts of posters and other material suitable for publication, e.g. large tables, colour pictures, program codes, etc.).

Acceptance

A project can only be accepted tentatively for publication, by both the editorial board and the publisher, following thorough examination of the material submitted. The book proposal sent to the publisher should consist of at least a preliminary table of contents outlining the structure of the book together with abstracts of all contributions to be included.
Final acceptance is issued by the series editor in charge, in consultation with the publisher, only after receiving the complete manuscript. Final acceptance, possibly requiring minor corrections, usually follows the tentative acceptance unless the final manuscript differs significantly from expectations (project outline). In particular, the series editors are entitled to reject individual contributions if they do not meet the high quality standards of this series. The final manuscript must be camera-ready, and should include both an informative introduction and a sufficiently detailed subject index.

Contractual Aspects

Publication in LNP is free of charge. There is no formal contract, no royalties are paid, and no bulk orders are required, although special discounts are offered in this case. The volume editors receive jointly 30 free copies for their personal use and are entitled, as are the contributing authors, to purchase Springer books at a reduced rate. The publisher secures the copyright for each volume. As a rule, no reprints of individual contributions can be supplied.

Manuscript Submission

The manuscript in its final and approved version must be submitted in camera-ready form. The corresponding electronic source files are also required for the production process, in particular the online version. Technical assistance in compiling the final manuscript can be provided by the publisher's production editor(s), especially with regard to the publisher's own LaTeX macro package which has been specially designed for this series.

Online Version/ LNP Homepage

LNP homepage (list of available titles, aims and scope, editorial contacts etc.):
http://www.springer.de/phys/books/lnpp/

LNP online (abstracts, full-texts, subscriptions etc.):
http://link.springer.de/series/lnpp/

S. Odenbach (Ed.)

Ferrofluids

Magnetically Controllable Fluids
and Their Applications

 Springer

Editor

Stefan Odenbach
ZARM
Universität Bremen
Am Fallturm
28359 Bremen, Germany

Cover picture: by Dr. S. Odenbach

Library of Congress Cataloging-in-Publication Data applied for.

Die Deutsche Bibliothek - CIP-Einheitsaufnahme

Ferrofluids : magnetically controllable fluids and their applications / S. Odenbach (ed.). - Berlin ; Heidelberg ; New York ; Barcelona ; Hong Kong ; London ; Milan ; Paris ; Tokyo : Springer, 2002
  (Lecture notes in physics ; 594)
  (Physics and astronomy online library)
  ISBN 3-540-43978-1

ISSN 0075-8450
ISBN 3-540-43978-1 Springer-Verlag Berlin Heidelberg New York

This work is subject to copyright. All rights are reserved, whether the whole or part of the material is concerned, specifically the rights of translation, reprinting, reuse of illustrations, recitation, broadcasting, reproduction on microfilm or in any other way, and storage in data banks. Duplication of this publication or parts thereof is permitted only under the provisions of the German Copyright Law of September 9, 1965, in its current version, and permission for use must always be obtained from Springer-Verlag. Violations are liable for prosecution under the German Copyright Law.

Springer-Verlag Berlin Heidelberg New York
a member of BertelsmannSpringer Science+Business Media GmbH

http://www.springer.de

© Springer-Verlag Berlin Heidelberg 2002
Printed in Germany

The use of general descriptive names, registered names, trademarks, etc. in this publication does not imply, even in the absence of a specific statement, that such names are exempt from the relevant protective laws and regulations and therefore free for general use.

Typesetting: Camera-ready by the authors/editor
Camera-data conversion by Steingraeber Satztechnik GmbH Heidelberg
Cover design: *design & production*, Heidelberg

Printed on acid-free paper
SPIN: 10886791   54/3141/du - 5 4 3 2 1 0

Preface

Magnetic control of the properties and the flow of liquids can be a challenging field for basic research as well as for applications. An important condition for a technically interesting magnetic control of fluids is that the requested magnetic fields should be as small as possible.

An excellent material fulfilling this condition are suspensions of magnetic nanoparticles - commonly called ferrofluids. These fluids exhibit coupled liquid and superparamagnetic properties, leading to the possibility to control their flows and properties with magnetic fields having a strength of about 10mT. After the first synthesis of stable ferrofluids in 1964, a rapid development of ideas for applications as well as of basic fluid mechanics investigations took place. Till now the research field experiences a strong ongoing growth, recently triggered by the establishment of a national research program on magnetic fluids in Germany and documented by numerous presentations of the 9th International Conference on Magnetic Fluids (ICMF9), held in Bremen in 2001.

One of the most challenging aspects of ferrofluid research is the interdisciplinarity of the field. Besides chemistry for the preparation of magnetic fluids, basic theoretical physics for the description of their properties and behavior, fluid physics and rheology for the investigation of flows and rheological properties under influence of magnetic fields are needed to cover the basic research interests. In addition engineering and medical applications contribute to the importance of ferrofluid research for everyday life.

Most of these aspects are covered in this issue of *Lecture Notes in Physics*, which is mainly based on a series of plenary talks given during ICMF9. Starting with a review of preparation techniques for ferrofluids the chemical aspect of production of stable magnetic colloids is highlighted, followed by two contributions showing different approaches for a quantitative characterization of ferrofluids.

Part two of the issue is devoted to basic questions concerning the behavior of ferrofluids in the presence of magnetic fields. Three contributions highlight different ways to describe the fluids theoretically. Macro- and microscopic approaches are shown together to enable a comparison of their advantages and problems. The section ends with a specific example of magnetic control - the influence of fields on heat and mass transfer in ferrofluids.

In the third part, the question of magnetic control of the fluids properties is addressed using the example of magnetoviscosity - i.e. the change of the viscous properties of magnetic fluids in a magnetic field. The combination of theoretical description and experimental investigations documents the complexity of the problem and gives a good overview on the actual state of knowledge in these problems. The part ends with a contribution on magnetorheological fluids, building the bridge to the application possibilities of magnetoviscous effects.

Finally, part four deals with question concerning application of magnetic fluids. Since the major progress in this respect is actually concentrated in the field of medical applications, a single contribution on cancer therapy using ferrofluids as a drug carrier is included in this issue.

In total, this issue of *Lecture Notes in Physics* provides the reader with all basic knowledge needed to enter the field of ferrofluid research as well as with the most recent developments achieved.

Bremen, May 2002 *Stefan Odenbach*

Table of Contents

Part I   Synthesis and Characterization

The Preparation of Magnetic Fluids
S.W. Charles 3

**Magnetic Spectroscopy
as an Aide in Understanding Magnetic Fluids**
P.C. Fannin 19

**Magnetic and Crystalline Nanostructures in Ferrofluids
as Probed by Small Angle Neutron Scattering**
A. Wiedenmann 33

Part II   Basic Theory

Basic Equations for Magnetic Fluids with Internal Rotations
R.E. Rosensweig 61

Ferrohydrodynamics: Retrospective and Issues
M.I. Shliomis 85

Ferrofluid Dynamics
H.W. Müller and M. Liu 112

Heat and Mass Transfer Phenomena
E. Blums 124

Part III   Rheological Properties

Statistical Physics of Non-dilute Ferrofluids
A. Zubarev 143

Magnetic Fluid as an Assembly of Flexible Chains
K.I. Morozov and M.I. Shliomis 162

Magnetoviscous Effects in Ferrofluids
S. Odenbach and S. Thurm 185

Magnetorheology: Fluids, Structures and Rheology
G. Bossis, O. Volkova, S. Lacis, and A. Meunier 202

Part IV Applications

Targeted Tumor Therapy with "Magnetic Drug Targeting":
Therapeutic Efficacy of Ferrofluid Bound Mitoxantrone
Ch. Alexiou, R. Schmid, R. Jurgons, Ch. Bergemann,
F.G. Parak, and W. Arnold 233

Magnetic Unit System 255

List of Contributors

Christoph Alexiou
Department of Otorhinolaryngology
Head and Neck Surgery,
Klinikum rechts der Isar
Technical University of Munich
81675 Munich
c.alexiou@lrz.tu-muenchen.de

W. Arnold
Physics-Department E 17
Technical University of Munich
81675 Munich
Germany

Christian Bergemann
Chemicell
10777 Berlin
chemicell@aol.com

Elmars Blums
Institute of Physics
University of Latvia
LV-2169, Salaspils
Latvia
eblums@tesla.sal.lv

Georges Bossis
LPMC
UMR 6622
University of Nice
Parc Valrose
06108 Nice Cedex 2
France
bossis@naxos.unice.fr

Stuart W. Charles
Department of Chemistry
University of Wales
Bangor Gwynedd LL57 2UW
UK
stuartcharles@uk.uumail.com

Paul C. Fannin
Department of Electronic and
Electrical Engineering
University of Dublin
Trinity College
Dublin 2
Ireland
pfannin@tcd.ie

Peter Hulin
Department of Otorhinolaryngology
Head and Neck Surgery,
Klinikum rechts der Isar
Technical University of Munich
81675 Munich
p.hulin@lrz.tum.de

R. Jurgons
Department of Otorhinolaryngology
Head and Neck Surgery,
Klinikum rechts der Isar
Technical University of Munich
81675 Munich

Sandris Lacis
Department of Physics
University of Latvia
Zellu str.8
LV-1586

Riga
Latvia
lacis@unice.fr

Mario Liu
Institut für Theoretische Physik
Universität Tübingen
D-72676 Tübingen
Germany
mario.liu@uni-tuebingen.de

Alain Meunier
LPMC
UMR 6622
University of Nice
Parc Valrose
06108 Nice Cedex 2
France
ameunier@unice.fr

Konstantin I. Morozov
Institute of Continuous Media
Mechanics
UB of Russian Academy of Sciences
614013 Perm
Russia
mrk@icmm.ru

Hanns Walter Müller
Max-Planck Institut für Polymerforschung
Ackermannweg 10
D-55128 Mainz
Germany
hwm@mpip-mainz.mpg.de

Stefan Odenbach
ZARM
University of Bremen
Am Fallturm
D-28359 Bremen
Germany
odenbach@zarm.uni-bremen.de

Fritz G. Parak
Physics-Department E 17
Technical University of Munich
81675 Munich
Germany
fritz.parak@ph.tum.de

Ronald E. Rosensweig
Consultant
Summit, NJ
U.S.A.
rerosen@comcast.net

Roswitha Schmid
Department of Otorhinolaryngology
Head and Neck Surgery,
Klinikum rechts der Isar
Technical University of Munich
81675 Munich
Rosi.Klein@lrz.tu-muenchen.de

A. Schmidt
Department of Otorhinolaryngology
Head and Neck Surgery,
Klinikum rechts der Isar
Technical University of Munich
81675 Munich

Mark I. Shliomis
Department of Mechanical Engineering
Ben-Gurion University of the Negev
P.O.B. 653
Beer-Sheva 84105
Israel
shliomis@netvision.net.il

Steffen Thurm
ZARM
University of Bremen
Am Fallturm
D-28359 Bremen
Germany
thurm@zarm.uni-bremen.de

Olga Volkova
LPMC
UMR 6622
University of Nice
Parc Valrose
06108 Nice Cedex 2
France
volkova@unice.fr

Albrecht Wiedenmann
Hahn-Meitner-Institut Berlin
Glienicker Strasse 100
D-14109 Berlin
Germany
Wiedenmann@hmi.de

Andrey Yu. Zubarev
Department of Mathematical Physics
Ural State University
Lenin Av., 51
620083 Ekaterinburg
Russia
Andrey.Zubarev@usu.ru

Part I

Synthesis and Characterization

The Preparation of Magnetic Fluids

Stuart W. Charles

Department of Chemistry, University of Wales, Bangor, Gwynedd LL57 2UW, UK
Liquids Research Limited, Mentec, Deiniol Road, Bangor, Gwynedd LL57 2UP, UK

Abstract. In any discussion of fluids, which have magnetic properties, it is convenient to divide them into the following categories, (A) ferrofluids; (B) magnetorheological fluids; (C) dispersions of micron-sized particles of a non-magnetic material containing magnetic nano-sized particles, and (D) fluids containing paramagnetic particles.

In this Chapter, discussion will be confined to categories (A) and (B). The preparation of particles, ferrofluids and magnetorheological fluids has been the subject of numerous patents and research publications. A comprehensive bibliography of this information can be found updated in each of Proceedings of the International Conference on Magnetic Fluids (ICMF), under the following authors Zahn et al. [1], Charles et al. [2], Kamiyama et al [3], Blums et al. [4], Cabuil et al. [5], Bhatnagar et al. [6], Vekas et al. [7]. Reviews of the subject have been given by Rosensweig [8–10], Charles et al. [11], Martinet [12] and Scholten [13].[1]

1 Ferrofluids

Ferrofluids are stable colloidal dispersions of nano-sized particles of ferro- or ferrimagnetic particles in a carrier liquid. A wide range of carrier liquids have been employed, and many ferrofluids are commercially available to satisfy particular applications, e.g. in rotary vacuum lead-throughs, it is essential that the carrier liquid has a very low vapor pressure. In other applications, temperature, either high, low or both, may be a critical consideration. Theoretically it should be possible to produce dispersions in any liquid thereby being able to tailor the requirements of viscosity, surface tension, temperature and oxidative stability, vapor pressure, stability in hostile environments, etc., to the particular application in mind, whether it be technological or biomedical.

When a ferrofluid is subjected to a magnetic field, magnetic field gradient and/or gravitational field, in order that the colloidal suspension remains stable the magnetic particles generally have to be of approximately 10 nm in diameter. Particles of this size, whether they be ferrite or metal, possess a single magnetic domain only, i.e., the individual particles are in a permanent state of saturation magnetization. Thus a strong long-range magnetostatic attraction exists between individual particles, the result of which would lead to agglomeration of the particles and subsequent sedimentation unless a means of achieving

---

[1] For further information and availability of Ferrofluids and Magnetorheological Fluids the reader is recommended to view www.liquidsresearch.com or contact Liquids Research Limited by e.mail info@liquidsresearch.com

a repulsive interaction can be incorporated. In order to achieve this repulsive mechanism, the particles can either be coated by a surfactant (surface active material) to produce an entropic repulsion, or the surface of the particles can be charged thereby producing an electrostatic repulsion. For dispersions in liquid metals, stability has not been achieved due to the lack of a method to produce a repulsive mechanism.

To discuss the preparation of ferrofluids it is convenient to divide the discussion into (1) the preparation of nano-sized magnetic particles and (2) the subsequent dispersion of such particles in various liquids.

2 Preparation of Nano-sized Magnetic Particles

2.1 Ferrite Particles

In most technological and biomedical applications of ferrofluids, the magnetic material may be one of a number of different ferrites. By far the most commonly used ferrites are magnetite (Fe_3O_4) and maghemite (γ-Fe_2O_3). Because magnetite can be oxidized to maghemite with only a relatively small reduction in moment, the actual structure of the particles in commercial and other ferrofluids usually involves the presence of both ferrites in an undefined ratio. Detailed reviews of the structure and magnetic properties of the normal and inverse spin structures of ferrites have been given [14–16]. The name spinel is given to those ferrites which have the formula MFe_2O_4 (where M is a divalent ion) and which have the cubic crystal structure of the mineral spinel ($MgAl_2O_4$). The oxygen atoms are arranged in layers in such a way that there are two types of interstitial sites, tetrahedral (A) sites and octahedral (B) sites. The net magnetic moment of each ferrite is determined by the moment of each cation, the arrangement of the cations in the A and B sites, and the interaction between cations. In magnetite, the moment of the two Fe^{3+} ions are split between an A site and a B site and are antiferromagnetically coupled so that the moments cancel, whereas the Fe^{2+} ion situated on the B site gives rise to the overall moment. This opposite and unequal arrangement of the moments gives rise to ferrimagnetism.

A discussion of the various ways of producing nano-sized ferrite particles now follows.

Wet-grinding. The original method of producing ferrofluids based on ferrites is attributed to [17]. The method involved wet-grinding ferrites in a ball-mill in the presence of a suitable surfactant until the ferrite is in a colloidal state. Centrifugation was usually employed to remove larger particles which could lead to agglomeration and sedimentation. However this process usually takes a very long time (1000 hours) and it is mainly for this reason that the process has been superceded by a rapid and simple method involving the co-precipitation of metal salts in aqueous solution using a base.

Co-precipitation method. The co-precipitation method [18] for the preparation of particles of magnetite, maghemite and substituted ferrites suitable for use in ferrofluids has been the subject of many patents and publications as given in the comprehensive bibliography referred to in the Introduction. This method is usually carried out between 0 and 100°C, and as a result the cation distribution on the A and B sites may not conform to the distribution in bulk annealed crystals of the same material. Nevertheless it is well established by X-ray diffraction that the particles formed are crystalline. Further, their magnetic characteristics with regard to magnetic crystalline anisotropy and Curie temperature are not that far removed from that of the bulk crystals.

Magnetite and maghemite particles

The co-precipitation process is an extremely versatile method of producing ferrite particles in that particles of different size (3-20 nm diameter) and magnetic properties may be prepared by simply controlling the experimental conditions. It has been shown [19,5] that control of the mole fraction ratio of Fe^{3+}:Fe^{2+}, their concentrations, nature and concentration of the alkali medium can enable particles to be prepared of the desired size. This work has been extended [20] to include studies of the effect of precipitation temperature on particle size, the effect of heating the precipitate in the alkaline medium, and the effect of addition of surfactant to the reaction mixture. They have shown that the particle size increases when uncoated particles are heated in the alkaline medium and that the extent of this increase depends on the temperature to which the particles are heated. For an increase in precipitation temperature (between 2 and 90°C) and without subsequent heating in the alkali media an increase in particle size is observed. The effect of changing the initial Fe^{3+}:Fe^{2+} molar ratio from 2:1 to 1.5:1 and also changing the bases (KOH, NaOH and NH_4OH) have been shown to have an effect on the particle size. In the case of magnetite, particles can be prepared from the co-precipitation of hydroxides from an aqueous solution of Fe^{3+} and Fe^{2+} in the mole ratio of approximately 2:1 using a base [18]. The reaction is complex and involves the conversion of the hydroxide particles to magnetite. Oxidation of Fe^{2+} leads to the stoichiometry of the particles not being purely magnetite which can be circumvented by adjusting the ratio of the Fe^{2+} and Fe^{3+} concentrations. Thus the particles produced may be effectively a mixture of magnetite and maghemite (berthollide). However, oxidation appears to be of little consequence in most technological applications as the oxidation has little effect, if any, on the colloidal stability and also results in only a relatively small decrease in the magnetic moment of the particles, at most 10%. To produce pure maghemite it is a relatively simple process to oxidize the particles of magnetite by heating the particles to 90°C for thirty minutes in a 0.34M solution of ferric nitrate [21].

Substituted ferrite particles. To produce substituted nano-sized ferrite particles suitable for use in ferrofluids the Fe^{2+} ion is simply replaced or partially replaced by another or combination of divalent metal ions such as Co^{2+}, Mn^{2+}, Ni^{2+}, Zn^{2+} etc. Ions of other valency, e.g., Li^+ can also be used to prepare substituted ferrites. Numerous papers and patents have been published on this subject. See the references given in bibliographies quoted earlier. The method of precipitation of the particles is essentially the same as that used to prepare magnetite. Thus it is not important that details of individual preparations be given here, except to say that in some cases the precipitate needs to be hydrothermally aged to facilitate the conversion of the precipitated hydroxides to the ferrite [15]. The reason for the interest in substituted ferrite particles stems from the wide-differences in the magnetic properties of these materials [14], [16]. Thus it is possible to tailor the properties of these particles and dispersions of such particles to satisfy various applications. For example, where particles are required with high magnetocrystalline anisotropy, cobalt ferrites can be employed [20]. Other ferrites, which have relatively low Neel temperatures (100^oC) and thus high thermomagnetic coefficients at these temperatures, have been used in studies of thermomagnetic convection and heat transfer [22].

Microemulsion techniques. In addition to simple straightforward precipitation, it is possible to produce particles using the microemulsion technique. Numerous papers have been published on the preparation of small nano-sized particles, some of which are ferri- or ferromagnetic, via the use of reverse micelles. A reverse micelle can be described as a water-in-oil microemulsion in which two immiscible liquids are stabilized by a surfactant. As it is a three-component system, a triangular phase diagram of the 3-component system is used to describe the various structures other than microemulsions that may be formed. For detailed information on micelles and particle formation the reader is recommended to study a review edited by Pileni [23]. The method of preparation of particles involves the preparation of two microemulsions, one containing an aqueous solution of a metal salt or mixture of metal salts and the other an aqueous solution of an alkali [24], and mixing the two in the appropriate ratio. Using pure surfactants, the micelle-size distribution is very narrow, with the result that the particles themselves also possess a narrow particle size. The drawback to this method is that the surfactant used in the preparation may not be compatible with the carrier required for a particular application. The only way to circumvent this problem is either to remove the surfactant from the particles and replace it with one compatible with the carrier or use a two-surfactant system, i.e. coat the initial surfactant with another surfactant compatible with the carrier.

2.2 Metal Particles

There are two major advantages of using ferrofluids based on metallic particles, such as cobalt and iron particles. Firstly, these metals have high saturation magnetizations compared with ferrites and secondly they can be produced easily with very narrow size distributions. However there is also a major drawback which has restricted their use in most commercial applications and that is their poor resistance to oxidation and subsequent loss of magnetic properties. Only by maintaining these fluids in an inert atmosphere can these fluids possess an extended life-time.

Decomposition of organo-metallic compounds. There are a number of methods of producing small magnetic metallic particles, the commonest of which for ferrofluid preparation being the decomposition of organo-metallic compounds. The decomposition of organo-metallic compounds will be discussed in most detail as it is the easiest and most versatile method of producing metal particles suitable for ferrofluids. In the 1960's several groups of workers [25,26] reported the production of colloidal-sized cobalt particles by the thermolysis of dicobalt octa-carbonyl in hydrocarbon solutions containing polymeric materials. The process is very simple in that it only involves refluxing a solution of the carbonyl usually in toluene, until the reaction is complete, with the formation of Co metal particles. Thomas [25] showed that the particles possess a very narrow size-distribution ideal for the production of stable colloids. The particles also possess an fcc structure instead of the bcc structure of bulk cobalt. The absence of a polymer in the thermolysis leads to the formation of large particles (>100 nm). A detailed investigation by Hess [26] of the presence of different copolymers of acrylonitrile-styrene in the reaction solution showed that the greater the concentration of polar groups along the polymer chain, the smaller the size of the particles formed. It was Smith [27] who proposed that the reaction is polymer-catalyzed which results in the formation of a polymer-carbonyl complex which then decomposes to form elemental cobalt. Papirer et al. [28] showed that the decomposition could also be catalyzed by the presence of surfactants. They investigated the evolution of carbon monoxide during the decomposition in the presence of diethyl-2-hexyl sodium sulphosuccinate (Manoxol-OT) and in its absence, both experiments carried out as a function of temperature. The results were consistent with the presence of 'micro-reactors' and a diffusion mechanism to control the growth. The higher the temperature used the smaller are the particles produced. During the process of thermolysis in toluene, the presence of $Co_4(CO)_{12}$ was identified as an intermediate. However the overall reaction mechanism is complex but it seems likely that a carbonyl complex and/or microreactors are formed which subsequently decompose to produce cobalt particles. Charles and Wells [29] investigated methods to control the size of the cobalt particles formed by the use of a variety of surfactants and combinations of surfactants. By taking aliquots of the reaction mixture as the reaction proceeded and making observations using electron microscopy they observed the presence of clusters or microreactors for the first half hour of a four hour reaction but no particles. Subsequently observations showed the presence of particles only. They were able to produce, consistently, particles with median diameters in the range 5 to 15 nm. The median diameter refers to a lognormal volume distribution.

In many of the papers quoted so far little information has been presented concerning the saturation magnetization of the particles compared to that of the bulk. Charles and Wells [29] found that the saturation magnetization of the particles to be consistently lower (25%) than that of the bulk value of 17900 Gauss (1.79 T). This result is in agreement with the work of Hess and Parker [26]. There are several possible reasons for this, namely oxidation, spin-pinning/canting or poor crystallization. Experiments where care had been taken to avoid oxidation still showed the reduced saturation magnetization. However evidence from X-ray diffraction indicated that the particles were poorly crystallized or amorphous with similar results being obtained for Fe particles produced by a similar process [30]. X-ray diffraction and Mössbauer studies indicated the presence of carbon in the particles which is thought responsible for the poorly crystallized state. The particles, when annealed showed an X-ray pattern consistent with α-Fe and χ-Fe_5C_2. Some researchers, using transmission electron diffraction, have reported that the particles produced by thermolysis have a crystalline structure. Whether this is a true indication of the structure of the particles or that annealing has taken place in the electron beam is not known. Measurements of the magnetic anisotropy of Co particles have been undertaken by several groups of workers [31,29].

Iron particles of colloidal dimensions can also be produced by thermolysis [30]; in this case of iron pentacarbonyl in decahydronaphthalene. Particle size is again controlled by the presence of surfactants. The actual structure of the particles is open to question. Some groups have reported a crystalline structure [32] whilst others [30] reported an amorphous structure which undergoes crystallization on annealing.

Hoon et al. [33] have reported the preparation of colloidal nickel particles by the irradiation of nickel carbonyl with ultraviolet light and by the reduction of nickelocene. The particles were reported to have an fcc structure from transmission electron diffraction.

Nakatani et al. [34,35] have prepared iron nitride particles by two methods. Firstly by a plasma CVD reaction of iron pentacarbonyl and nitrogen gas and secondly by heating a solution of iron pentacarbonyl in kerosene containing a surfactant through which a constant stream of ammonia is passed. The crystal structure observed is ϵ-Fe_3N.

It is also possible to prepare alloy particles of the transition metals by the thermolysis of mixed-metal organometallic compounds in the presence of surfactants. Lambrick et al. [36,37] prepared organo-metallic compounds from which particles of Fe-Co and Ni-Fe alloys with median diameter 8 nm were obtained. The saturation magnetizations of the particles are >84% of that of the bulk values.

Inverse micro-emulsion techniques. The stimulus for much of this work has been an interest in the production of small-particle catalysts. The method is similar to that described for the production of ferrite particles by the same technique except that the microemulsions containing alkali are replaced by microemulsions containing a reducing agent such as sodium borohydride on sodium hypophosphite. As with ferrite production by simply mixing the two in the appropriate ratio, metal particles, in which some boron or phosphorus is usually incorporated, are produced within the micelles. Because of the incorporation of boron or phosphorus, the particles often have an amorphous structure. Using pure surfactants, the micelle size distribution is very narrow, with the result that the particles themselves also possess a narrow particle size. It has been shown by Lopez-Quintela et al. [38] that it is possible to subsequently coat the metal particles formed with another metal using a further microemulsion. This process is of particular interest in that by the use of an appropriate second metal it may be possible to overcome the recurring problem of oxidation of ferromagnetic particles of iron, cobalt etc.

Reduction of metal salts in aqueous solution. Particles produced by the reduction of metal salts in aqueous solution using reducing agents such as sodium borohydride or sodium hypophosphite are typically micron-sized. This precluded their use in ferrofluids because of their large magnetic moments and consequent strong magnetostatic interactions particularly in the presence of a magnetic field. However nano-sized particles have been produced, the formation of which is induced by the addition of $PdCl_2$ to the solution as a nucleating agent. It has been shown by Akashi et al. [39] that by the introduction of water-soluble viscosity-increasing materials that particle size can be reduced sufficiently to produce a stable colloidal suspension. Harada et al. [40] have reported the formation of Co-P particles, using sodium hypophosphite as reducing agent, in aqueous solutions containing protein or synthetic polypeptides. Similar studies on Fe-P particles have also been undertaken. There have been a few other reports of the formation of such particles.

As in the case of the microemulsion preparations, the at.% of P or B in the particles is very dependent on the conditions used. Above a critical concentration of B, P and indeed C, the particles possess an amorphous structure very similar to structures produced by rapid-quench techniques. The incorporation of P, B and C is a disadvantage in one respect, in that the saturation magnetization of the particles may be significantly reduced.

Small acicular particles of Co particles have been prepared using this method performed in the presence of a uniform magnetic field [41].

Miscellaneous methods. Fine particles of most of the metallic elements can be produced by evaporation of the metals in an inert gas atmosphere. The size of the particles, so produced, is very dependent on the gas used and its pressure. See 3.2 covering the preparation of magnetic fluids using miscellaneous methods.

2.3 Particle-Size Distribution

In order to achieve stable colloidal dispersions reproducibly, it is of paramount importance to monitor the particle size for each preparation. It is an obvious advantage to have a narrow particle-size distribution since large particles, which may adversely affect the performance of the fluid, are absent.

Nano-sized particles of ferrites produced by the methods described almost invariably have a rough, very approximate spherical shape, see Fig. 1(a). For the co-precipitation method the particle-size distribution is relatively wide in that the standard deviation of the lognormal size distribution is usually about 0.4. In the case of metal particles, the particles are near-spherical having a smooth appearance, see Fig. 1(b). Further the metals invariably have a narrow particle-size distribution with a standard deviation of <0.2.

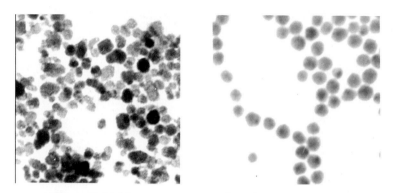

Fig. 1. (left) Magnetite particles (right) Cobalt particles

Particle-size distribution is best monitored by using transmission electron microscopy. However, another method based on measurement of the magnetization curve of the fluid is particularly convenient [42]. To obtain the particle size and standard deviation, the saturation magnetization of the particles is needed. Use of the bulk value may lead to erroneous values, particularly in the case of ferromagnetic particles where values of the saturation magnetization of the particles and bulk can differ greatly. The method also assumes that the particle-size distribution can be described by a log normal volume distribution [43]. This distribution has been found to be a good representation for most systems studied whether they be based on metallic or ferrite particles.

X-Ray diffraction and use of the Debye-Scherrer expression is a satisfactory pointer to the particle size for well-crystallized particles. This method is obviously of no value for particles, such as some of the metal particles, which have an amorphous structure.

3 Preparation of Ferrofluids

For most applications it is absolutely essential that the ferrofluids be very stable with regard to temperature and in the presence of magnetic fields. The presence of agglomeration of particles must be avoided at all costs. The literature contains numerous papers where phenomena can be attributed to the presence of agglomerates and yet their presence has been omitted in any discussion of the phenomena.

For most applications a low viscosity, low vapor pressure and chemical inertness are desirable for the carrier liquid and surfactant. Scholten [44] gives a review of the chemical and physical problems associated with these parameters and lists the advantages and disadvantages associated with different carrier liquids.

3.1 Fluids Containing Ferrite Particles

Surfactant stabilized colloids. Various methods of coating particles prior to dispersion in a carrier liquid have been described but all are basically the same with small differences in procedure. Fig. 2 shows an example of one method.

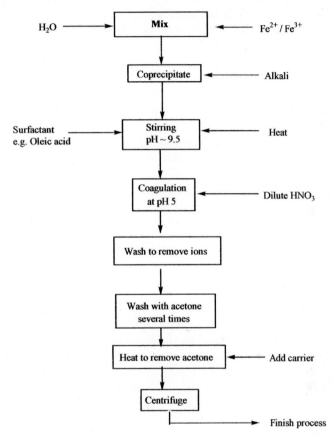

Fig. 2. Typical process for preparation of surfactant stabilized ferrofluids.

Magnetite can be coated with oleic acid by adding the acid to the precipitated phase in alkaline solution at pH 9.5. The mixture is usually left stirring for approximately 1 hour after which time it is heated to 95°C to facilitate the conversion of hydroxides formed to ferrite. After cooling, the product is acidified to pH 5 using nitric acid. The oleate ion produced chemisorbs strongly on to the magnetite particles, and the precipitate coagulates and falls out of solution. The coagulation may also be accompanied by the appearance of a thin film of oleic acid on the surface of the supernatant liquid which is unadsorbed oleate ions reconverted to oleic acid in acid solution. Sufficient oleic must be added to form a monolayer coverage [20]. The supernatant liquid can be decanted and the agglomerated hydrophobic precipitate of coated particles washed several times to remove salts, such as nitrates, chlorides and sulphates. Water and any physisorbed oleic acid can be removed by washing with acetone. By adding the appropriate carrier liquid to the acetone-wet slurry and gently heating, the acetone and any residual water can be removed leaving a colloidal dispersion of magnetite. If any aggregates are present it may be necessary to centrifuge the fluid or subject the fluid to a high-gradient magnetic field separator (HGMS) [45]. If the washing procedure with water has not been carried out carefully the presence of small quantities of micron-sized particles of salt may be present, which must be filtered.

Using this method with the appropriate surfactant it is possible to produce stable colloids in low vapor pressure carriers such as diesters [46], polyphenyl ethers [47], silicone oils [48], hydrocarbons [18], perfluorocarbons, perfluoropolyethers etc.

Stable water-based fluids can be produced by this method by the addition of a variety of secondary surfactants to oleate-coated particles [49].

As with fluids prepared from metal particles, the limiting value of saturation magnetization of ferrite-based fluids depends on the desired viscosity of the fluids. Fluids with magnetization up to approximately 1000 Gauss (0.1 T) can be prepared [50] which represents approximately 25 volume % of ferrite. Most ferrites have saturation magnetizations at 20°C up to a maximum value of 5000 Gauss (0.5 T).

Ionically-stabilized colloids. For the production of ionically-stabilized colloids particles of ferrites are produced by methods similar to that described in 2.2.

The particles produced by these methods are macro-ions in which the electric changes are due to specific adsorption of the amphoteric hydroxyl group. In an alkaline medium, the particles are negatively charged and in an acidic medium, positively charged. These particles can be stabilized in water by the presence of low-polarizing counter-ions such as $N(CH_3)_4^+$ in alkaline medium, or ClO_4^- in acidic medium. Strongly polarizing counter-ions lead to flocculation. For the -OH ligand, the point-of-zero-charge (PZC) occurs at approximately pH = 7.5. The small surface-charge density in this region of pH precludes the formation of a stable colloidal suspension. It was Massart (1981) developed the method of producing stable colloidal dispersions in alkaline and acidic aqueous media. He and his co-workers have subsequently made extensive and detailed studies of such dispersions, e.g., [51,21].

Using OH^- ligands, stable suspensions are restricted to pH values outside the range 6-10. Massart and co-workers [51] overcame this problem by modifying the nature of the surface charges by substituting the -OH surface ligands by citrate ligands. In so doing they were able to synthesize stable fluids in the pH range 3-11.

3.2 Fluids Containing Metal Particles

As already pointed out in 3.1, a drawback exists in the production of metal particles by the thermal decomposition of organo-metallic compounds, in that where a surfactant is chosen to produce particles of a particular size that surfactant may not be compatible with the carrier liquid needed for a particular application. Although the surfactants used in the preparation can be removed prior to dispersion in another carrier liquid containing an appropriate surfactant, this is not always an easy option. Thus sometimes a compromise has to be made. To achieve compatibility between the surfactant and carrier liquid, one may have to be flexible in one's choice of particle size which is dictated by the surfactant used in the preparation of the particles. Alternatively, sometimes it is possible to achieve a stable colloid by the addition of a secondary surfactant. Transfer of the coated particles to the carrier of choice is best facilitated by rotary evaporation under reduced pressure of the solvent used in the preparation, in the presence of the carrier liquid. The volume of the latter dictates the value of the saturation magnetization of the ferrofluid. This value will in turn be dictated by the requirements of viscosity for particular applications. For particles of Fe, Co and Fe_3N, saturation magnetizations of 2000 Gauss (0.2 T) are feasible [50], representing a volume fraction of the metal particles between 10 and 20%.

Similar problems occur to the above in the production of metal particles by the inverse-microemulsion technique, because the presence of a surfactant to produce the microemulsion may not be compatible with the carrier fluid required. This is also the case for the production of metal particles by the reduction of metal salts in aqueous solution.

Ferrofluids containing particles of Fe, Co or Ni have also been prepared by the vacuum evaporation of metals on to a running oil substrate containing a surfactant [52]. The particles produced by this method are very small (2 nm in diameter). On heating the product to $270^\circ C$ in an argon atmosphere for a few minutes, the particle size can be increased. Fluids produced by this method are gravitationally stable.

This method has also been used to produce dispersions of particles of Gd and transition metal alloys in mercury [53,54]. Unfortunately these fluids are not colloidally stable despite attempts to stabilize the dispersions with metallic additives [55].

Spark erosion has been used to produce ferrofluids [56]. A wide range of particle sizes is produced, the smaller particles of which are suitable for dispersion to form a ferrofluid. The method suffers from disadvantages in that the process is slow and the particle distribution is very wide. However it does have an advantage in that novel particles with similar properties to the bulk can be produced, e.g., TbFeB which possess a large magnetostriction.

3.3 Assessment of Colloidal Stability

As has already been pointed out, it is of paramount importance that ferrofluids are well-characterized, i.e. the particle-size distribution be defined and the presence of agglomerates be eliminated.

In many applications of ferrofluids, such as exclusion seals, loudspeakers, the fluids are subjected to large magnetic fields and field gradients and thus the presence of agglomerates or large particles will also certainly have a detrimental effect on the performance of the device. This problem also extends to the interpretation of certain phenomena observed in ferrofluids.

A review of the role that aggregation plays in determining the properties of magnetic fluids has been presented by Charles [57] as well as the various techniques by which aggregation can be studied, both experimentally and theoretically. The stability of a fluid can be best monitored by simply measuring the saturation magnetization of the fluid and the magnetic particle size [42] before and after subjecting the fluid to a strong magnetic field gradient. An effective and simple way of studying the effect of a magnetic field gradient on colloidal stability is by the use of a high gradient magnetic field separator (HGMS) [45]. Other methods to monitor the presence of agglomerates and large particles include ac susceptibility measurements [58], light scattering [59], or small-angle neutron scattering, SANS [60].

A simple method to undertake routine and reproducible measurements of stability and thereby make meaningful comparisons between different fluids is by means of a system based on a Colpitts oscillator [61]. In this method a long (15 cm) narrow tube is filled with ferrofluid. Measurements of inductance along the length of the tube enable variations in the concentration of the particles to be monitored. Application of a magnetic field gradient allows a comparison of the concentration profile to be made before and after exposure.

4 Magnetorheological Fluids

All magnetic fluids, of whatever nature, exhibit magnetorheological effects to a greater or lesser degree. The magnetorheological (MR) fluids to be discussed here will be confined to those that exhibit very strong effects only. MR fluids usually contain micron-sized particles (5 μm) of ferromagnetic materials, as opposed to the nano-sized particles used in ferrofluids. Even under saturating magnetic fields stable ferrofluids exhibit relatively modest magneto-viscous effects. For example, viscosities of the fluids may change by a factor of 2 or so. Larger effects have been observed but almost invariably these effects can be attributed to 'poor-quality' ferrofluids, i.e. fluids with poor colloidal stability namely those containing large particles or aggregates of particles.

As stated previously, nano-sized particles (10 nm) are magnetically single-domain and are in a permanent state of magnetization, i.e. they always possess a magnetic dipole even in the absence of an applied field. Micron-sized particles, on the other hand, are magnetically polydomain. If the material used in these micron-sized particles is magnetically 'soft', i.e. does not retain a remanent magnetization after exposure to an external applied magnetic field, then the particles possess zero overall moment and magnetic interactions between particles are minimal. However, the application of even modest external magnetic fields (say 500 Oe $\equiv$ 40 kAm^{-1}) will result in the particles attaining large magnetic moments resulting in strong magnetic interaction leading to chaining. The effect on the MR fluid is immediate and pronounced. The MR fluid passes from a liquid state to an extremely highly viscous 'solid-like' state, i.e. the magnetic particles are capable of imparting a high yield stress capability to these fluids. Control of the magnitude of the applied magnetic field allows easy control of the viscosity of the fluid, and for this reason these fluids and the analogous electrorheological fluids are often referred to as *smart* fluids. Because of these properties, which will be discussed in detail in Chapter 4 , great interest has been shown in these fluids, some of which have been exploited commercially in a wide variety of applications, e.g. in dampers, clutches etc.

Numerous patents have been published (see bibliography references given in the Introduction to this Chapter) describing the preparation of MR fluids and development of the various additives used in these fluids. Generally these MR fluids contain micron-sized particles of magnetically 'soft' ferromagnetic materials dispersed in a variety of carrier fluids. In principle dispersions can be made in any fluid. Probably the most commonly used magnetic material to date, but by no means the only one, is carbonyl iron, which is readily available commercially in a wide range of particle sizes. Shulmann et al. [62], Bibik [63] and many others have studied the properties of MR fluids based on carbonyl iron. Alloys of iron with other transition metals [64] and carbonyl nickel and electrolytic nickel [65] have also been used. MR fluids containing particles with a high specific gravity (e.g. iron = 7800 kgm^{-3}) in a low specific gravity carrier (e.g. silicone oil 950 kgm^{-3} or a perfluoropolyether 2000 kgm^{-3}) suffer from excessive gravitational settling, the result of which can interfere with the magnetorheological activity of the fluid due to the non-uniform distribution of particles. Brownian movement is of no importance in this system unlike ferrofluids. However, in many applications large shear forces are exerted which can render the fluid relatively homogeneous in a short time, thereby minimizing disturbance of the magnetorheological activity. To overcome this problem, metallic soap-like surfactants such as lithium stearate, aluminium stearate etc, have been used in the past to reduce particle settling, but they suffer to a certain degree from water retention limiting the operating temperature of the fluid. Ideally the MR fluid should exhibit minimal particle-settling over a broad temperature range. The problem of gravitational settling can be achieved by the introduction of thixotropic agents, of which a large number are presently available, the use of which has also been the subject of a number of patents. The thixotropic agent creates a network of the particles that at low shear rates form a loose structure, which imparts a small degree of rigidity to the MR fluid, thereby reducing particle settling. On application of a shearing force, this structure is easily disrupted, but is rapidly reformed on removal of the shear force. Where hydrogen-bonded thixotropic agents have been used, colloidal metal-oxide additives have been used to facilitate the thixotropic network [66].

The actual preparation of these systems is relatively simple and considerably less complicated than the preparation of ferrofluids. Generally, it is sufficient to mix the ingredients of the MR fluid, i.e. particles, carrier, thixotropic agents etc, by mechanical or ultrasonic means. The relative amounts of the ingredients and carrier are dependent on the particular application in mind. Typically MR fluids contain up to 70 wt% of metal particles. Because of the simplicity of the preparation, in contrast to ferrofluids, and the fact that the components to produce MR fluids are readily commercially available, it is not considered necessary to describe any one particular procedure, other than to say the choice of dispersing agent/thixotropic materials depends on the nature of the carrier used. The bibliography/patents on the preparation of MR fluids makes available information concerning the best combination of components to use.

References

1. Zahn M, Shenton K E 1980 Magnetic fluids bibliography. *IEEE Trans. Mag.* MAG-16(2): 387-415.
2. Charles, S W, Rosensweig R E 1983 Magnetic fluids bibliography. *J. Magn. Magn. Mat.* 39: 190-220
3. Kamiyama S, Rosensweig R E 1987 Magnetic fluids bibliography. *J. Magn. Magn. Mat.* 65: 401-39.
4. Blums E, Ozols R, Rosensweig R E 1990 Magnetic fluids bibliography. *J. Magn. Magn. Mat.* 85: 303-82
5. Cabuil V, Neveu S, Rosensweig R E 1993 Magnetic fluids bibliography. *J. Magn. Magn. Mat.* 122: 437-82
6. Bhatnagar S P, Rosensweig R E 1995 Magnetic fluids bibliography. *J. Magn. Magn. Mat.* 149: 198-232
7. Vekas L, Sofonea V, Balau O 1999 Magnetic fluids bibliography. *J. Magn. Magn. Mat.* 39: 454-89
8. Rosensweig R E 1979 Fluid dynamics and science of magnetic liquids. In: *Advances in Electronics and Electron Physics*. Vol. 48: Martin M (ed.) Academic Press, New York, pp. 103-99
9. Rosensweig R E 1985 *Ferrohydrodynamics*. Cambridge University Press, Cambridge, UK
10. Rosensweig R E 1988 Special Issue on Ferrofluids. *Chem. Eng. Comm.* 67: 1-340
11. Charles S W, Popplewell J 1980 Ferromagnetic liquids. In: Wohlfarth E P (ed.) *Ferro-magnetic Materials* Vol. 2: North-Holland, Amst. pp. 509-59
12. Martinet A 1983 The case of ferrofluids. In: *Aggregation Processes in Solution*, Elsevier, New York, Ch. 18, pp. 1-41
13. Scholten P C 1978 In: Berkovsky B M (ed.) *Thermomechanics of magnetic fluids*. Hemisphere, Washington D C, pp. 1-26
14. Smit J, Wijn H P J 1959 *Ferrites*. Wiley, New York
15. Schuele W J, Deetscreek V D 1963 *Ultrafine particles*. Kuhn W E (ed.) Wiley, New York
16. Dormann J-L, Nogues M 1990 Magnetic structure of substituted ferrites. *J. Phys: Condens. Mat.* 2: 1223-37
17. Papell S S 1965 *Low viscosity magnetic fluid obtained by the colloidal suspension of magnetic particles*. U S Patent 3,215,572
18. Khalafalla S E, Reimers G W 1973 *Magnetofluids and their manufacture*. U S Patent 3,764,540
19. Massart R 1981 Preparation of aqueous magnetic liquids in alkaline and acidic media. *IEEE Trans. Mag.* MAG-17: 1247-8

20. Davies K J, Wells, S, Charles S W 1993 The effect of temperature and oleate adsorption on the growth of maghemite particles. *J. Magn. Magn. Mat.* 122: 24-8
21. Bee A, Massart R, Neveu S 1995 Synthesis of very fine maghemite particles. *J. Magn. Magn. Mat.* 149: 6-9
22. Nakatsuka K, Hama Y, Takahashi J 1990 Heat transfer in temperature-sensitive magnetic fluids. *J. Magn. Magn. Mat.* 85: 207-9
23. Pileni M P 1989 *Structure and Reactivity of Reverse Micelles*, Elsevier, Amst.
24. Gobe M, Kon-No K, Kitahara A 1984 Preparation of magnetite superfine sol in w/o microemulsion. *J. Coll. Int. Sci* 93: 293-5
25. Thomas J R 1966 Preparation and magnetic properties of colloidal cobalt particles. *J. Appl. Phys.* 37(7): 2914-15
26. Hess P H, Parker P H 1966 Polymers for stabilization of colloidal cobalt particles. *J. Appl. Polymer Sci* 10: 1915-17
27. Smith T W 1981 *Surfactant-catalyzed decomposition of dicobalt octacarbonyl*. U S Patent 4,252,673
28. Papirer E, Horny P, Balard H, Anthore R, Petipas R, Martinet A 1983 The preparation of a ferrofluid by the decomposition of dicobalt octacarbonyl. *J. Coll. Int. Sci.* 94: 207-20
29. Charles S W, Wells S 1990 Magnetic properties of colloidal suspensions of cobalt. *Magnetohydrodynamics* 26: 288-92
30. Wonterghem J van, Mørup S, Charles S W, Wells S, Villadsen J 1986 Formation of a metallic glass by thermal decomposition of $Fe(CO)_5$. *Phys. Rev. Lett.* 55: 410-13
31. Chantrell, R W, Popplewell J, Charles S W 1980 The coercivity of a system of single domain particles with randomly oriented easy axes. *J. Magn. Magn. Mat.* 15-18: 1123-4
32. Kilner M, Hoon S R, Lambrick D R, Potton J A, Tanner B K 1984 Preparation and properties of metallic iron ferrofluids. *IEEE Trans. Mag.* MAG-20: 1735-7
33. Hoon S R, Kilner M, Russell G J, Tanner B K 1983 Preparation and properties of nickel ferrofluids. *J. Magn. Magn. Mat.* 39: 107-10
34. Nakatani I, Furubayashi T 1990 Iron-nitride magnetic fluids prepared by plasma CVD technique. *J. Magn. Magn. Mat.* 85: 11-3
35. Nakatani I, Hijikata M, Ozawa K 1993 Iron-nitride fluids prepared by vapor-liquid reaction. *J. Magn. Magn. Mat.* 122: 10-4
36. Lambrick D B, Mason N, Harris N J, Russell G J, Hoon S R, Kilner M 1985 An iron-cobalt alloy magnetic fluid. *IEEE Trans. Mag.* MAG-21: 1891-3
37. Lambrick D B, Mason N, Hoon S R, Kilner M 1987 Preparation and properties of Ni-Fe magnetic fluids. *J. Magn. Magn. Mat.* 65: 257-60
38. Lopez-Quintela M A, Rivas J 1992 *Covering of magnetic particles to produce stable magnetic fluids*. Spanish Patent 9,201,984 (see also 1994, Structural and magnetic characterisation of Co particles coated with Ag. *J. Appl. Phys.* 76: 6564-6)
39. Akashi G, Fujyama M 1969 *Process for the production of magnetic substances*. US Patent 3,607,218
40. Harada S, Yamanashi T, Ugaji M 1972 Preparation and magnetic properties of cobalt alloy particles. *IEEE Trans. Mag.* MAG-8: 468-72
41. Charles S W, Issari B 1986 The preparation and properties of small acicular particles of cobalt. *J. Magn. Magn. Mat.* 54-57: 743-4
42. Chantrell R W, Popplewell J, Charles S W 1978 Measurements of particle-size distribution parameters in ferrofluids. *IEEE Trans. Mag.* MAG-14: 975-7
43. O'Grady K, Bradbury A 1983 Particle-size analysis in ferrofluids. *J. Magn. Magn. Mat.* 39: 91-4

44. Scholten P C 1988 Some material problems in magnetic fluids. *Chem. Eng. Comm.* 67: 331-40
45. O'Grady K, Stewardson H R, Chantrell R W, Fletcher D, Unwin D, Parker M R 1986 Magnetic filtration of ferrofluids. *IEEE Trans. Mag.* MAG-22: 1134-6
46. Wyman J E 1984 *Ferrofluid composition and method of making the same.* U S Patent 4,430,239
47. Bottenberg W R, Chagnon M S 1982 *Low vapor-pressure ferrofluids using surfactant-containing polyphenyl ether.* U S Patent 4,315,827
48. Chagnon M S 1982 *Stable ferrofluid compositions.* U S Patent 4,356,098
49. Shimoiizaka, J, Nakatsuka K, Chubachi R 1978 In: *Thermomechanics of magnetic fluids.* Berkovsky B M (ed.) Hemisphere, Washington DC, pp. 67-76
50. Scholten P C 1983 How magnetic can a fluid be? *J. Magn. Magn. Mat.* 39: 99-106
51. Bacri J C, Perzynski R, Salin D, Cabuil V, Massart R 1990 Ionic ferrofluids: a crossing of chemistry and physics. *J. Magn. Magn. Mat.* 85: 27-32
52. Nakatani I, Furubayashi T, Takahashi T, Hanaoka H 1987 Preparation and magnetic properties of colloidal ferromagnetic metals. *J. Magn. Magn. Mat.* 65: 261-4
53. Shepherd P G, Popplewell J, Charles S W 1970 A method of producing a ferrofluid with Gd ferromagnetic particles. *J. Appl. Phys. D* 3: 1985-7
54. Shepherd P G, Popplewell J, Charles S W 1972 Ferrofluids containing Ni-Fe alloy particles in mercury. *J. Phys D* 5: 2273-82
55. Keeling L, Charles S W, Popplewell J 1984 The prevention of diffusional growth of cobalt particles in mercury. *J. Phys. F* 14: 3093-3100
56. Berkowitz A E, Walter J L 1983 Ferrofluids produced by spark erosion. *J. Magn. Magn. Mat.* 39: 75-8
57. Charles S W 1988 Aggregation in magnetic fluids and magnetic fluid composites. *Chem. Eng. Comm.* 67: 145-80
58. Fannin P C, Scaife B K P, Charles S W 1987 The study of the complex susceptibility of ferrofluids and rotational Brownian motion. *J. Magn. Magn. Mat.* 65: 279-81
59. Beresford J, Smith F M 1973 In: Parfitt G D (ed.) *Dispersion of Powders in Liquids.* Publ. Appl. Science.
60. Cebula D J, Charles S W, Popplewell J 1983 Aggregation in ferrofluids studied by neutron small angle scattering. *J. Physique* 44: 207-13
61. Bissell P R, Chantrell R W, Spratt G W D, Bates P A, O'Grady K 1984 Long term stability measurements on magnetic fluids. *IEEE Trans. Mag.* MAG-20: 1738-40
62. Shulman Z P, Kordonsky V I, Zaltsgendler E A, Prokhorov I V, Khusid B M and Demchuk S A 1986 Structure, physical properties and dynamics of magnetorheological suspensions. *Int. J. Multiphase Flow* 12: 935-55
63. Bibik E E 1981 *Rheology of Dispersed Systems.* Leningrad State Univ. Press. 171
64. Carlson J D, Weiss KD 1995 *Magnetorheological materials based on alloy particles.* U S Patent 5,382,373
65. Shulman Z P 1991 Magnetorheological systems and their application. European Advanced Short Course of UNESCO. Minsk, USSR
66. Weiss K D, Nixon D A, Carlson J D, Margida A J 1994 *Thixotropic magnetorheological materials.* W O Patent 9,410,693

Magnetic Spectroscopy as an Aide in Understanding Magnetic Fluids

Paul C. Fannin

Department of Electronic and Electrical Engineering, University of Dublin, Trinity College, Dublin 2, Ireland.

Abstract. Accurate data on the frequency-dependent, complex susceptibility, $\chi(\omega) = \chi'(\omega) - i\chi''(\omega)$ of magnetic fluids are vital for an understanding of the dynamic behavior of these colloidal suspensions. They enable relaxation mechanisms, both Brownian and Néel, as well as ferromagnetic resonance to be identified and investigated. They also provide a convenient means of determining the macroscopic and microscopic properties of the fluids, such as the mean particle radius, $\bar{r}$, the mean value of anisotropy field, $\bar{H}_A$, the gyromagnetic constant, γ, and the damping constant, α. Also, through the medium of $\chi(\omega)$ one can also investigate the frequency dependence of the loss tangent (tanδ) and power factor (sinδ) of such particulate systems. In this Chapter the above-mentioned topics together with the relevant theory are presented and by means of suitable examples it is demonstrated how the appropriate equations may be employed to determine such properties from experimental data.

1 Relaxation Mechanisms

The major advances of recent times in the investigation of the relaxation mechanisms of ferrofluids are in no small way due to the application of dielectric formalism in the representation of ferrofluid data [1-5]. There are two distinct mechanisms by which the magnetisation of ferrofluids may relax after an applied field has been removed: either by rotational Brownian motion of the particle within the carrier liquid, with its magnetic moment, m, locked in an axis of easy magnetisation, or by rotation of the magnetic moment within the particle.

$$m = M_s v \qquad (1)$$

where v is the volume of the particle and M_s its saturation magnetisation. The time associated with the rotational diffusion is the Brownian relaxation time τ_B [6] where

$$\tau_B = 3V\eta/kT \qquad (2)$$

V is the hydrodynamic volume of the particle and η is the dynamic viscosity of the carrier liquid. In the case of the second relaxation mechanism, the magnetic moment may reverse direction within the particle by overcoming an energy barrier, which for uniaxial anisotropy, is given by Kv, where K is the anisotropy constant of the particle. The probability of such a transition is $\exp(\sigma)$ where σ is the ratio of anisotropy energy to thermal energy (Kv/kT). This reversal time is characterised by a time τ_N, which is referred to as the Néel relaxation time [7], and given by the expression,

$$\tau_N = \tau_0 \exp(\sigma) \quad (3)$$

τ_0 is a damping or extinction time having an often quoted approximate value of 10^{-8} to 10^{-10}s [8-10]. According to Brown [11], for high and low barrier heights,

$$\tau_N = \tau_0 \sigma^{-1/2} \exp(\sigma), \qquad \sigma > 2 \quad (4)$$
$$= \tau_0 \sigma, \qquad \sigma \ll 1$$

Other workers [12,13] have subsequently derived a single expression to cater for a continuous range of σ, however because of the difficulty in characterising small magnetic particle systems it is perfectly adequate to use the expressions of equation (4). A distribution of particle sizes implies the existence of a distribution of relaxation times, with both relaxation mechanisms contributing to the magnetisation. They do so with an effective relaxation time τ_{eff} [14], where, for a particular particle,

$$\tau_{eff} = \tau_N \tau_B / (\tau_N + \tau_B) = \tau_{\parallel} \quad (5)$$

The mechanism with the shortest relaxation time being dominant. The theory developed by Debye to account for the anomalous dielectric dispersion in dipolar fluids has been successfully used [15,16] to account for the analogous case of magnetic fluids. Debye's theory holds for spherical particles when the magnetic dipole-dipole interaction energy, U, is small compared to the thermal energy kT. According to Debye's theory the complex susceptibility, $\chi(\omega)$ has a frequency dependence given by the equation,

$$\chi(\omega) = \chi_\infty + (\chi_0 - \chi_\infty)/(1 + i\omega\tau) \quad (6)$$
$$= \chi_\infty + (\chi_0 - \chi_\infty)(1/1 + \omega^2\tau^2) - i\omega\tau/(1 + \omega^2\tau^2)) \quad (7)$$

where

$$\chi_0 = nm^2 / 3kT\mu_0 \quad (8)$$

is the static susceptibility and

$$\tau = 1/\omega_{max} = 1/2\pi f_{max}. \quad (9)$$

f_{max} is the frequency at which $\chi''(\omega)$ is a maximum, n is the particle number density and χ_0 and χ_∞ indicate susceptibility values at f=0 and at very high frequencies.

Fig. 1 shows a plot of the Debye equation (7) which clearly indicates how $\chi'(\omega)$ falls monotonically and that the maximum possible value of the absorption component, $\chi''(\omega)$, is equal to half that of the $\chi'(\omega)$ component. Here $\tau = \tau_{eff} = \tau_\parallel$.

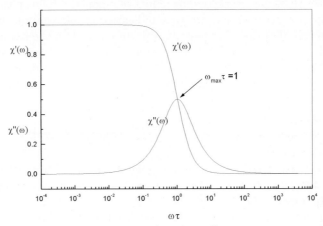

Fig. 1. Debye plot of $\chi'(\omega)$ and $\chi''(\omega)$ against $\omega\tau$

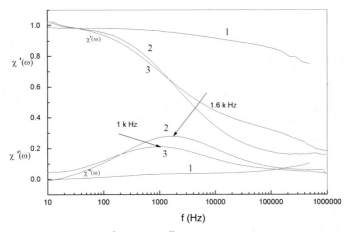

Fig. 2. Normalised plot of $\chi'(\omega)$ and $\chi''(\omega)$ against f(Hz) for samples 1, 2 and 3

The susceptibility plots of Fig 2 over the approximate frequency range, 10 Hz to 1 MHz, for three different samples, (samples 1, 2 and 3), obtained by means of the toroidal technique [15], clearly illustrates that the loss peak can lie in any part of the frequency spectrum. Sample 1 consists of a suspension of magnetite in isopar M whilst samples 2 and 3 are suspensions of magnetite in water; the corresponding mean particle radii as determined by electron microscopy, are 5 nm, 5.5 nm and 7 nm respectively. It is apparent that the plots of samples 2 and 3 have Debye-type profiles with a maximum in $\chi''(\omega)$ occurring at 1 kHz and 1.6 kHz; from equation (1) and using a viscosity of 10^{-4}Nsm^{-2}, corresponding hydrodynamic radii of 80 nm and 69 nm are obtained. These values are greater than the corresponding values of magnetic radius plus surfactant thickness(approx 2-4 nm) and are thus indicative of aggregation. In contrast, the profile of sample 1 does not have an absorption peak in the frequency range up to 1 MHz and corresponds to the profile of a finely dispersed sample. For the case of Brownian relaxation, a distribution of particle volumes corresponds to a distribution of relaxation times where $\chi(\omega)$ may be expressed in terms of a distribution function, $f(\tau)$ giving

$$\chi(\omega) = \chi_\infty + (\chi_0 - \chi_\infty) \int_0^\infty f(\tau) d\tau / (1 + i\omega\tau) \qquad (10)$$

$f(\tau)$ may be represented by a Normal, Log-normal, Frohlich or a Cole-Cole distribution function. In the Cole-Cole case where the complex susceptibility data fits a depressed circular arc, the relation between $\chi'(\omega)$ and $\chi''(\omega)$ and their dependence on frequency, $\omega/2\pi$, can be displayed by means of the magnetic analogue of the Cole-Cole plot [2]. In the Cole-Cole case the circular arc cuts the $\chi'(\omega)$ axis at an angle of $\alpha_c \pi/2$; α_c is referred to as the Cole-Cole parameter and is a measure of the particle-size distribution . The magnetic analogue of the Cole-Cole circular arc is described by the equation

$$\chi(\omega) = \chi_\infty + (\chi_0 - \chi_\infty)/[(1 + (i\omega\tau_0)^{1-\alpha_c}], \quad 0 < \alpha < 1 \qquad (11)$$

which for $\alpha_c = 0$, reduces to that of equation (6). The Cole-Cole distribution function [17] is described by,

$$G(\ln \tau) = (1/2\pi)(\sin\pi\alpha_c/(\cosh\{(1-\alpha_c)\ln(\tau_0/\tau)\} - \cos\pi\alpha_c)) \qquad (12)$$

For the Frohlich [3,17] distribution function for which it is assumed that the distribution of relaxation times is confined to a certain range from τ_1 to τ_2; $f(\tau)d\tau = G(\ln \tau)d(\ln \tau)$ where

$$G(\ln \tau) = 1/\ln(\tau_1/\tau_2) \qquad \tau_2 < \tau < \tau_1 \qquad (13)$$
$$= 0 \qquad 0 < \tau < \tau_2, \tau_1 < \tau < \infty$$

and as τ_1 and τ_2 are proportional to the cube of the maximum and minimum hydrodynamic radii respectively, then $(r_{max}/r_{min}) = (\tau_1/\tau_2)^{1/3}$

As an example of obtaining an estimate of the aggregate size distribution in the case of sample 2, Fig 3 shows the result of fitting the measured susceptibility plots to profiles generated by the Debye equations modified by Frohlich and Cole-Cole distribution functions. The fits obtained from the two fitting functions are found to be similar with the Frohlich distribution function giving the value for the particle radii of from 37 nm to 174 nm whilst the Cole-Cole fit was realised with $\alpha_c = 0.31$.

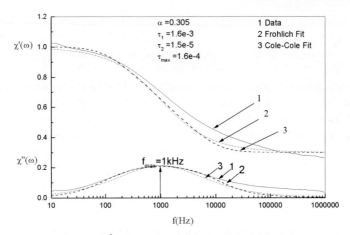

Fig. 3. Normalised plot of $\chi'(\omega)$ and $\chi''(\omega)$ against f(Hz) for samples 2. Curve 1 represents the measured data whilst curves 2 and 3 represent Frohlich and Cole-Cole fits

2 The Use of the Hilbert Transform in Generating Complex Susceptibility Data

On occasions it may only be possible to obtain good quality data for $\chi'(\omega)$ and not $\chi''(\omega)$. However $\chi'(\omega)$ and $\chi''(\omega)$ are a Hilbert transform pair so in knowing $\chi'(\omega)$ one can determine $\chi'(\omega)$ [18-21]. The Hilbert transform of a function $x(t)$ is defined as

$$\hat{x}(t) = 1/\pi \int_{-\infty}^{\infty} x(\tau)/(t-\tau)d(\tau) \tag{14}$$

$$= (1/\pi t) * x(t). \tag{15}$$

where $*$ represents a convolution operation. This operation corresponds to phase-shifting all frequency components of $x(t)$ by $90°$ and it may be represented by a linear system with an impulse response of

$$h(t) = 1/\pi t \tag{16}$$

and transfer function

$$H(f) = -i \ \text{sgn(f)}. \qquad (\text{f} = \omega/2\pi) \tag{17}$$

As convolution in the time domain is equivalent to multiplication in the frequency domain, equation (15) may be written as

$$\hat{X}(f) = H(f)X(f) \tag{18}$$

where $\hat{X}(f)$, $H(f)$ and $X(f)$ are the Fourier transforms of $\hat{x}(t)$, $h(t)$ and $x(t)$ respectively. $\hat{X}(f)$ and $\hat{x}(t)$ are related by the expression,

$$\hat{x}(t) = \int_{-\infty}^{\infty} \hat{X}(f)e^{j2\pi ft}df \tag{19}$$

and from equation (18), assuming that $X(f)$ is an even function it follows that

$$\hat{x}(t) = \int_0^\infty X(f)\sin(2\pi f)df \qquad (20)$$

As an illustration of the accuracy of the technique is initially applied to theoretical Debye curves where it is known what the resultant Hilbert transform should be. Initially the technique is applied to equation (7) with $\chi_\infty = 0$; the corresponding plots of $\chi'(\omega)$ and $\chi''(\omega)$ and its Hilbert transform $\hat{\chi}'(\omega)$ are shown in Fig.4; the Hilbert transform, proves to be a good approximation to $\chi''(\omega)$ with some error existing in the low-frequency region of the plots. The number of data points used was N=8192. To account for the situation where a χ_∞ component exists, the technique is shown to be equally successful when applied to case where $\chi_\infty = 0.2$ and N= 8192 ; the resultant plots of $\chi''(\omega), \chi'(\omega)$ and its Hilbert transform $\hat{\chi}'(\omega)$ are shown in Fig.5.

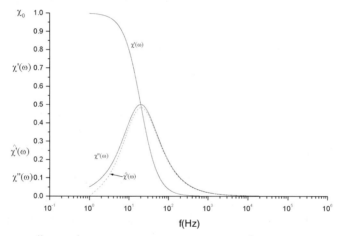

Fig. 4. Plot of $\chi''(\omega)$, $\chi'(\omega)$ and its Hilbert transform $\hat{\chi}'(\omega)$, against f(Hz), for the Debye case with $\chi_\infty = 0$ and N= 8192 points

In reference [22] the technique is applied to a dynamic situation, and this example illustrates the usefulness of the Hilbert transform in determining complex susceptibility components once one component is known. Of course, as already mentioned, the technique can also be used to generate data on $\chi'(\omega)$ from a knowledge of $\chi''(\omega)$, simply by taking the inverse Hilbert transform of $\chi''(\omega)$. This is due to the fact that the Hilbert transform of a Hilbert transform returns the original signal with a change in sign [23].

3 Resonance

In the GHz frequency range the character of the dispersion changes from relaxation to one of resonance and it is convenient to describe $\chi(\omega)$ in terms of its

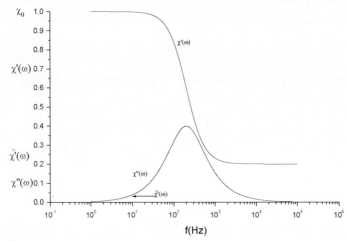

Fig. 5. Plot of $\chi''(\omega)$, $\chi'(\omega)$ and its Hilbert transform $\hat{\chi}'(\omega)$, against f(Hz), for the Debye case with $\chi_\infty = 0.2$ and N= 8192 points

parallel (relaxational) $\chi_\parallel(\omega)$ and perpendicular (resonant) $\chi_\perp(\omega)$, components, with,

$$\chi(\omega) = \frac{1}{3}\left(\chi_\parallel(\omega) + 2\chi_\perp(\omega)\right) \qquad (21)$$

with corresponding relaxation times $\tau_\parallel$ and $\tau_\perp$, respectively.

$\chi_\perp(\omega)$ is associated with resonance which is indicated by a change in sign of the value of $\chi'(\omega)$ at an angular frequency, ω_{res}; where for a small polar angle,

$$\omega_{res} = \gamma H_A = \gamma 2K/M_s. \qquad (22)$$

The transverse or resonant component of the susceptibility, $\chi_\perp(\omega)$, may be described by equations derived by Raikher and Shliomis [24] and modified by Coffey et al. [25] with,

$$\chi_\perp(\omega) = \chi_\perp(0)\frac{1 + i\omega\tau_2 + \Delta}{(1 + i\omega\tau_2)(1 + i\omega_\perp) + \Delta} \qquad (23)$$

where $\tau_\perp$, Δ and τ_2 are as given in reference [25] and $\chi_\perp(0)$ is the static transverse susceptibility.

When combined with equation (6), the overall frequency-dependent susceptibility is given as,

$$\chi(\omega) = \frac{1}{3}\left[\frac{\chi_\parallel(0)}{1 + i\omega\tau_\parallel} + 2\chi_\perp(0)\frac{1 + i\omega\tau_2 + \Delta}{(1 + i\omega\tau_2)(1 + i\omega\tau_{\perp eff}) + \Delta}\right] \qquad (24)$$
$$= \chi'(\omega) - i\chi''(\omega)$$

In equation (24) $\tau_\perp$ has been replaced by $\tau_{\perp eff} = \tau_\perp \tau_B/(\tau_\perp + \tau_B)$ in order to take account of the effects of Brownian relaxation. Fig 6 shows high-frequency susceptibility measurements obtained, by means of the transmission

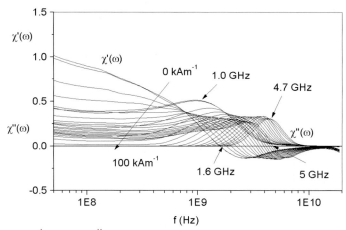

Fig. 6. Plot of $\chi'(\omega)$ and χ'' over the frequency range 50 MHz-18 GHz for 17 values of polarising field over the range 0-100 kA/m, for sample 4

line technique [26,27], for a 760 G suspension (sample 4) of magnetite in isopar M subjected to a polarising field over the range 0-100 kAm^{-1}. From the figure it is seen that for the unpolarised case, resonance, indicated by the $\chi'(\omega)$ component changing sign, occurs at a frequency f_{res}= 1.6 GHz whilst the maximum $\chi''(\omega)$ of the loss-peak is also shown to occur at a frequency of f_{max} =1.0 GHz. Variation of the polarising field, H, over the stated range results in f_{res} and f_{max} increasing up to a frequencies of 5.0 GHz and 4.7 GHz , respectively. The loss-peak can be attributed to the presence of two different processes, namely the Néel relaxation of the smallest particles in the distribution and resonance. The application of H to the sample effectively results in an increase in the barrier Kv (and hence $\sigma = Kv/kT$) which the magnetic moment of the particles must overcome. This increase in H results in a reduction in spontaneous flipping of the magnetic moments (Néel relaxation) leading to an increasing dominance from $\chi_\perp(\omega)$ to $\chi(\omega)$. The effect is to shift f_{max} from 1.0 GHz to 4.7 GHz and f_{res} from 1.6 GHz to 5.0 GHz. The result of this is that the value of f_{max} corresponding to the maximum in the value of $\chi''(\omega)$ (loss peak) approaches the value of f_{res} as resonance becomes the dominant process. A further increase in H would move the value of f_{max} closer to that of f_{res}, however it is of interest to note that in references [29] it has been reported that the condition $f_{max} = f_{res}$ is not realisable.

A plot of f_{res} against H is shown in Fig 7. and as $\omega_{res} = 2\pi f_{res} = \gamma(H+\bar{H}_A)$, the value of $\bar{H}_A$, a mean value of the anisotropy field, is determined from the intercept of Fig 7. as being equal to 41 kA/m. This corresponds to a mean value of anisotropy constant, $\bar{K}$, at room temperature and bulk M_s of 0.4 T, of $8.2 \cdot 10^3$ J/m^3. From the slope of Fig 7 the magneto-mechanical ratio, γ, is found to be $2.32 \cdot 10^5$ s^{-1}A^{-1}m.

By determining one further parameter, namely the damping constant α, one can evaluate the magnetic viscosity, η_m. Here an estimate of α is obtained by

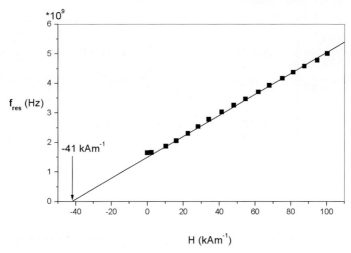

Fig. 7. Plot of f_{res} against H with $\bar{H}_A = 41$ kA/m

fitting [19] the original susceptibility data to theoretical susceptibility profiles generated by equation (24), suitably modified to cater for a distribution of particle size, r, and anisotropy constant, K.

Table 1. Fluid parameters

Fluid Mag. [G]	Sat.Mag. Ms [T]	$\bar{d}$ [nm]	$\bar{H}_A$ kA/m	$\bar{K}$ [10^3 J/m^3]	γ [m/As]	α	η_m [10^{-6} Ns/m^2]	τ_o [10^{-9} s]
760	0.4	9.0	41	8.2	2.32	0.1	2.9	1.06

Fig 8 shows the fit obtained for a Normal distribution of K with a standard deviation of 6.10^3 and a mean $\bar{K} = 1 \cdot 10^4$ J/m^3 together with a Nakagami distribution of radii, r, with a width factor $\beta = 4$ and a mean particle radius, $\bar{r} \approx 4.5$nm and a saturation magnetisation of 0.4 T. For the fit shown, α was found to be $\alpha = 0.1$, a figure within the range of values normally quoted for α.

Thus having obtained data on the macroscopic and microscopic properties of the particles one is in a position to determine the 'magnetic viscosity', η_m, where $\eta_m = M_s/6\alpha\gamma$, the exponential prefactor (τ_0) of Néels expression for τ_N, where $\tau_0 = M_s/2\alpha\gamma K$. The results obtained are shown in Table 1 with τ_0 found to have a value of $0.92 \cdot 10^{-9}$ s, which is within the range of values predicted for τ_0.

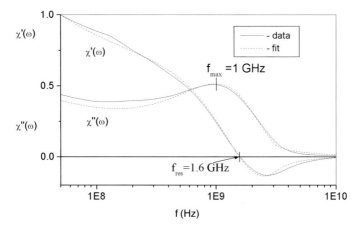

Fig. 8. Plot of $\chi'(\omega)$ and $\chi''(\omega)$, and corresponding fits, for unpolarised plot of Fig 6

4 Time Frequency Transformation

By placing an ensemble of small magnetic particles in a large dc magnetic field, thereby aligning the moments of the particles, the magnetisation decay, or after-effect function, $b(t)$, can, on removal of the field, be observed over time scales of seconds and longer. However, the equivalent measurements over a much shorter time range cannot be realised using this conventional method and thus an alternative indirect technique, such as the transformation of the frequency-dependent magnetic susceptibility, $\chi(\omega)$. Frequency-dependent χ and the time-dependent decay function are a Fourier transform pair and this fact enables the utilisation of the susceptibility data in the investigation of $b(t)$, of magnetic fluids. $b(t)$ represents the decay of the magnetisation after the removal of an external field and according to Scaife [19], is related to the frequency dependent complex susceptibility, $\chi(\omega)$, via the following equation:

$$\chi(\omega) = \chi_0 - i\omega \int_0^\infty b(t) \left[\cos \omega t - i \sin \omega t\right] dt \qquad (25)$$

where χ_0 is the static susceptibility.

The complex susceptibility is defined as $\chi(\omega) = \chi'(\omega) - i\chi''(\omega)$ and writing the integral in parts gives,

$$\chi'(\omega) - i\chi''(\omega) = \chi_0 - \omega \int_0^\infty b(t) \sin \omega t\, dt - i\omega \int_0^\infty b(t) \cos \omega t\, dt. \qquad (26)$$

Since $b(t)$ is a real function we can see that

$$\chi''(\omega) = \omega \int_0^\infty b(t) \cos \omega t\, dt. \qquad (27)$$

Now $b(t) \cos \omega t$ is an even function so that,

$$\frac{\chi''(\omega)}{\omega} = \frac{1}{2} \int_{-\infty}^\infty b(t) \cos \omega t\, dt \qquad (28)$$

which may be written as,

$$\frac{\chi''(\omega)}{\omega} = \frac{1}{2}\Re\left\{\int_{-\infty}^{\infty} b(t)e^{-i\omega t}dt\right\} \qquad (29)$$

Equation (29) is known as the fluctuation dissipation theorem. The component within the brackets of eqn (29) is the Fourier transform of $b(t)$; thus $b(t)$ can be obtained by carrying out an inverse Fourier transform on $\chi''(\omega)/\omega$ with,

$$b(t) = 2\Re\left\{\frac{1}{2\pi}\int_{-\infty}^{\infty} \frac{\chi''(\omega)}{\omega} e^{i\omega t}d\omega\right\}$$

$$= 2\Re\left[F^{-1}\left\{\frac{\chi''(\omega)}{\omega}\right\}\right] \qquad (30)$$

where F^{-1} denotes the inverse Fourier transform.

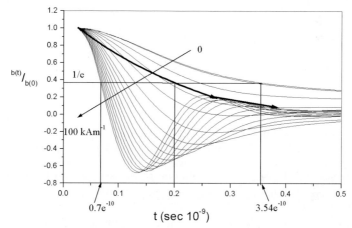

Fig. 9. Plot of normalised after-effect function, $b(t)/b(0)$, for data in Fig 6

A plot of the normalised after-effect functions, $b(t)/b(0)$, for the data in Fig 6, is shown in Fig 9. Here it is seen that the effect of the polarising field is to change $b(t)/b(0)$ from an exponential type decay for $H = 0$, with a time-constant, or $1/e$ value, of $2.5 \cdot 10^{-10}$ s to a damped oscillatory type waveform having a time-constant of $6.9 \cdot 10^{-11}$s at $H = 100$ kA/m. Thus there is an approximate fourfold increase in the rate of decay over the polarising field range.

As previously mentioned χ can be described in terms of parallel, $(\chi_\parallel)$, and perpendicular, $(\chi_\perp)$, components (Eq. (21)) with the corresponding relaxation times $\tau_\parallel$ and $\tau_{\perp eff}$. However, as the effect of the biasing field is to cause $\chi(\omega)$ to be dominated by the $\chi_\perp(\omega)$ component, the decay time at $H = 100$ kA/m corresponds approximately to that of $\tau_{\perp eff}$. This decrease in the time-constant with increasing polarising field arises because a ferrofluid has particles of different sizes and shapes giving rise to a distribution of precession or resonant frequencies. Thus, when an external field is removed the magnetic moments of the particles precess back to the directions of the natural easy axes with different precession frequencies, resulting in the transverse component of the magnetisation, $\chi_\perp(\omega)$, decaying rapidly to zero. Here $\tau_{\perp eff}$ is approximately $2 \cdot 10^{-10}$s.

5 Magnetic Losses

In the context of device application, losses in magnetic fluids are of particular interest. The permeability of a magnetic fluid, $\mu(\omega) = \mu'(\omega) - i\mu''(\omega)$, is a complex quantity which expresses the loss of energy which occurs as the magnetisation alternates. The loss mechanisms cause the flux density, B, to lag behind the applied alternating field, H, by a phase angle δ, and in this context two important relations are the loss tangent, $\tan \delta$, which is also known as the dissipation factor [19,28] and $\sin \delta$, which is referred to as the power factor [19]. The dissipated energy per cm^3 of the ferrofluid sample is directly proportional to both $\tan \delta$ and $\sin \delta$ where in direct analogy with the dielectric case, $\tan \delta = \mu''(\omega)/\mu'(\omega)$ and $\sin \delta = \mu''(\omega)/|\mu(\omega)|$ [19].

As $\mu''(\omega) = \chi''(\omega)$, and $\mu'(\omega) = \chi'(\omega) + 1$, the corresponding equations for the loss factor and power factor are,

$$\tan \delta = \frac{\chi''(\omega)}{(1 + \chi'(\omega))} \tag{31}$$

and

$$\sin \delta = \frac{\chi''(\omega)}{\left((1 + \chi'(\omega))^2 + \chi''(\omega)^2\right)^{1/2}} \tag{32}$$

Thus measurement of $\chi(\omega)$ enables one to investigate the frequency dependence of the loss tangent ($\tan \delta$) and power factor ($\sin \delta$) as a function of particle packing fraction and hence fluid saturation magnetization where, Fig 10 shows the high frequency dependence of $\tan \delta$ and $\sin \delta$ for a 200 G fluid (sample 5). Here the profile of $\tan \delta$ is identical to that of $\sin \delta$ with both differing from the $\chi''(\omega)$ profile.

In [28] such measurements are reported for four magnetic fluids with corresponding saturation magnetizations of 720 G, 360 G, 175 G and 75 G. The results show that, with reducing fluid magnetisation, the profiles of both $\tan \delta$ and $\sin \delta$ gradually become closer to that of the $\chi''(\omega)$ profile, until, in the case of a 75 G Fluid, the frequency dependent profiles become almost identical. Thus, apart from the results giving an indication of the measure of fluid magnetisation (and hence packing fraction) required for the above condition to occur it is of significance to note that the results also enable one to conclude that, in the case of the most dilute fluid, the high frequency dependent loss tangent and power factor can be modelled by the imaginary component of equation (23), with,

$$\tan \delta = \sin \delta = \chi_\perp''(\omega) = \chi_\perp'(0) \frac{\omega^2 \tau_2^2 \tau_\perp + \omega \tau_\perp (1 + \Delta)}{(1 - \omega^2 \tau_2 \tau_\perp + \Delta)^2 + \omega^2 (\tau_\perp + \tau_2)^2}. \tag{33}$$

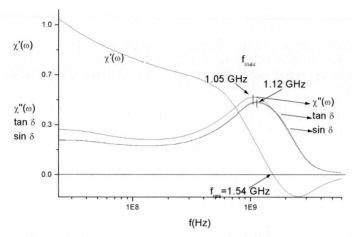

Fig. 10. Normalised plot of $\chi'(\omega), \chi''(\omega), \tan\delta$ and $\sin\delta$, against f in Hz, for sample 5

Finally, with the advent of magnetoelectronics and the possible use of nanoparticles as devices, such as switches, one is interested in the signal-to-noise(SNR) [29] ratio of such systems. Here again the usefulness of the availability of complex susceptibility data come to the fore since the SNR for a field driven magnetisation of a nanoparticle is a function of the parameters $\tau_0, \alpha, \gamma, \eta$ and K; all of which can be determined from susceptibility measurements as demonstrated in this Chapter.

Acknowledgements

Acknowledgement is extended to B.K.P.Scaife for useful discussions and to the Irish Higher Education Authority and Prodex for funding this work.

References

1. P. DEBYE. Polar Molecules, *The Chemical Catalog Company*, New York (1929).
2. K.S. COLE, R.H. COLE. *J. Chem. Phys.*, vol. 9, (1941), pp. 341.
3. H. FROHLICH. Theory of Dielectrics, *2 nd edn (Oxford: Clarandon)*, (1958).
4. D.W. DAVIDSON, R.H. COLE. *J. Chem. Phys.*, vol. 9, (1951), pp. 341.
5. S. HAVRILIAK, S. NEGAMI. *Polymer*, vol. 8 (1967), pp. 161.
6. W.F. BROWN. 5.*Appl.Phys.*, vol. 34 (1963), pp. 1319.
7. L. NÉEL. *Ann. Géophys.*, (1949), pp. 99.
8. D.P.E. DICKSON, N.M.K. REID, C.A. HUNT, H.D. WILLIAMS, M. EL-HILO, K. O'GRADY. *J. Magn. Magn. Mater.*, (1993), pp. 345.
9. V. SCHUNEMANN, H. WINKLER, H.M. ZIETHEN, A. SCHILLER, A.X. TRAUTWEIN. Magnetic Properties of Fine Particles *Elsevier Science Publishers, Netherlands*, (1992), pp. 371.
10. E. KNELLER, E.P. WOHLFAHRT. *J.Appl.Phys.*, vol. 37, (1966), pp. 4816.
11. W.F. BROWN. *Phys. Rev.*, vol. 130, (1963), pp. 1677.

12. L. BESSAIS, L.B. JAFFEL, J.L. DORMANN. *Phys, Rev B.*, vol. 45, (1992), no. 14, pp. 7805.
13. A. AHARONI. *Phys. Rev. B*, vol. 46, (1992), no. 9, pp. 5434.
14. M.I. SHLIOMIS. *Sov. Phys.-Usp.*, vol. 17, (1974), pp. 53.
15. P.C. FANNIN, B.K.P. SCAIFE, S.W. CHARLES. *J. Phys. E.Sci. Instrum*, (1986), pp. 238.
16. M.M. MAIROV. *Magnetohydrodynamics (Trans. of Magnitnia Hidrodinamika)*, vol. 2, (1979), no. 21, pp. 53.
17. C.J.F. BOTTCHER, P. BORDEWIJK. Theory of Electric Polarisation *Amsterdam, Elsevier Scientific Publishing Company)*, vol. 2, (1978).
18. V.V. DANIEL. Dielectric Relaxation *Academic press, London*, (1967).
19. B.K.P. SCAIFE. Principles of Dielectrics *rev.edn (London, Oxford Science Publications)*,(1998).
20. L. LUNDGREN, P. SVEDLINDH, O. BECKMAN. *J. Magn. Magn. Mater*, vol. 25 (1986), pp. 33.
21. C.C. PAULSEN, S.J. WILLIAMSON, H. MALETTA. *J. Magn. Magn. Mater.*, vol. 54-57, (1986), pp. 209.
22. P.C. FANNIN, A. MOLINA, S.W. CHARLES. *J. Magn. Magn. Mater.*, vol. 26, (1993), pp. 194.
23. M.S. RODEN. Analog and Digital Communication Systems *Prentice Hall International, London*, (1991).
24. Y.L. RAIKHER, M.I. SHLIOMIS. *Sov.Phys.JETP*, vol. 40, (1975), pp. 526.
25. W.T. COFFEY, J.P. KALMYKOV, E.S. MASSAWE. Modern Nonlinear Optics *Adv.Chem.Phys.*, vol. 85 part 2, (1993), pp. 667, Wiley Interscience, New York.
26. S. ROBERTS, A.R. VON HIPPEL. *J. Appl. Phys.*, vol. 17, (1946), pp. 610.
27. P.C. FANNIN, T. RELIHAN, S.W. CHARLES. *J. Phys. D; Appl. Phys.*, vol. 28, (1995), pp. 2003.
28. E.C. SNELLING, A.D. GILES. Ferrites for Inductors and Transmission Lines *Wiley, New York*,(1983).
29. P.C. FANNIN, S.W. CHARLES. Investigation of the frequency dependent loss tangent and power factor of colloidal suspension of nano-particles *J. Magn. Magn. Mater.* , vol. 226, (2001), pp. 1887.

Magnetic and Crystalline Nanostructures in Ferrofluids as Probed by Small Angle Neutron Scattering

Albrecht Wiedenmann

Hahn-Meitner-Institut Berlin, Glienicker Strasse 100, D-14109 Berlin, Germany

Abstract. We present a newly developed technique of nuclear and magnetic contrast variation by using polarised neutrons in Small Angle Neutron Scattering (SANSPOL) which allows density, concentration and magnetisation fluctuations in magnetic liquids to be analysed simultaneously. Diluted Ferrofluids based on different magnetic materials (Co, Magnetite, Ba-ferrite) and stabilized by charges or surfactants in different carrier liquids have been investigated. In such polydisperse systems several constituents of similar sizes have been identified by this technique: Magnetic core-shell composites, magnetic aggregates and free surfactants. The corresponding size distributions, compositions and magnetic moments have been determined. In more concentrated Co-FF the nature of field induced particle arrangements has been determined.

1 Introduction

Small angle neutron scattering (SANS) is known to be well suited for studying density- and concentration fluctuations on a length scale between 0.5 nm and 300 nm which corresponds to typical sizes of microstructural features in nanoscaled (n-) materials [1,2]. This technique allowed different types of inhomogeneities to be identified in crystalline, amorphous and liquid materials. In addition, magnetisation fluctuations in domain like structures resulting from strong intergranular correlations have been monitored by SANS in compacted n-Fe, n-Co, or n-Ni alloys [3,4]. In some diluted systems such as n-Fe_3O_4 embedded in a glass ceramic [5] and n-Fe_3Si in an ferromagnetic amorphous Fe-Si-B-Nb-Cu matrix [6,7] magnetic single domain behaviour of the nanocrystalline grains and the nature of interfaces have been evaluated.

Here we'll focus on the determination of the crystalline and magnetic microstructures of magnetic colloids. Magnetic colloids are stable dispersions of ferromagnetic materials. In such "Ferrofluids" (denoted as FF) nanoscaled magnetic particles are stabilized against coagulation either by electrostatic repulsion or by coating the core with organic chain molecules acting as surfactants [8]. Currently great effort is undertaken to prepare new bio-compatible Ferrofluids for potential biomedical applications [9]. Such applications are based on the super-paramagnetic behaviour of nanosized particles, which disappears when aggregation takes place as the consequence of an inefficient screening [10,11]. Therefore a precise knowledge of the microstructural parameters is a pre-requisite for the interpretation of macroscopic phenomena and for a tailored fabrication of FF.

Only few methods give access to these parameters. From macroscopic techniques such as magnetisation $M(H,T)$ measurements [12] or magneto-viscous effects [13] average parameters of concentration, size and arrangements of magnetic particles are derived. However, the actual values obtained depend strongly on the basic assumptions for the underlying processes [14]. Wide angle and small angle X-ray scattering as well as transmission electron microscopy [15] are sensitive to the particle core only, since the light elements of the organic shell give no sufficient contrast. The advantages of neutrons in such liquids are twice: First, the strong scattering power of hydrogen contained in the organic surfactants gives access to the shell and second, the interaction of the neutron spin with magnetic moments allows to visualise the magnetism of the core. Some SANS studies have been performed mainly on concentrated systems FF, which allowed the stability phase diagram to be established [16,17]. However, complications arise in poly-disperse systems when different constituents are present. Then we face the problem that weak magnetic scattering signals have to be analysed beside strong nuclear contributions from other sources or *vice versa* which can lead to considerable inconsistencies in the interpretation of results.

In our SANS investigations we intended to study the mesoscopic contituents of such poly-disperse FF using polarised neutrons as a labelling technique. Size, distribution and composition of the particles as well as magnetic nanostructures are determined systematically depending on the magnetic materials, surfactants and carrier liquids and on the particle concentration.

The paper is organised as follows: In section 2 we recall the basic concept of SANS and present the newly developed technique of nuclear and magnetic contrast variation by using polarised neutrons. We will show how this combination allows constituents of magnetic liquids to be identified. Results of our recent investigations on Co-FF, Fe_3O_4-FF and Ba-hexaferrite-FF are presented in the third section. It will be shown that this technique provides additional information about the nature of the nanocrystalline magnetic core and magnetic correlation as well as of the nonmagnetic shell which are not available by other techniques. Some common features of the different materials investigated are summarized in section 4.

2 Small Angle Neutron Scattering Technique

2.1 Basic Concept

Scattering of neutrons result from both nuclear and magnetic interactions between neutron and the nucleus and with the magnetic moment, respectively. The formalism is described in detail in various monographs e.g. [18]. In this section we recall some basic terms which are important for the understanding of the present subject.

The concept and experimental set-up for SANS is very simple and schematically shown in Fig. 1. From the "white" spectrum of the neutron source (reactor or spallation source) a small band of wavelengths $\Delta\lambda$ is filtered out, collimated and directed to the sample. The transmitted beam is absorbed in a beam stop whereas neutrons scattered around the primary beam are registered in a detector. The intensity distribution I(Q) has to be analysed as a function of the scattering vector Q(reciprocal space) which by Fourier transform gives access to correlation functions in real space i.e. to size, composition and magnetization of inhomogeneities present in the material. Small angle neutron scattering (SANS) is a special regime of low values of Q between typically 10^{-2} nm^{-1} and 5 nm^{-1} which allows to probe fluctuations on a length scale D$\propto 2\pi/$Q ranging from 1 nm to 500 nm.

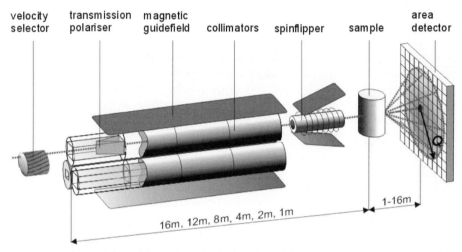

Fig. 1. Sketch of a typical SANS instrument (V4 at HMI Berlin) with the option for polarised neutrons (SANSPOL): The velocity selector picks out a small wavelength band $\Delta\lambda$ from the reactor spectrum. The beam is focussed in the collimator and directed to the sample where part of the neutrons are scattered elastically at nanosized inhomogeneities in an angle 2θ. The scattering intensity around the primary beam is counted in a position sensitive area detector an analysed a function of the scattering vector **Q**. For SANSPOL polarised neutrons are produced in the transmission polariser. Two scattering intensities are measured subsequently with neutron spin (n+) and (n-) after reversal in a spin flipper.

The signal is obtained by summing up the scattering amplitudes of all atoms weighted by the phase shift at each atomic position. Elastic scattering of neutrons with wavelength λ at an angle 2Θ leads to a **momentum transfer** Q according to

$$Q = 4\pi \sin\Theta/\lambda \tag{1}$$

and hence to a phase shift at an atomic position **r** of exp(i**Qr**). In the SANS regime of low spatial resolution, the discrete atomic scattering amplitude b can be replaced by a locally averaged **scattering length density** $\eta(r)$. A nuclear scattering length density is defined by

$$\eta_N = \Sigma c_i b_i / \Omega_i \qquad (2)$$

where b is the nuclear scattering length, c the concentration and Ω the atomic volume of the species i. η is easily calculated from the tabulated values of each element e.g. for water and heavy water η (H$_2$O,D$_2$O) = ρ N$_A$ (2b$_{H,D}$ + b$_O$)/MW which, with the mass density ρ, the Avogadro number N$_A$ and the molecular weight MW yields values of -0.56 10^{10} cm^{-2} and 6.33 10^{10} cm^{-2}, respectively.

For the interaction between the neutron spin S=1/2 with an assembly of magnetic moments a magnetic scattering amplitude is defined similarly by

$$\eta_M = \left(e^2\gamma/2mc^2\right) \Sigma c_i \boldsymbol{M}_i^\perp / \Omega_i \qquad (3)$$

where only the projection of the magnetic moment $\boldsymbol{M}_i{}^\perp$ onto a plane perpendicular to the scattering vector **Q** contributes to the interaction. $\boldsymbol{M}_i{}^\perp$ is given in units of Bohr magnetons μ_B and $(e^2\gamma/2mc^2) = 0.27 \cdot 10^{-12}$cm.

Throughout this paper we restrict our formulations to two phase-systems where particles are embedded in a homogeneous matrix. The total scattering amplitude of the particle is called **"form factor"** and defined by

$$F(QR) = \int dr^3 \Delta\eta \exp(i\boldsymbol{Q}\boldsymbol{r}_j) = \Delta\eta V_p f(QR) \qquad (4)$$

where the **contrast** $\Delta\eta$ is the difference between scattering length densities of particle and matrix, i.e. $\Delta\eta = \eta_p - \eta_{matrix}$. Note that the magnetic contrast $\Delta\eta_M$ and magnetic form factor $\boldsymbol{F}^\perp{}_M(\boldsymbol{Q})$ are vectors and depend on the orientation of the moment with respect to **Q**. The factor f(QR) depends only on the shape of the particle and is known analytically for various simple geometrical units. E.g. for spherical particles of radius R, f(QR) is given by the oscillating function

$$f(QR) = 3(\sin(QR) - QR\cos(QR))/(QR)^3 \qquad (5)$$

The differential scattering cross section in diluted systems, where N_p identical particles scatters independently, is given by

$$d\sigma(Q)/d\Omega = N_p F_j^2(QR) \qquad (6)$$

For such simple mono-disperse systems the scattered intensity follows general approximations which are independent of the shape of the particle: At low Q ($QR < 1$) the scattered intensity $I(Q)$ is described by the Guinier law

$$I(Q \to 0) = \Delta\eta^2 N_p V_p^2 \exp(-Q^2 R_g^2/3) \quad \text{for} \quad (QR < 1), \qquad (7)$$

from which a radius of gyration R_g can be derived. When the shape is known the particle dimensions are obtained from R_g, e.g. for spheres of radius R, $R_g = (5/3)^{0.5} R$. At large Q, scattering arises from the total surface of the particle S according to the Porod approximation

$$I(Q \to \infty) 2\pi (\Delta\eta^2 S) Q^{-4}. \qquad (8)$$

On the other hand the integrated intensity is known as "invariant"

$$\int (d\sigma/d\Omega)Q^2 dQ \propto N_p V_p \Delta\eta^2. \tag{9}$$

Combining Eq. (7 - 9) one obtains some indications of size, shape and concentration of the particle. However, in most of materials the two extreme regions are in practice not well defined due to additional scattering contributions, such as grain boundaries, surface defects, dislocations or strong background.

In this low-Q regime of SANS there is no access to distances between atoms or molecules; e.g. in simple liquids scattering in the SANS regime is independent of Q and given by

$$I(Q) = I(0) = N <b>^2 k_B T \chi(T) + I_{inc}. \tag{10}$$

The first term results from thermodynamic fluctuations, where $\chi(T)$ is the isothermal compressibility and $<b>$ is the average atomic scattering length, the second term results from the incoherent cross section of each atom which in some cases might be very high such as for protons where I_{inc} is 80 barn/atom.

2.2 Polydisperse Multiphase Systems

Complications arise in polydisperse multiphase systems when different types of particles j of different shape $F_j(QR)$ and size distributions $N_j(R)$ coexist which all will contribute to the scattering signal according to

$$d\sigma(Q)/d\Omega = \int_j N_{pj} F_j^2(QR) N_j(R) dR. \tag{11}$$

$N_j(R)dR$ is the incremental volume fraction of particles of type j in the radius interval between R and $R + dR$ and N_{pj} is the total particle number. $N_j(R)$ can be evaluated in principle from the scattering curves by the Indirect Fourier transform method [20] or by model fitting using analytical functions. However, if the particles of different type are of similar size their scattering contributions superimpose within the same Q-range. In addition, all oscillations of $I(Q)$ from the individual form factors are then smeared out due to the size distributions leading to more or less broadened signals.

A second complication appears in concentrated systems where interactions between particles can no longer be neglected. The simplest case of excluded volume interaction results from steric effects, i.e. the volume occupied by one particle can not be occupied by a second one. Such an assembly of interacting particles embedded in a homogenous matrix can be considered as a liquid, for which the particle positions are determined by the interaction potential $V(\boldsymbol{r})$. $V(\boldsymbol{r})$ is related to the pair correlation function $g(r)$ which defines the probability to find a particle at a distance r from a first particle at the origin. The Fourier transform of $g(r)$ is measured in scattering experiments as the so called "**structure factor**" $S(\boldsymbol{Q})$ given by

$$S(\boldsymbol{Q}) = 1 + N_p \int [g(\boldsymbol{r}) - 1] \exp(i\boldsymbol{Q}\boldsymbol{r}) d\boldsymbol{r} \tag{12}$$

The scattered intensity therefore depends on the form factors $F(Qr)$ and partial structure factors $S_{ij}(\mathbf{Q})$ of all particles of type i and j. This general case is difficult to treat for which different approximations have been proposed [19-21]. Analytical solutions are available only for mono-disperse systems; for spherical particles of identical size it simplifies to the product

$$I(Q) = N_p F^2(QR) S(Q) \tag{13}$$

The structure factor S(Q) will be unity when particles are fully uncorrelated which is the case only for a perfect gas. In disordered condensed matter the scattering at large values of Q is determined mainly by the particle form factor alone, where S(Q)=1. For a repulsive potential such as excluded volume effects or electrostatic repulsion S(Q) decreases at low Q and oscillates with Q when the particles form a pseudo-periodic arrangement. Attractive interaction potentials will give rise to an increase of the intensity at low Q which in practice might also be assigned to polydispersity or aggregates.

2.3 Magnetic and Nuclear Scattering with Unpolarised Neutrons

In a classical SANS experiment, where the incoming monochromatic neutron beam is unpolarised i.e. neutron spins are distributed at random, the scattering intensity is the sum of the squared amplitudes (Eq.6) from individual magnetic and nuclear contrasts. Due to the vector nature of $\mathbf{F}^{\perp}{}_M(\mathbf{Q})$ the scattering profile from a magnetic sample in an external magnetic field $\mathbf{H}$ is anisotropic and described by

$$d\sigma(Q)/d\Omega = A(Q) + B(Q)\sin^2\alpha \tag{14}$$

where A(Q) and B(Q) are the isotropic and anisotropic terms, respectively and α is the azimuth angle between the direction of the magnetic field $\mathbf{H}$ and the scattering vector $\mathbf{Q}$. For full alignment of all moments along the magnetic field, $A(Q) = F_N{}^2$ is solely of nuclear origin and $B(Q) = F_M{}^2$ represents the magnetic contribution resulting from the squared difference of the spontaneous magnetisation between particle and matrix.

In this cases the intensities measured perpendicular to the external magnetic field $I(Q\perp H)$ gives the sum of nuclear and magnetic contributions A(Q)+B(Q) whereas $I(Q//H)$ yields the nuclear contribution A(Q). When an area detector is used the validity of Eq. 14 can be checked by analysing the 2D scattering pattern $I(\alpha)$ which then yields both contributions with higher precisions. This commonly used techniques has two limitations. First, when the magnetic contrast is weak compared to the nuclear one (or vice versa, e.g. 10%) the anisotropy of the 2D scattering signal will be very weak (1 %) due to the squaring of the amplitudes. Such weak effects are very hard to be detected precisely in practice. Second, since the signs of the contrast is lost in the squaring of the individual nuclear and magnetic amplitudes the compositions, densities and magnetisation of matrix and particles cannot be obtained in absolute scale.

2.4 Polarised Neutrons

For polarised neutrons where the neutron spins are aligned antiparallel denoted by (+) or parallel (-) to a magnetic field **H**, four types of scattering process are to be distinguished, two for conserving the neutron spin $|F^{++}|^2$ and $|F^{--}|^2$ (spin non-flip scattering: snf) and two with reversal of the spin by the scattering $|F^{+-}|^2$ and $|F^{-+}|$ (spin-flip scattering: sf) [22,23]. Whereas coherent nuclear scattering is nsf, differences in the absolute values of the magnetic moments give rise to magnetic nsf and fluctuations of the magnetisation away from perfect alignment yields sf scattering.

When the polarisation of the scattered neutrons is not analysed, the intensity collected in the detector contains both snf and sf contributions and depends on the polarisation state of the incident neutrons. We denote this technique as SANSPOL. For simplicity we treat first the case of a mono-disperse system with non-interacting particles where the magnetic moment is *fully aligned* [24]. When magnetic moments and neutron polarisation are both parallel to an external magnetic field the intensities for the two states are given by

$$I^+(Q) = <|F^{++}|^2> + <|F^{+-}|^2>$$
$$= F_N^2 + \{F_M^2 - 2P(1-2f^+)F_N F_M\}\sin^2\alpha$$
$$I^-(Q) = <|F^{--}|^2> + <|F^{-+}|^2>$$
$$= F_N^2 + \{F_M^2 - 2P(1-2f^-)F_N F_M\}\sin^2\alpha. \quad (15)$$

P is the degree of polarisation defined by $P = (n^+ - n^-) / (n^+ + n^-)$ where n^+ are the numbers of neutrons with spin antiparallel, and n^-, parallel to H, respectively. The reversal of the polarisation direction from (+) to (-) is achieved with a spin-flipper of efficiency $f^{\pm}$, which is close to unity ($f^{(-)} \approx 1$) when the flipper is on, and zero when the flipper is off ($f^{(+)}=0$).

The anisotropy of both scattering signals is again described by a relation similar to Eq. (14),

$$I^{\pm}(Q) = A(Q) + B^{\pm}(Q)\sin^2\alpha \quad (16)$$

where the nuclear scattering $A(Q) = F_N^2$ is independent of the neutron polarisation. For polarised neutrons the anisotropic part $B^{\pm}(Q)$ is different for the two polarisation states and given by

$$B^{\pm}(Q) = F_M^2 - 2P(1-2f^{\pm})F_N F_M. \quad (17)$$

The scattering intensity for polarised neutrons contains a magnetic-nuclear cross term which is obtained from the difference between the intensities of the two polarisation states according to

$$I^-(Q) - I^+(Q) = 4Pf^{(-)}F_N F_M \sin^2\alpha = B_{int}\sin^2\alpha \quad (18)$$

The cross term is linear in the magnetic amplitude and hence makes it possible to determine the absolute value of the magnetic contrast with respect to the nuclear contrast, i.e. magnetic moment and compositions of particles and matrix.

The ratio $\gamma = F_M / F_N$ between the magnetic and nuclear form-factors is directly obtained from the intensity ratio of both polarisation directions ("flipping-ratio" FR) for Q perpendicular to H ($\alpha=\pi/2$) by solving Eq. (19)

$$FR = I^-(Q \perp H)/I^+(Q \perp H) = (1 - 2P(1 - 2f)\gamma + \gamma^2)/(1 - 2P\gamma + \gamma^2) \quad (19)$$

or from the ratio

$$R_2 = B^-(Q)/B^+(Q) = (\gamma - 2P(1 - 2f))/(\gamma - 2P) \quad (20)$$

The ratios FR and R_2 can be measured very precisely which allows the magnitude of γ to be determined much more accurately than for nonpolarised neutrons using Eq. (6), for which the modulus of γ is given by $|\gamma| = [B(Q)/A(Q)]^{1/2}$.

Note that the average of the intensities for both neutron polarisations is given by

$$[I^+(Q) + I^-(Q)]/2 = F_N^2(Q) + [F_M(Q)^2 + 2P(f - 1)F_n F_M]\sin^2\alpha \quad (21)$$

which corresponds to the intensity of non-polarised neutrons (Eq. 14) since the second term in brackets vanishes for $f \to 1$. For the paramagnetic case where all moment directions are equally probable the scattering is fully isotropic and independent of the polarisation and given by

$$I(Q)_{H=0} = F_N^2(Q) + 2/3 F_M(Q)^2 \quad (22)$$

Now we consider the case of *non perfect alignment* of the magnetic moments of a ferromagnetic single domain particle of saturation magnetization $M_s{}^p$ embedded in a nonmagnetic matrix [25]. For such particles superparamagnetic behaviour is expected, where the orientation distribution of the magnetic moments as a function of an effective magnetic field H_{eff} and temperature follows the Langevin statistics

$$L(x) = \coth(x) - 1/x \quad (23)$$

where $x = M(R')H_{eff}/k_B T$.

The total moment, $M(R')$, depends on the radius R' of the particles according to

$$M(R') = 4\pi R'^3 m_0/3\Omega_{at} \quad (24)$$

where m_0 is the saturation value of the atomic magnetic moment. Using the formalism presented in [6] SANSPOL cross sections for the case of a dilute system of non-interacting particles are again described by Eq. (16). If we include steric interaction effects via phenomenological structure factors $S(Q//H)$ and $S(Q\perp H)$ the intensities parallel and perpendicular to the magnetic field are then given by

$$\begin{aligned} I^\pm(Q, //H) &= A(Q) = F_N^2 S(Q//H) + F_M^2 2L(x)/x \\ I^\pm(Q \perp H) &= F_M^2[1 - L^2(x) - L(x)/x] \\ &\quad + S(Q \perp H)[F_M^2 L^2(x) + F_N^2 - 2P(1 - 2f^\pm)F_N F_M L(x)] \end{aligned} \quad (25)$$

The isotropic term as obtained from the intensity parallel to H is independent of the polarization state. Beside the nuclear term $F_N^2\ S(Q\|H)$ it contains an additional term of magnetic origin, $F_M^2 2L(x)/x$, which vanishes for complete alignment of all moments along **H**, when $L(x)/x = 0$. The anisotropic part $B^\pm(Q)$ is different for the two polarization states. Since the cross-term $B_{int}(Q)$ as obtained from the intensity difference

$$B_{int}(Q, H) = 4PfF_N F_M L(x) S(Q \perp H) \tag{26}$$

is linear in the magnetic amplitude the field variation of the magnetic moment according to the Langevin statistics can be precisely determined from the cross-term. Note that in Eq. (25) inter-particle correlations parallel to the field affect only the nuclear part but correlations perpendicular to H influences nuclear, magnetic and cross terms. We will see later that these features of Eq. (25) will help to distinguish between inter-particle correlations and particular form-factors which both may give rise to characteristic peaks in the SANS curves at low Q.

2.5 Combined Contrast Variation

The interpretation of SANS signals in polydisperse multiphase systems is generally not unique and requires special labelling techniques in order to identify and separate individual contributions. One conventional technique consists in the substitution of some atoms in the particle by isotopes with different scattering lengths. E.g. subsequent replacement of hydrogen by deuterium ("Triangulation") [26] allowed the localisation of sub-units in biological macro-molecules [27]. Similarly, in some cases the isotope composition of the matrix can be adjusted in such a way that the contrast vanishes (contrast matching), i.e. the particle (or some part of it) will be "transparent" for neutrons. However, in this kind of labelling techniques *different* samples have to be used which may present unknown sample specific characteristics. Secondly, the original isotope distribution may change after preparation. E.g. hydrogen atoms contained in the organic molecules can be partly exchanged by deuterium from the contrast matching isotope mixture in the solvent.

These ambiguities are removed when labelling is due to an intrinsic physical variable which is independent of the sample. The neutron-spin in SANSPOL is such a variable for magnetic systems [28]. This is illustrated in Fig. 2 for a composite particle, where the different grey scales represent the scattering length densities of a central magnetic core (η_1), a nonmagnetic outer part (η_2) and a nonmagnetic liquid as matrix η_{matrix}. Since $\eta_1^{(\pm)} = \eta_1^{nuc} \pm \eta_1^{mag}$ depend on the polarisation states (+) or (-) (black and tiled in Fig. 2) the scattering contrasts between core and shell as well as between core and matrix are different for both polarisation states. The contrast between nonmagnetic shell and matrix $\Delta\eta_2 = \eta_2^{nuc} - \eta_{matrix}$ is independent of the neutron polarisation. Two scattering curves $I^+(Q)$ and $I^-(Q$ are measured alternatively on the same sample by switching the neutron polarisation from (+) to (-). Both curves may then be different solely as a result of the different magnetic contrasts. Further, the difference

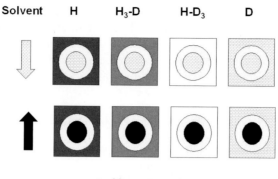

$$\Delta \eta^{(\pm)} = \eta^{nuc} \pm \eta^{mag} - \eta_{solvent}$$

Fig. 2. Illustration of the labelling technique for SANSPOL combined with contrast matching as a grey scale representation of the scattering length densities η in a composite particle: $\eta_1^{(\pm)}$ of the central magnetic depends on the neutron polarisation. The contrast between the nonmagnetic shell and matrix is adjusted using different isotope mixtures of the solvent.

$I^+(Q)$-$I^-(Q)$ is proportional to the product $\eta_1^{mag} \cdot \eta_1^{nuc}$ of *one and the same particle* whereas nuclear and magnetic contrasts from *different* particles do not contribute to the cross-term. This allows additional nonmagnetic particles to be distinguished from magnetic particles even when they are of the same size.

In addition, the "colour" of the surrounding matrix can be adjusted using the above mentioned contrast matching with different mixtures of isotopes in the solvent.

Both scattering curves $I^+(Q)$ and $I^-(Q)$, the difference $I^+(Q)$-$I^-(Q)$ and $F_N^2(Q)$ und $F_M^2(Q)$ as obtained from the averaged signal of Eq. 14 in all isotope mixtures must be adjusted using the same structural model. This combined contrast variation thus reduces considerably the number of possible structural models and ensures the consistence and accuracy of the constrained model parameters.

2.6 Experimental

SANS measurements have been performed at the instrument V4 installed at the BERII reactor of HMI, Berlin [29](Fig. 1). A horizontal magnetic field of strength up to 1.1 T was applied at the sample position, oriented perpendicular to the incoming neutrons. Polarized neutrons (n^+) are provided by a transmission polarizing super-mirror cavity [30] which is contained in one of the three tubes of the first segment of the collimator (Fig. 1). The polarisation direction is conserved in a guide field downstream to the spin flipper. In this device the polarisation can be reversed by injection of a radio-frequency superposed to a magnetic field gradient. The two scattering intensities I(+) and I(-) are collected alternatively with flipper off and on. The SANSPOL option, which can be set without any modification of the instrument alignment, is characterized by a high neutron flux of more than 30% of non-polarized neutrons, a high degree of polarization ($P=95\%$) and by the high efficiency of the spin flipper ($f(-) \approx 98\%$) at the wavelength $\lambda=0.6$ nm used in this experiments. The reliability of this option has been demonstrated by comparing the results of a SANSPOL study on magnetic glass ceramics to those of a conventional SANS study [28]. By choosing three distances between the sample and detector a Q range of 0.08 nm^{-1} to 3 nm^{-1} could be accessed which allows particles to be detected with sizes ranging from 0.5 to 80 nm.

We investigate ferrofluids where the nanoscaled magnetic particles are stabilized against coagulation either by electrostatic repulsion or by coating the core with organic chain molecules acting as surfactants. Magnetic moments of the core materials (Co, Magnetite Fe_3O_4, and Ba-Hexaferrite), shell forming surfactants (mono- and bi-layers) and carrier liquids (Water, organic solvents) have been systematically varied in order to study the influence on the microstructures. Co- and Magnetite ferrofluids have been prepared by Berlin Heart AG [41] and Ba-Hexaferrite by R. Müller of IPHT, Jena Germany [31].

3 Results

3.1 Cobalt Ferrofluides

Diluted Co-FF. Colloidal solutions of ferromagnetic Cobalt can be prepared by coating the nanosized particles with organic chain molecules C_{21}-H_{39}-N-O_3. The hydrophilic head groups are considered to be in contact with the metal surface and the hydrophobic tail pointing to the non-polar toluene as carrier liquid. We performed a SANS study on dilute solutions with a nominal concentration of 0.5 vol. % of Cobalt as derived from saturation magnetisation measurements. Four different mixtures of protonated and deuterated toluene, denoted as AF1-AF4 (Table 1) have been studied in a magnetic field of 1.1 T in which full alignment of the magnetic moments is expected.

Table 1. Scattering length densities as obtained from a fit to a shell model using the bulk value of $\eta_1(nuclear) = 2.53 \cdot 10^{10} cm^{-2}$ for Co.

Sample	Content C_7D_8	η (solvent) $10^{10}cm^{-2}$	η_1 (mag) $10^{10}cm^{-2}$ SANSPOL	η (mag) $10^{10}cm^{-2}$ SANS	η_2 (shell) $10^{10}cm^{-2}$
AF2	100%	5.65	4.29	4.38	0.56
AF4	43%	3.0	4.5	5.3	0.3
AF3	14%	1.6	1.52	1.62	0.9
AF1	0%	0.97	0.94	1.24	0.8

A first conventional SANS study showed that the scattering on of the sample AF4 is dominated in the high Q range by a high level of the incoherent background (mainly from hydrogen contained in the solvent and surfactant). Nuclear and magnetic scattering contributions have been extracted and analyzed

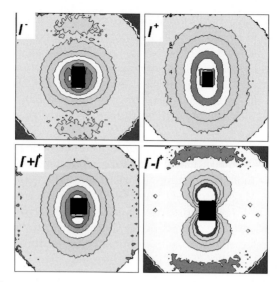

Fig. 3. SANSPOL patterns in Co-ferrofluid AF4 for neutron spins antiparallel (I^-) and parallel (I^+) to the horizontal field of strength H=1.1 T. The arithmetic mean $[(I^-) + (I^+)]/2$ corresponds to the 2D pattern of non-polarized neutrons. The difference (I^-)-(I^+) yields the interference term [equation (18)] which presents negative values in the center.

in terms of number-distributions $N(R)$ of non-interacting spherical particles. For the magnetic contribution a maximum of $N_{mag}(R)$ appeared at R =3.3 nm whereas from the nuclear contribution, a bimodal distribution had to be assumed with maxima at R_1=1.12 nm and R_2=4.9 nm, respectively.

The 2-dimensional SANSPOL pattern for the same sample AF4 shown in Fig. 3 are highly anisotropic with a dramatic change of the aspect ratios for the two polarization states. We emphasize the particularity that only for the polarization state I^- a maximum appears in the outer part of the patterns. The difference signal $(I^- - I^+)$ of Fig. 3, where all background contributions are cancelled out, shows the angular dependence as expected from Eq. (16) with negligible intensity along the direction of the magnetic field. Note that in the inner part the intensities are negative and positive values in the outer part. SANSPOL intensities perpendicular to the applied field $I^\pm(\mathbf{Q}\perp\mathbf{H})$ as obtained by an adjustment of the 2-d pattern to Eq. (16) are compared for the different solvents in Fig. 4a.

In the fully (AF2) and partly deuterated solvents (AF4, AF3) pronounced maxima occur only for $I^-(\mathbf{Q}\perp\mathbf{H})$ whereas $I^+(\mathbf{Q}\perp\mathbf{H})$ decreases always continuously with increasing Q. The position of the maximum of $I^-(\mathbf{Q}\perp\mathbf{H})$ shifts to lower Q for increasing content of hydrogen. The cross-term B_{int} reverses the sign from negative values below Q=0.61 nm^{-1} to positive values above, with a maximum at Q=0.8 nm^{-1}(Fig. 3). For Q larger than 1.35 nm^{-1}, B_{int} is zero, which proves that there is no magnetic contribution left. The ferrofluid AF1 in a fully

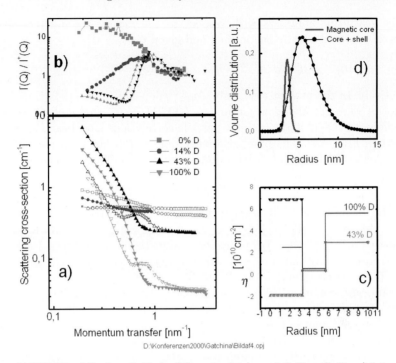

Fig. 4. SANSPOL of Co-ferrofluids in different mixtures of $C_7D_8 - C_7H_8$: a) Intensities $I^+(\mathbf{Q}\perp\mathbf{H})$ (solid symbols) and $I^-(\mathbf{Q}\perp\mathbf{H})$ (open symbols). Solid lines: Fit according to a shell model using the parameters Table 1. b) Flipping ratio according to Eq. (19), c) Scattering length density profiles for two isotope mixtures of the solvents. d) volume distribution of a composite particle as a function of the outer radius (SANSPOL). The results from an unpolarised SANS study could have been described by an artificial bimodal distribution of spheres of radius R

protonated solvent shows only a splitting of $I^+(\mathbf{Q}\perp\mathbf{H})$ and $I^-(\mathbf{Q}\perp\mathbf{H})$ below $Q=1$ nm^{-1}, where $I^-(\mathbf{Q}\perp\mathbf{H}) > I^+(\mathbf{Q}\perp\mathbf{H})$. The flipping ratio FR derived from Eq. (19) (Fig. 4b) are different for each solvent at low Q but follow nearly the same Q dependence at high Q. This indicates that the surface of the magnetic Co-particles must be surrounded by a layer of almost constant density different from that of the solvent. While from non-polarized neutrons size distributions corresponding to individual nuclear and magnetic units are derived, polarized neutrons show that both contributions must result from a "composite" particle built up by a magnetized core of Co atoms surrounded by a nonmagnetic surface layer. As the simplest description of such a "composite" we use a shell model consisting of a sphere with an inner core radius R' surrounded by a concentric shell of radius R. The form factor is given by

$$F_{shell}(Q) = [(\Delta\eta_1 - \Delta\eta_2)f_{sph}(QR') + \Delta\eta_2 f_{sph}(Q(R))]V_p \qquad (27)$$

with the shape function for spheres as given by Eq. (5), i.e. $f_{sph}(x) = 3[\sin(x) - x\cos(x)]/x^3$.

The scattering contrasts with respect to the matrix are different for the magnetic core and non-magnetic shell and given by

$$\Delta\eta_1^{(\pm)} = \eta_1^{nuc} \pm \eta_1^{mag} - \eta_{matrix} \text{ and } \Delta\eta_2 = \eta_2^{nuc} - \eta_{matrix} \qquad (28)$$

respectively. In the present case, only $\Delta\eta_1^{(\pm)}$ depends on the polarization. The intensities were calculated according to $I(\mathbf{Q}\perp\mathbf{H}) = N_p \int F_{shell}^2(Q,R) N(R') dR$, where for the diluted case inter-particle correlations have been neglected i.e $S(Q)=1$ in Eq. (25). The number distribution $N(R')$ has been parameterized using a log-normal number distribution of the core radius R' according to

$$N(R')1/(\sqrt{2\pi R\sigma})\exp[-\ln^2(R'/R'_0)/2\sigma^2] \qquad (29)$$

where $\mathbf{R'_0}$ denotes the median and s the width of the distribution.

The thickness of the shell D is assumed to be constant, i.e. R=R'+D. The parameters η_1^{nuc}, η_1^{mag}, η_2^{nuc} have been adjusted in a non-linear least square fitting routine using the constraints of Eq. (28) for the contrasts $\Delta\eta_1^{(\pm)}$ and $\Delta\eta_2$. The scattering length density of the solvent η_{matrix} was known for the H/D ratio and confirmed by the incoherent background level. The parameters N_p, R', D and σ have been constrained to be identical for both polarization states. The solid lines in Fig. 4 represent the calculated intensities $I^\pm(\mathbf{Q}\perp\mathbf{H})$. It turned out that for all solvents this model function lead to one set of parameters which are consistent with the intensities $F_N(Q)^2$ and $F_M(Q)^2$ as calculated for the nuclear and magnetic contributions from the averaged signal according to Eq. (14).

The distribution $N(R')$ was found to be rather sharp (Fig. 4d) corresponding to a volume weighted average of the core radius of $<R'> = 3.7$ nm and a constant thickness of the shell of $D= 2.47$ nm. We emphasize that from non-polarized neutrons alone it would not have been possible to derive the shell model due to the very low nuclear contrast of the shell beside the high incoherent background. The first maximum of the size distribution of spheres as obtained from classical SANS is therefore identified as artificial and reflects in fact half of the thickness of the shell.

The scattering length densities resulting from fits of both polarization states using the same model function are included in Table 1. The values of η_2 for the shell of the same order of magnitude as calculated for densely packed surfactant molecules ($\eta_2 = 0.33 \ 10^{10} \text{cm}^{-2}$) and do not depend significantly on the solvent composition. This supports strongly the conclusion that the shell of thickness of about 2.4 nm formed by organic surfactants is nearly impenetrable for the solvent in i.e. the Co-core is not in direct contact with the solvent. The values of $\eta_1(\text{mag})$ in AF2 and AF4 experimentally derived at $H = 1.1$ T correspond closely to $\eta_1(\text{mag}) = 4.14 \ 10^{10} \text{cm}^{-2}$ expected from Eq. (3) for bulk Co ($m_0=$ 1.715 μ_B/ atom , $\Omega_{at}(\text{Co}) = 0.01099 \ \text{nm}^3/\text{at.}$) Discrepancies appeared in the samples AF1 and AF3 where much lower values for $\eta_1(\text{mag})$ were found. The reason for such anomalous low values is not yet clear; it might result from a lower particle density, to some oxidation , or to non-magnetic amorphous surface layers as reported for magnetic glass ceramics [32].

Variation of the magnetic field. For the solvent composition AF4 the cross-term $B_{int}(Q,H)$ has been evaluated from the difference $I^-(Q) - I^+(Q)$ for different values of the external magnetic field and plotted in Fig. 5. All additional non-magnetic contributions are eliminated in this difference [25].

$B_{int}(Q,H)$ could be well fitted according to Eq. (26) by the product $F_N F_M L(x)$ alone showing that for this diluted sample no significant inter-particle correlations are present or induced by the magnetic field, i.e. $S(Q\perp H)=1$ [38]. The ratio $B_{int}(Q,H) / B_{int}(Q,H=1.1T)$ was found to be constant for all values of Q. The H-dependence of this ratio is plotted in the insert of Fig. 5. The experimental points follow closely a Langevin function corresponding to spherical Co-particles with average core radius $<R'> = 3.7$ nm. This is in good agreement with the prediction of Eq. (24) and with $<R'>$ as derived from the direct model fit [33].

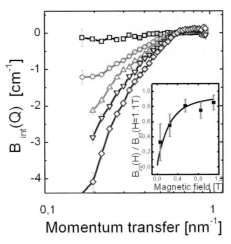

Fig. 5. Cross-term $B_{int}(Q)$ for Co-ferrofluid AF4 as a function of Q at $H = 1.1$ T ($\Diamond$), 0.85T(∇), 0.25T(Δ), 0.06T(O) and 0 T ($\Box$). Inset: Field variation of the ratio $B_{int}(Q,H))/B_{int}$ $(Q, H =1.1$ T) (symbols) follows the Langevin behavior for non-interacting single domains particles of $<R>=3.7$ nm

Concentrated Co-FF. The setup of inter-particle correlation $S(Q)$ under the influence of magnetic field was studied in a second series of higher concentrated systems with nominal Co-concentrations as about 1 vol.% (D1) and 5 vol. % (D5) ("MF 239") [34]].

The azimutally averaged intensities for unpolarised neutrons $<I(Q)>$ plotted in Fig. 6 illustrate how the scattering behaviour changes as a function of Co-concentration and magnetic field. For random orientation $<I(Q)_{H=0}> = [F_N^2 + 2/3F_M^2]S(Q)$ is expected at zero field while for full alignment in strong magnetic fields $<I(Q)>$ is given by $<I(Q)_{Hmax}> = [F_N^2 + 1/2F_M^2]S(Q)$. For the sample D1% the scattering observed at zero field scales fully with that at

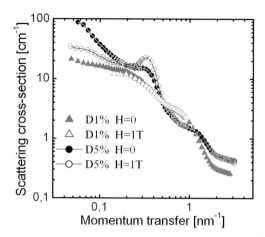

Fig. 6. Radial averaged SANS intensity of 1-vol% (triangles) and 5-vol% (circles) Co-Ferrofluids in C_7D_8 at $H = 0$ (full symbols) and $H = 1.1$ T (open symbols).

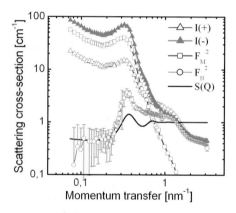

Fig. 7. SANSPOL intensities $I^+(\mathbf{Q}\perp\mathbf{H})$ (solid triangles) and $I^-(\mathbf{Q}\perp\mathbf{H})$ (open triangles) of 5vol% Co-Ferrofluids in C_7D_8 and magnetic (squares) and nuclear (circles) contributions from non-polarised neutrons. The solid line represents the common structure factor S(Q) used in the model fits.

$H = 1$ T, i.e. there is no change of the particle arrangement in the external field. In the concentrated sample D5% a pronounced peak occurs at $Q_1 = 0.32$ nm^{-1} when the magnetic field is turned on indicating the set-up of correlation between the Co particles. This is confirmed by the SANSPOL results shown in Fig. 7. In addition to the high Q features characteristic for the core-shell profile for both polarisation states a sharp peak occurs at $Q_1 = 0.32$ nm^{-1} in the concentrated case D5% only i.e. the magnetic-nuclear cross term is clearly peaked resulting from the structure factor $S(Q)$. This example shows clearly that SANSPOL allows definitely to distinguish between scattering maxima arising from interparticle correlations $S(Q)$ and from the form-factor F(QR). Compared to Fig. 4 an additional isotropic nonmagnetic scattering contribution is superimposed to I^+ and I^- at high Q corresponding to a small amount (1-5%) of spherical particles with $R \approx 1.8 \pm 0.4$ nm. This contribution results from excess surfactant molecules present in the solutions and which might form spherical micelles. Taking into account this additional contribution, all curves $I^\pm(\mathbf{Q}\perp\mathbf{H})$, $F_M{}^2$ and $F_N{}^2$ could be adjusted simultaneously by the same core-shell model as described above with very similar values for the structural parameters. In the concentrated sample D5% the set-up of strong inter-particle-correlation between the Co-particles in an external magnetic field are clearly reflected by a structure factor $S(Q)$. All curves were well described using the same $S(Q)$ as given by the hard sphere model of Percus-Yevick [19]. The resulting $S(Q)$ plotted in Fig.7 corresponds to volume fraction η=0.3 of hard spheres with a radius R_{hc}=8.65 nm. R_{hc} is by a factor of about 1.5 larger than the total radius $R' + dR$ of the particles which implies a rather dense packing of the particles under the influence of the magnetic field. It compares favourably with the hydrodynamic radius as experienced from field induced magneto-viscous damping [15]. Dipole interaction between the particle moment should favour the formation of chains with the moment and chain directions along the magnetic field. Such a chain formation should give rise to stronger inter-particle correlation along H i.e. the structure factor $S(\mathbf{Q}\|\mathbf{H})$ should be different from $S(\mathbf{Q} \perp \mathbf{H})$ [35]. However, comparing the intensities along H and perpendicular to H did not reveal any evidence for such an anisotropy of the structure factor. $S(\mathbf{Q}\|\mathbf{H}) \approx \mathbf{S}(\mathbf{Q} \perp \mathbf{H})$ seems to reflect the rather dense packing of the core-shell particles in the concentrated systems which is induced by an external magnetic field. A detailed study of the field induced correlations are currently performed as a function of concentration and core sizes.

3.2 Magnetite Ferrofluids

Ferrofluids based on Magnetite are available in aqueous solution which makes them interesting as biocompatible materials for medical applications. Nanosized Magnetite particles, Fe_3O_4, have been prepared by co-precipitation of ferric salt mixtures with concentrated ammonium hydroxide [41]. The magnetite cores are charge stabilized (denoted as ELEC) which optionally can be coated by different surface active organic molecules. Size, composition and structural arrangement of the molecules in this coating are almost unknown. In the first example, denoted as LM, coating should consists on a bi-layer of dodecanoic acid (inner layer) and C_{12} ethoxylated alcohol with 9 mol/mol ethoxy groups. The ferrofluid (DEX) has a single dextrane shell, whose chains are believed to be entangled by subsequent heat treatment. In a first series we investigated the structural parameters as a function of the stabilizing surface materials with no organic shell (ELEC) and two different layer materials (LM and DEX) [36]. For SANS contrast requirements the content of D_2O with respect to H_2O in the carrier liquid is larger than 90%. The volume fraction of magnetite was nominally about 1 vol.% in all 3 samples. In the second series a combined contrast variation study have been performed for LM in three different isotope mixtures H_2O/D_2O denoted as $H_{0.2}D_{1.8}O$, $H_{0.6}D_{1.4}$, DHO [37].

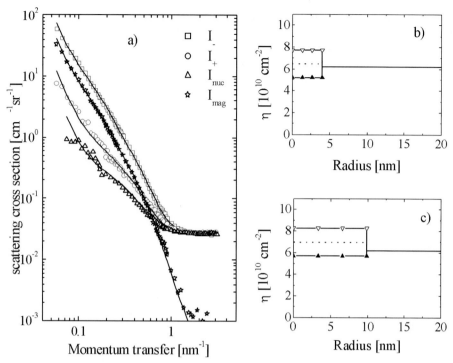

Fig. 8. Charge stabilized Magnetite FF (ELEC) in $H_{0.2}D_{1.8}O$,: a) SANSPOL intensities $I^+(\mathbf{Q}\perp\mathbf{H})$ and $I^-(\mathbf{Q}\perp\mathbf{H})$ where solid lines corresponds to a fit with using a bimodal distribution of spheres with b) scattering length density profiles of small and c) large spherical particles.

The SANSPOL intensities $I^\pm(\mathbf{Q}\perp\mathbf{H})$ as obtained by an adjustment of the 2-d pattern to Eq. (16) are compared to nuclear and magnetic contributions as derived from Eq. (14) for the different samples in Figs. 8a-10a. In DEX and LM samples $I^-(\mathbf{Q}\perp\mathbf{H})$) is lower than $I^+(\mathbf{Q}\perp\mathbf{H})$ at high values of Q and exhibit a crossover around Q=0.2 nm below which $I^-(\mathbf{Q}\perp\mathbf{H})) < I^+(\mathbf{Q}\perp\mathbf{H})$. For the ELEC no such crossover occurs where $I^-(\mathbf{Q}\perp\mathbf{H})) > I^+(\mathbf{Q}\perp\mathbf{H})$ for all values of Q.

The crossover phenomena observed in the polarized neutron data of DEX and LM is a characteristic feature of "composite" particle similar to that observed in Co-ferrofluids which is expected to be built up by a magnetized core of Fe_3O_4 atoms surrounded by a nonmagnetic surface layer. In fact the solid lines in Fig. 9-10 for LM and DEX represent the intensities $I^\pm(\mathbf{Q}\perp\mathbf{H})$, I(nuc) and I(mag) calculated with a lognormal distribution of the core-shell particles model as described above. In ELEC the absence of a crossover of the intensities $I^-(\mathbf{Q}\perp\mathbf{H})$ and $I^+(\mathbf{Q}\perp\mathbf{H})$ indicates clearly that in this FF the particles are not coated. The scattering is therefore described by a lognormal distribution of spherical particles (solid lines in Fig. 8). In all samples the scattering curves of I(mag) and I(nuc) revealed some shoulder at low Q which is a strong indication of the presence of a second fraction of much larger particles. This second contribution had to be included in the fit. A spherical shape and a lognormal size distribution have been assumed for these aggregates. In LM the combined contrast variation revealed an additional nonmagnetic contribution resulting from free organic surfactants, which had to be added in excess for stability reasons during the preparation process.

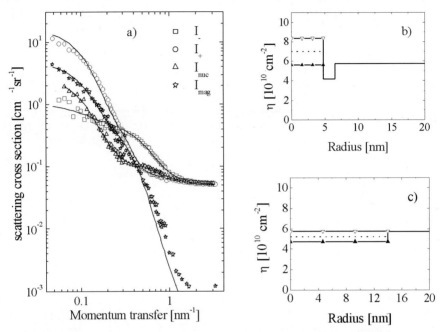

Fig. 9. Magnetite FF with DEX coating in $H_{0.2}D_{1.8}O$,: a) SANSPOL intensities $I^+(\mathbf{Q}\perp\mathbf{H})$ and $I^-(\mathbf{Q}\perp\mathbf{H})$ where solid lines corresponds to a fit with a distribution of small core-shell particles and larger spherical aggregates with the corresponding scattering length density profiles shown in b) and c). The low value of η_1 (nuc) in the aggregates reflects inclusion of free surfactant molecules.

It turned out that for all samples this model function lead to consistent parameters. The volume weighted size distributions are compared in Fig. 11. For the small particles a rather sharp distribution $N(R')$ corresponding to a volume weighted average of the core radius of $<R'>$=6.73nm (LM), 4.8 nm (DEX) and 4.0 (ELEC) a constant thickness of the shell of $D = 2.61$ nm and 2.4 nm (DEX). For the second fraction the volume averaged radius is by a factor of 2.5-3.5 times larger than that of the core.

The scattering length densities resulting from simultaneous model fits of both polarization states as well as I_{nuc} and I_{mag} using the same model function are presented in Fig. 8-10 (b,c).

From SANSPOL it turned out that for the small particles the scattering length density of the core (η_1) is clearly higher than $\eta_{solvent}$ and higher than that of the shell (η_2) for the samples DEX and LM. The magnetic moment for Fe_3O_4 as derived from the fit is very close to the bulk value. In addition, the resulting scattering length density (Fig. 10) for the shell in LM was found to be independent on the H/D composition of the solvent, which indicates that the organic surfactant is nearly impenetrable for the solvent.

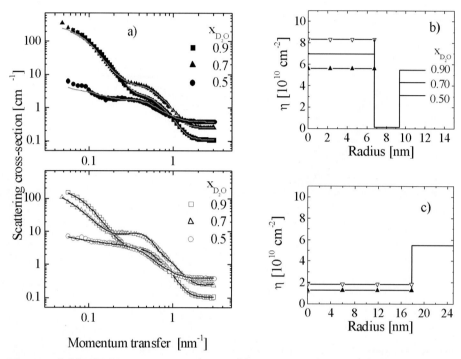

Fig. 10. a) SANSPOL intensities $I^-(\mathbf{Q} \perp \mathbf{H})$(open symbols) and $I^+(\mathbf{Q}\|\mathbf{H})$ (full symbols) of Magnetite FF with LM-coating in different isotope mixtures of D_2O/H_2O and b) corresponding scattering length density profiles of small bi-layer coated composites particles and c)of spherical aggregates.

For the large particles η_1 is found to be lower than $\eta_{solvent}$ in LM and DEX but higher in ELEC. Since the scattering intensity for the large particles depends on the polarisation, they must also contain magnetic materials. However, the observed ratio $\eta^{(-)}/\eta^{(+)}$ was much smaller than that expected for pure magnetite. It corresponds to a magnetite content of only 22%. These result suggests that the large particles consist of a loose and mixed arrangement of magnetite together with surfactant molecules forming a larger aggregate. Size and volume fraction of these aggregates seem to depend on the preparation conditions (see Table 2). In ELEC where surfactants are absent the magnetic contrasts of both fractions are very similar which indicates that also the aggregates presents the ferrimagnetic ordering of magnetite.

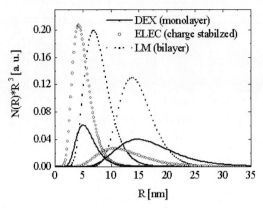

Fig. 11. Volume weighted size distribution of magnetic particles in charge stabilized ELEC (without shell), monolayer coated DEX and bilayer coated LM. The size of the central core are very similar while that of the aggregates depends on the preparation conditions.

In any case the presence of aggregates indicate that steric or electrostatic screening is not fully efficient in these magnetite based materials.

3.3 Barium-Hexaferrite Fluids

Magnetic nano-particles of Ba-hexaferrite $BaFe12-2xTi_xCo_xO19$ (x=0.8) have been prepared [31] by a glass crystallisation method. Magnetic liquids are prepared with oleic acid $CH_3(CH_2)_7CH=CH(CH_2)_7COOH$ as surfactant and dodecane as carrier liquid. Due to the high uniaxial crystalline anisotropy a large magneto-viscous effect for barium ferrite Ferrofluids (FF) was expected which is of potential technical interest. Samples with a particle concentration between 1 and 4 vol % were diluted in mixtures of protonated and deuterated dodecane, denoted by H(content of 100% H), HD (64% H) and DD (33% H) [39].

The two-dimensional SANSPOL intensities for both polarisation states are slightly anisotropic and consequently $I^+(\mathbf{Q} \perp \mathbf{H})$ and $I^-(\mathbf{Q} \perp \mathbf{H})$ show only small differences at low Q values (Fig. 12). Therefore, the magnetic contrast is very weak compared to the strong nuclear contrast. Again the simultaneous fit of two SANPOL curves together with I(nuc) in all three isotope mixtures allowed three contributions to be distinguished: i) magnetic "composite" particles ii) free surfactant molecules giving an additional (non-magnetic) contribution at high Q while iii) at low Q an additional magnetic component of aggregates occurs. The final fits (solid lines in fig. 12a) correspond to the scattering length densities of core-shell particles and aggregates of Fig. 12b and to the volume size distributions of Fig. 12c. The magnetic particles have an averaged core radius of 4.1 nm with a shell thickness of 1.8 nm. Since in all H/D mixtures the same (theoretical) value of the contrast of the oleic acid shell was found we conclude that the shell is impenetrable for the carrier liquid. The low value of the scattering length densities found for the large particles with a radius of 9.8 nm

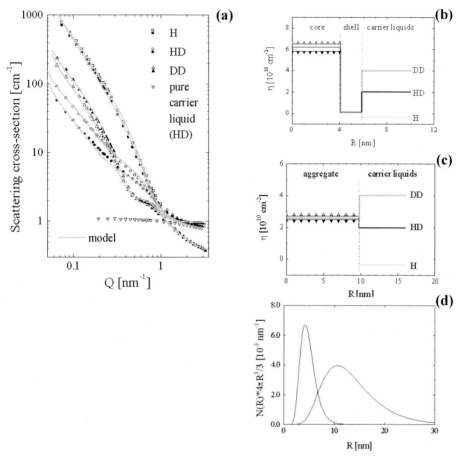

Fig. 12. SANSPOL cross sections $I^-(\mathbf{Q} \perp \mathbf{H})$ (open symbols) and $I^+(\mathbf{Q} \perp \mathbf{H})$ (full symbols) for Ba-ferrite-FF in different H/D mixtures 100 % H (H), 64 % H (HD) and 33 % H (DD) of the dodecane a carrier liquid (a). The grey lines represent model curves with the scattering length density profiles of core-shell particles (b) and of aggregates (c) and the corresponding volume distributions shown in of d).

is attributed to aggregates of Ba-ferrite containing a large amount of surfactant molecules (the overall of the magnetization is only 1/3 of the small particles.

The magnetic scattering contrast derived from SANSPOL to about 43% of bulk magnetization in agreement with the measured magnetization of the pure particles used in the preparation of the FF [40]. Non-stoichiometry or a non-magnetic surface layers might be responsible for the observed low values of magnetization. Volume fractions f were calculated from the size distributions. About 1/2 of the total amount (12 vol %) of surfactants remains as free molecules in the carrier liquid.. The volume fractions of the small ferrite cores were 2.3% (H), 2.6% (HD) and 1.6% (DD) which correspond to about 2/3 of the value from magnetization measurements i.e. 2/3 of all ferrite particles are separated and encapsulated with surfactants. The remaining 1/3 of ferrite particles are in the aggregates which have volume fractions of about 4.5% to 7%.

4 Summary

The use of polarized neutrons in Small Angle Neutron Scattering is a powerful labeling technique for investigations of magnetic materials. In the present study of magnetic colloids we combined this technique with the conventional contrast variation using different isotope mixtures of the carrier liquid. This combination allows magnetic and nonmagnetic particles to be distinguished and density-, composition-, and magnetization profiles to be precisely determined. The microstructure parameters have been evaluated in polydisperse multiphase systems where magnetic materials (Co, Magnetit, Ba-Hexaferrite), shell forming surfactants (mono- and bi-layers) and carrier liquids (Water, organic solvents) have been systematically varied. The results are compared in Table 2 and summarized as follows:

As a common feature three different components were identified in the magnetic liquids, magnetic composite particles, free organic shell molecules and magnetic aggregates. In diluted Ferrofluids which are stabilized by surfactants composite particles are well described by a magnetic core and an organic shell of constant thickness. The size of the core depends on material and preparation conditions while the thickness of the shell is characteristic for the surfactant materials. The shell was found to be of homogeneous density and almost impenetrable for the solvent. No shell structure was found in charge stabilized samples where solvent molecules are in touch with the magnetic core. The magnetic ordering of diluted systems corresponds to non-interacting ferromagnetic single domain particles. Small amounts of aggregates with lower densities and magnetization and a typical size of 2-4 times the core radius were identified. This indicates that screening is incomplete in some systems leading to aggregates in which surfactant molecules are included. Depending on preparation conditions nonmagnetic contributions have been found which are ascribed to free organic surfactants or micelles.

The set-up of strong field induced inter-particle correlations was detected in Co- Ferrofluids with higher concentrations. The observed structure factor $S(Q)$ corresponds to an interaction of hard spheres with a radius which is by a factor of about 1.5 larger than the total radius of the composite particles. Surprisingly, no evidence for an anisotropic structure factor was found which implies a dense packing of the particles rather than formation of chains as predicted for dipole interactions.

The application of the combined contrast variation technique will enable a more detailed insight in the nature of more complicated core-shell structures, which are generally not accessible from other microscopic methods. This knowledge is a necessary pre-requisite for potential biomedical applications of ferrofluids, where immunoassays or therapeutic agents should be connected to the surfactant shell.

Table 2. Structural parameters resulting from SANSPOL investigations in Ferrofluids with different core and surfactant materials. Average radii $<R>$, shell thickness D and volume fractions f of core in composites, aggregates (agg) and micelles (mic). f_{magn} denotes the total volume fraction of magnetic particles as obtained from magnetization measurements.

Sample	Shell	f_{magn} vol %	$<R_{core}>$ [nm]	D_{shell} [nm]	f_{core} vol %	$<R_{Agg}>$ [nm]	f_{Agg} vol %	$<R_{Mic}>$ [nm]	f_{Mic} vol %
Co-	Single-layer	0.5	3.7	2.4	0.3	–	–	–	
Oleyl Sarkosyl	$C_{21}H_{39}NO_3$	1	3.7	2.2	0.6	–	–	1.8	2
in Toluol		5	3.7	2.2	3.2	$R_{hc}=8.7$	–	1.8	5
Fe_3O_4 DEX in $H_{0.2}D_{1.8}O$	Dextran entangled single layer	1	4.8	2.4		16.8			
Fe_3O_4 LM in	Double layer								
$H_{0.2}D_{1.8}O$	$C_{12}H_{25}COOH$	1	6.62		1.06	13.4	0.14	3.05	0.5
$H_{0.4}D_{1.6}O$	/$C_{30}H_{62}O_{10}$	1	6.7	2.7	0.24	17.8	0.07	2.62	0.7
$H_{0.6}D_{1.4}O$		2	6.7	2.7	1.04	22.8	0.32	2.62	1.2
HDO		1	6.7	2.7	0.26	12.3	0.28	2.62	1
Ba-Hexaferrit in $C_{12}H_{26}$	Single layer $C_{17}H_{33}COOH$								
H		3.6	4.1	1.8	2.3	9.8	1	0.8	5.8
H:D = 67:33		3.9	4.1	1.8	2.6	9.8	1	0.8	8.7
H:D = 36:64		2.5	4.1	1.8	1.5	9.8	1	0.8	5.8

Acknowledgements

The author acknowledges the fruitful cooperation with his colleagues Dr. A. Heinemann, Dr. A. Hoell, M. Kammel, and Dr. U. Keiderling of the SANS team at HMI Berlin. Many thanks are given for sample preparation to Dr. N. Buske and Dr. M. Gansau from Berlin Heart AG, and to Dr. R. Müller from IPHT Jena, Germany.

This work is supported by the German DFG priority program "Colloidal magnetic fluids", Grant No WI-1151/2.

References

1. J. Loeffler, W. Wagner, H. Van Swygenhoven, & A. Wiedenmann, Nanostruct. Mat., **9** (1997), 331-334.
2. A. Wiedenmann , J. Appl. Cryst., **30** (1997), 580-585.
3. J. F. Loeffler, H. B. Braun, W. Wagner, G. Kostorz, A. Wiedenmann Materials Science and Engineering A, **304-306, 31** (2001), 1050-1054.
4. A. Michels,, J. Weissmüller, A. Wiedenmann, J.S. Pedersen, J.G. Barker Philosoph. Magazine Letters **80** (2000), 785-792.
5. A. Wiedenmann, U. Lembke, A Hoell,., R. Müller, & W. Schüppel Nanostruct. Mat., **12** (1999), 601-604.
6. J. Kohlbrecher, A. Wiedenmann, H. Wollenberger, Z. Physik, **104**(1997),1-4.
7. A. Heinemann, H. Hermann, A. Wiedenmann, N. Mattern, K. Wetzig J. Applied Cryst. **33**(2000),1386-1392.
8. S.W. Charles, JMMM (2002).
9. P.P.Gravina et al., JMMM (2002).
10. E. Dubois et.al, Langmuir **13**(2000),5617.
11. G. A.van Ewijk, PhD Thesis Utrecht (2001).
12. B. Huke et al., Phys. Rev. **E 62**(2000),6875.
13. S. Odenbach, K.Raj, Magnetohydrodynamics **36**(2000),379-386.
14. B. U. Felderhof Phys. Rev. **E 62**(2000),3848.
15. J.P.Embs et al. Phys. Rev. **E**(2001), preprint
16. E. Dubois, V. Cabuil, F. Boué, R. Perzyski, J.Phys. Chem. **111**(1999),7147-7160.
17. F. Cousin, V. Cabuil, Prog. Colloid Polym. Sci. 2000),77-83.
18. G. Kostorz in Treatease on Materials Science and Technology, Acad. Press (1979) ed. G. Kostorz, 227-290.
19. J.K. Percus, G.J. Yevick Phys. Rev **110**(1959),1-13.
20. J. Brunner-Popela, O. Glatter J. Appl. Cryst **30**(1997),431-442.
21. J.S. Pedersen J. Appl. Cryst. **27**(1994),595-608.
22. R.M. Moon, T. Riste, & W.C. Koehler, Phys. Rev.(1969) **181**, 920-931.
23. R. Pynn, &J. B.Hayter. Phys. Rev. Letters , **51**, 710-713.
24. A. Wiedenmann, Mat. Science Forum, **312-314**(1999) 315. J. Metastable and Nanocrystalline Materials, 2-6 (1999), 315-324.
25. A. Wiedenmann, Physica B 297 (2001), 226-233.
26. R. P. May and K. Ibel and J. Haas, J. Appl. Cryst., **15** (1982), 15-19.
27. R. Stegmann, E. Manakova, , M. Rössle, H. Heumann, S. Axmann, A. Plückthun, T. Hermann, R. May, A. Wiedenmann. J. Structural Biology **121**(1998),30-30.
28. A. Wiedenmann, J. Appl. Cryst. 33 (2000), 428-432.
29. U. Keiderling & A. Wiedenmann, A. Physica B, **213&214**(1995). 895-897.

30. T. Keller, T. Krist, A. Danzig, U. Keiderling, F. Mezei, A. Wiedenmann, J. Nuclear Instruments **A451**(2000), 474-479.
31. R. Müller et al, J. Magn. Magn. Mat. **201**(1999), 34.
32. A. Hoell, A. Wiedenmann, U. Lembke, Physica **B 276-278**(2000), 886-887.
33. A. Wiedenmann, Magnetohydrodynamics **37**(2001),236-242.
34. A. Wiedenmann, A. Hoell, M. Kammel, JMMM (2002) in press.
35. J. B.Hayter. J. Appl. Cryst., **21**(1988), 737-742.
36. M. Kammel, A. Hoell, A. Wiedenmann, Scripta Materialia **44**(2001), 2341-2345.
37. M. Kammel, A. Wiedenmann, A. Hoell, JMMM (2002) in press.
38. A. Heinemann, Tatchev, A. Hoell, A. Wiedenmann J.Appl. Cryst.to be published
39. A. Hoell, M. Kammel, A. Wiedenmann, JMMM (2002) in press
40. A. Hoell, R. Müller, A. Wiedenmann, W. Gawalek, JMMM (2002) in press.
41. N. Buske, DE patent .197 58 **350** (1997)

Part II

Basic Theory

Basic Equations for Magnetic Fluids with Internal Rotations

Ronald E. Rosensweig

Consultant, Summit, NJ (U.S.A.),
Former Blaise Pascal Professor, ENS Fondation, Paris

Abstract. Several authors have attempted with varying success to derive a complete set of basic equations for magnetic fluids having internal rotations. In this work a complete set of governing equations is derived on the basis of dynamic balance relationships with the dissipation function derived from thermodynamic consideration. The magnetization relaxation equation is thereby determined from requirement of positive entropy production along with a complete set of well-accepted constitutive laws. The analysis employs the Minkowski expression of electromagnetic momentum and assumes that the product of Maxwell stress and velocity for polarized media contributes to the energy balance on the same footing as contact stresses of pressure and viscous origin.

1 Introduction

Magnetic fluids, also known as ferrofluids, are stable suspensions of subdomain ferrimagnetic or ferromagnetic fine particles in a liquid carrier. Early theory of magnetic fluids assumed that the magnetization is colinear with the magnetic field at all times [1]. More generally, the magnetic particles lag the changing field and the description of the flow becomes more complex. The work of Dahler, Scriven, and Condiff [2] exposed the role of asymmetric stress in this circumstance, although the relaxation equation was postulated rather than derived. Independently the work of Shliomis [3] has been influential in the development of the science, albeit the relaxation law is derived separately [4] from the main theory [5].

Irreversible thermodynamics provides a framework for the consistent description of dissipative processes. In their well-known monograph deGroot and Mazur [6] arrive at a plausible expression for entropy production of a polarized medium only after introducing an artificial definition of electromagnetic energy density. Also, their work did not treat the case when the direction of the magnetization differs from the direction of the field. Felderhof and Kroh [7] building on the DeGroot and Mazur treatment present a theory that yields a plausible form of the relaxation relationship when magnetization and field are not colinear. Yet the derivation is not transparent to us for the reason cited, and is incomplete; e.g., the asymmetric hydrodynamic stress tensor is postulated rather than deduced.

A main objective of the present work is to circumvent the infelicities cited above while employing the irreversible thermodynamic framework. We utilize with modifications the formulation of Shizawa and Tanahashi [8] with its methodology based on integral formulation of equations governing the balance of linear momentum, angular momentum, and energy. In the latter the electromagnetic work is formulated as the product of electromagnetic stress and the velocity of the fluid. So far as we know, it is the only prior work that attempts that formulation. However, their analysis produced an unsatisfactory result for magnetization relaxation. In the present work this deficiency is overcome using a strict formulation of entropy production rate

In addition, a recent criticism has arisen concerning the mentioned use of the generalized Maxwell stress tensor in formulating the balance of energy [9]. The present work shows that incorporation of the stress in the energy formulation in fact leads deductively to a self-consistent exposition yielding a complete set of governing equations and constitutive relationships including the Shliomis relaxation equation. Furthermore, there has been uncertainty concerning whether the Minkowski expression $\boldsymbol{D} \times \boldsymbol{B}$ for electromagnetic momentum or the Abraham expression $\epsilon_0 \mu_0 \boldsymbol{E} \times \boldsymbol{H}$ is the appropriate one. The present work is based on the Minkowski expression, and helps validate its use.

In this work we essentially trace the development in [8] up to the point of formulating entropy change in the fluid dynamic system. A detailed exposition is employed in developing the well-known balances to set the pattern for consistent treatment of unfamiliar balances and the work treats conductive magnetic fluid. The SI system of units is employed throughout, and the conventions employed in the dyadic notation are defined in the Appendix.

2 Equations of Electromagnetic Field

The well-known system of equations governing the electromagnetic field in the presence of matter is the set of four Maxwell equations.

$$\frac{\partial \boldsymbol{B}}{\partial t} = \nabla \times \boldsymbol{E} \tag{1}$$

$$\frac{\partial \boldsymbol{D}}{\partial t} + \boldsymbol{j} = \nabla \times \boldsymbol{H} \tag{2}$$

$$\nabla \cdot \boldsymbol{B} = 0 \tag{3}$$

$$\nabla \cdot \boldsymbol{D} = \rho_e \tag{4}$$

3 Law of Conservation of Mass

A constant mass of fluid in a flow field occupies a volume V_m that deforms with time. Expressed in integral form this may be written as

$$\frac{\mathrm{d}}{\mathrm{d}t} \int_{V_m} \rho \mathrm{d}V = 0 \tag{5}$$

where $\frac{d}{dt} = \frac{\partial}{\partial t} + \mathbf{v} \cdot \nabla$ is the material derivative and ϱ is the mass density. The Reynolds transport theorem for any function Ψ of position and time (scalar, vector or tensor component) is the kinematic relationship having the following visualizable form:

$$\frac{d}{dt}\int_V \Psi dV = \int_V \frac{\partial \Psi}{\partial t} dV + \oint_S (\boldsymbol{n} \cdot \boldsymbol{v}\Psi) dS \qquad (6)$$

where **n** is the unit outward facing normal to the surface. In words, the total time rate of change equals the time rate of change within the volume plus the net convection out of the volume through the surrounding surface. Applying the Gauss divergence theorem to the surface integral term transforms it to a volume integral yielding the transport theorem in alternate form as

$$\frac{d}{dt}\int_{V_m} \Psi dV = \int_{V_m} \left[\frac{\partial \Psi}{\partial t} + \nabla \cdot (\boldsymbol{v}\Psi)\right] dV \qquad (7)$$

The theorem applies with Ψ a vector or a tensor as may be demonstrated by summing the components.

Identifying Ψ with ϱ and invoking the arbitrariness of the volume yields the well-known equation of continuity in differential form.

$$\frac{\partial \rho}{\partial t} + \nabla \cdot (\rho \boldsymbol{v}) = 0 \qquad (8)$$

This transforms easily to a corollary form

$$\frac{1}{\rho}\frac{d\rho}{dt} = -\nabla \cdot \boldsymbol{v} \qquad (9)$$

A corollary of the Reynolds transport theorem is obtained from Eq. (7), substituting $\rho\Psi$ and employing the continuity equation in the form of Eq. (9). This yields

$$\frac{d}{dt}\int_V \rho\Psi dV = \int_V \rho\frac{d\Psi}{dt} dV \qquad (10)$$

4 Balance Equation of Linear Momentum

The continuum extension of Newton's fundamental law states that a change in momentum of a fixed mass system with respect to time is equal to a sum of surface and body forces. Defining the system as the contents of a volume V_m enclosed by surface S_m the balance equation of momentum is written in the form

$$\left(\frac{d}{dt}\right)\int_{V_m} (\rho\boldsymbol{v} + \boldsymbol{g}) dV = \int_{S_m} \boldsymbol{t}_n dS + \int_{V_m} \rho\boldsymbol{b} dV \qquad (11)$$

where $\boldsymbol{g}$ is the electromagnetic momentum vector, $\boldsymbol{t}_n$ the surface stress vector representing the combination of pressure, viscous and electromagnetic effects, and $\boldsymbol{b}$ body force density excluding electromagnetic effects. A definite expression for $\boldsymbol{g}$ is not required until a later point in the development. Strictly speaking we consider that any apparent body force arises due to transmission via field of surface stresses from afar. However, when the presence of matter in the system perturbs the field to a negligible extent, it is convenient to formulate body force as $\boldsymbol{b}$ in the above. The body force of gravity furnishes an example of $\boldsymbol{b}$ for problems we consider.

From the Reynolds transport theorem of Eq. (10),

$$\left(\frac{d}{dt}\right) \int_{V_m} (\rho \boldsymbol{v}) dV = \int_{V_m} \rho \frac{d\boldsymbol{v}}{dt} dV \qquad (12)$$

From Eq. (7) with $\Phi = \boldsymbol{g}$,

$$\left(\frac{d}{dt}\right) \int_{V_m} \boldsymbol{g} dV = \int_{V_m} \left[\frac{\partial \boldsymbol{g}}{\partial t} + \nabla \cdot (\boldsymbol{v}\boldsymbol{g})\right] dV \qquad (13)$$

Next, rewriting the first term of the right side of Eq. (11) using Cauchy's fundamental theorem: $\boldsymbol{t}_n = \boldsymbol{n} \cdot \tilde{\mathbf{T}}$ where $\tilde{\mathbf{T}}$ is the total stress vector (sum of viscous, pressure, and electromagnetic components), transforming the surface integral into a volume integral by applying Gauss' theorem, collecting terms and noting that the volume of integration is arbitrary, the equation of motion in Cauchy (unconstituted) form is obtained.

$$\rho \frac{d\boldsymbol{v}}{dt} + \frac{\partial \boldsymbol{g}}{\partial t} = \nabla \cdot \mathbf{T} + \rho \boldsymbol{b} \qquad (14)$$

where

$$\mathbf{T} \equiv \tilde{\mathbf{T}} - \boldsymbol{v}\boldsymbol{g} \qquad (15)$$

5 Balance Equation of Angular Momentum

Fundamental law for angular momentum states that a change of angular momentum of a system with respect to time is equal to a sum of moments due to surface and body forces. In integral form the balance equation of angular momentum is expressed as follows:

$$\left(\frac{d}{dt}\right) \int_{V_m} [\boldsymbol{r} \times (\rho \boldsymbol{v} + \boldsymbol{g}) + \rho \boldsymbol{s}] dS = \int_{S_m} (\boldsymbol{r} \times \boldsymbol{t}_n + \boldsymbol{\lambda}_n) dS + \int_{V_m} (\boldsymbol{r} \times \rho \boldsymbol{b} + \rho \boldsymbol{l}) dV \qquad (16)$$

where $\boldsymbol{r}$ is the position vector of a fluid element relative to an arbitrary fixed origin, $\boldsymbol{s}$ angular momentum density (spin) per unit mass, $\boldsymbol{\lambda}_n$ surface couple density, and $\boldsymbol{l}$ volume couple density transmitted from afar. The remarks above concerning $\boldsymbol{b}$ apply as well to $\boldsymbol{l}$.

The various terms of Eq. (16) transform as follows:

$$\left(\frac{d}{dt}\right) \int_{V_m} r \times \rho v \, dV = \int_{V_m} \rho \frac{d}{dt}(r \times v) \, dV$$

$$= \int_{V_m} \rho \left(\frac{dr}{dt} \times v + r \times \frac{dv}{dt}\right) dV$$

$$= \int_{V_m} \left(r \times \rho \frac{dv}{dt}\right) dV \qquad (17)$$

where the first equality results from the Reynolds corollary of Eq. (10), and the last equality from the circumstance that $dr/dt = v$.

Multiplying and dividing the integrand of the next term in Eq. (16) by ϱ, using the Reynolds transport theorem of Eq. (10) and then the equation of continuity of Eq. (9), and noting that $dr/dt = v$, the term transforms as follows.

$$\left(\frac{d}{dt}\right) \int_{V_m} (r \times g) \, dV = \int_{V_m} \rho \frac{d}{dt}\left(r \times \frac{g}{\rho}\right) dV$$

$$= \int_{V_m} \left[(v \times g) + r \times \left(\frac{dg}{dt} + g \nabla \cdot v\right)\right] dV$$

$$= \int_{V_m} \left[-\epsilon : vg + r \times \left\{\frac{\partial g}{\partial t} + \nabla \cdot (vg)\right\}\right] \qquad (18)$$

Using the Reynolds transport theorem of Eq. (10) the spin term on the left side of Eq. (16) becomes

$$\frac{d}{dt} \int_{V_m} \rho s \, dV = \int_{V_m} \rho \frac{ds}{dt} dV \qquad (19)$$

Using the Cauchy theorem $t_n = n \cdot \tilde{T}$ and Gauss' divergence theorem the first term on the right side of Eq. (16) transforms as follows.

$$\int_{S_m} (r \times t_n) \, dS = -\int_{S_m} \left(n \cdot \tilde{T} \times r\right) dS$$

$$= -\int_{V_m} \nabla \cdot \left(\tilde{T} \times r\right) dV \qquad (20)$$

Introducing the identity $-\nabla \cdot \left(\tilde{T} \times r\right) = r \times \left(\nabla \cdot \tilde{T}\right) - \epsilon : \tilde{T}$, see proof in [16, p. 247], Eq. (20) becomes

$$\int_{S_m} (r \times t_n) \, dS = \int_{V_m} \left[r \times \left(\nabla \cdot \tilde{T}\right) - \epsilon : T\right] dV \qquad (21)$$

From the definition of the couple stress vector, $\boldsymbol{\lambda}_n = \boldsymbol{n} \cdot \boldsymbol{\Lambda}$, where $\boldsymbol{\Lambda}$ is the couple stress tensor, and thus

$$\int_{S_m} \boldsymbol{\lambda}_n \mathrm{d}S = \int_{S_m} (\boldsymbol{n} \cdot \boldsymbol{\Lambda}) \mathrm{d}S$$
$$= \int_{V_m} (\nabla \cdot \boldsymbol{\Lambda}) \mathrm{d}V \qquad (22)$$

Substituting the collection of transformed terms into the angular momentum balance of Eq. (16) yields

$$\int_{V_m} \left[\rho \frac{\mathrm{d}\boldsymbol{s}}{\mathrm{d}t} + \boldsymbol{r} \times \left\{ \rho \frac{\mathrm{d}\boldsymbol{v}}{\mathrm{d}t} + \frac{\partial \boldsymbol{g}}{\partial t} - \nabla \cdot \left(\tilde{\mathsf{T}} - \boldsymbol{v}\boldsymbol{g} \right) - \rho \boldsymbol{b} \right\} \right] \mathrm{d}V$$
$$= \int_{V_m} \left[\rho \boldsymbol{l} + \nabla \cdot \boldsymbol{\Lambda} - \boldsymbol{\epsilon} : \left(\tilde{\mathsf{T}} - \boldsymbol{v}\boldsymbol{g} \right) \right] \mathrm{d}V \qquad (23)$$

Substituting the definition of $\mathbf{T}$ from Eq. (15), noting that the bracketed term in the cross product with the radius vector disappears by virtue of the equation of motion of Eq. (14), and invoking the arbitrariness of the volume element yields the balance equation of internal angular momentum.

$$\rho \frac{\mathrm{d}\boldsymbol{s}}{\mathrm{d}t} = \rho \boldsymbol{l} + \nabla \cdot \boldsymbol{\Lambda} - \boldsymbol{\epsilon} : \mathbf{T} \qquad (24)$$

For a monodispersion of spherical particles spin can be expressed as $\boldsymbol{s} = I\boldsymbol{\Omega}/\varrho$ so that spin rate is given by

$$\rho \frac{\mathrm{d}\boldsymbol{s}}{\mathrm{d}t} = I \frac{\mathrm{d}\boldsymbol{\Omega}}{\mathrm{d}t} \qquad (25)$$

where I is moment of inertia per unit mass and $\boldsymbol{\Omega}$ is average angular velocity of particles about their own center.

6 Equilibrium Thermodynamic Relationships

Thermodynamics deals with changes in states of matter, which for polarized matter depend on $\boldsymbol{P}$ and $\boldsymbol{M}$, the electric and magnetic polarizations. The polarizations are defined in terms of the field variables appearing in Maxwell's equations.

$$\boldsymbol{D} = \epsilon_0 \boldsymbol{E} + \boldsymbol{P} \qquad (26)$$

and

$$\boldsymbol{B} = \mu_0 \boldsymbol{H} + \boldsymbol{M} \qquad (27)$$

Here we have chosen to absorb the factor of μ_0 into the $\boldsymbol{M}$, a deviation from standard SI nomenclature, so that $\boldsymbol{M}$ and $\boldsymbol{P}$ appear on a symmetrical basis and $\boldsymbol{M}$ is expressed in units of tesla. The volumetric density of polarization work is given by $\boldsymbol{E} \cdot \mathrm{d}\boldsymbol{D}$ and $\boldsymbol{H} \cdot \mathrm{d}\boldsymbol{B}$ [10]. The combined first and second law expression (Gibbs equation) written for the contents of a fixed unit volume of space and incorporating these work terms is absent a pressure term and, in its place contains a term proportional to the chemical potential of the medium, considered to be of constant composition. Later in the present study the Gibbs equation is required in a form in which internal energy and entropy are expressed on the unit mass basis. Conversion from the unit volume basis to unit mass basis yields the following expression [11]

$$\mathrm{d}u^{(e)} = T\mathrm{d}s^{(e)} + p^{(e)}\mathrm{d}\rho/\rho^2 + \boldsymbol{E}^{(e)} \cdot \mathrm{d}\left(\boldsymbol{D}^{(e)}/\rho\right) + \boldsymbol{H}^{(e)} \cdot \mathrm{d}\left(\boldsymbol{B}^{(e)}/\rho\right) \quad (28)$$

where superscript (e) denotes equilibrium value. $u^{(e)}$ is internal energy per unit mass and $s^{(e)}$ entropy per unit mass. The same expression is derived by Chu [12] based on a roundabout treatment, and adopted by Tarapov [13] and Blums et al. [14]; our development in [11] shows that the result follows from classical thermodynamics alone. The $p^{(e)}$ denotes the value of pressure in the presence of the fields and it differs from the pressure $p_0(\varrho, T)$ that attains in the absence of fields.

The analysis of [11] relates the two pressures as follows:

$$p^{(e)} = p_0(\rho, T) + \left(\epsilon_0 E^{(e)2} + \mu_0 H^{(e)2}\right)/2$$
$$+ \int_0^E (\partial v P/\partial v)^{(e)}_{T,H,E} \mathrm{d}E^{(e)} + \int_0^H (\partial v M/\partial v)^{(e)}_{T,H,E} \mathrm{d}H^{(e)}$$
$$\equiv p + \left(\epsilon_0 E^{(e)2} + \mu_0 H^{(e)2}\right)/2 \quad (29)$$

where $p_0(\varrho, T)$ is ordinary pressure and $v = 1/\rho$ is specific volume. Naturally, when $E^{(e)}$ and $H^{(e)}$ are absent, $p^{(e)}$ reduces to $p_0(\varrho, T)$. Also, from Eq. (29), the pressure-like variable p is given by

$$p = p_0(\rho, T) + \int_0^E (\partial v P/\partial v)^{(e)}_{T,H,E} \mathrm{d}E^{(e)} + \int_0^H (\partial v M/\partial v)^{(e)}_{T,H,E} \mathrm{d}H^{(e)} \quad (30)$$

Substituting for $p^{(e)}$ from Eq. (29) into Eq. (28) and expanding the electromagnetic work terms yields the following form of the Gibbs equation after rearrangement and substitution from Eqs. (26,27).

$$\rho \mathrm{d}u^{(e)} = \rho T \mathrm{d}s^{(e)}$$
$$+ \left(p - \epsilon_0 E^{(e)2}/2 - \mu_0 H^{(e)2}/2 - \boldsymbol{P}^{(e)} \cdot \boldsymbol{E}^{(e)} - \boldsymbol{M}^{(e)} \cdot \boldsymbol{H}^{(e)}\right) \mathrm{d}\rho/\rho$$
$$+ \boldsymbol{E}^{(e)} + \boldsymbol{H}^{(e)} \cdot \mathrm{d}\boldsymbol{B}^{(e)} \quad (31)$$

This form of the Gibbs equation with each term expressed on the per-unit-volume basis is employed subsequently below.

7 Balance Equation of Energy

The balance equation of energy is derived from the following integral equation expressing time rate of change of total energy in a system of fixed mass and variable volume as the sum of work done on the system and energy supplied to the system.

$$\left(\frac{d}{dt}\right) \int_{V_m} \rho \left(u + v^2/2 + I\Omega^2/2\right) dV = \int_{S_m} \boldsymbol{t}_n \cdot \boldsymbol{v} dS + \int_{V_m} \rho \boldsymbol{b} \cdot \boldsymbol{v} dV$$
$$+ \int_{S_m} \boldsymbol{\lambda}_n \cdot \boldsymbol{\Omega} dS + \int_{V_m} \rho \boldsymbol{l} \cdot \boldsymbol{\Omega} dV - \int_{S_m} \boldsymbol{n} \cdot \boldsymbol{q} dS$$
$$- \int_{S_m} \boldsymbol{n} \cdot (\boldsymbol{E}^* \times \boldsymbol{H}^*) dS + \int_{V_m} \rho R dV \quad (32)$$

The u is internal energy per unit mass, q is the heat flux vector and R is volumetric heat release rate, e.g., due to radioactive decay. The integrand of the first term on the right side treats all components (viscous, pressure and electromagnetic) of the stress vector on equal basis in expressing the work rate as the scalar product with the velocity vector [1].

$$\boldsymbol{t}_n = \boldsymbol{n} \cdot \tilde{\mathsf{T}} \quad (33)$$

$$\mathsf{T} = \tilde{\mathsf{T}} - \boldsymbol{v}\boldsymbol{g} \quad (34)$$

The integrand of the sixth term on the right side is the Poynting vector expressing electromagnetic energy flux density. Since the integration is over a closed material surface, the Poynting vector is evaluated in the moving medium as indicated by the asterisks suffixed to the vector fields. The electric field vector $\boldsymbol{E}^*$ and magnetic field vector $\boldsymbol{H}^*$ are related to the components of vectors $\boldsymbol{E}$ and $\boldsymbol{H}$ observed in a stationary coordinate system as follows [10].

$$\boldsymbol{E}^* = \boldsymbol{E}_{//} + \gamma(\boldsymbol{E}_\perp + \boldsymbol{v} \times \boldsymbol{B}) \quad (35)$$

$$\boldsymbol{H}^* = \boldsymbol{H}_{//} + \gamma(\boldsymbol{H}_\perp - \boldsymbol{v} \times \boldsymbol{D}) \quad (36)$$

where

$$\gamma \left[1 - (v/c)^2\right]^{-1/2} \quad (37)$$

In Eqs. (35,36), $\boldsymbol{v} \times \boldsymbol{B}$ and $\boldsymbol{v} \times \boldsymbol{D}$ are respectively the apparent electric field and magnetic field produced when an observer crosses $\boldsymbol{B}$ and $\boldsymbol{D}$ with velocity $\boldsymbol{v}$. Since the speed v of the suspensions can be assumed much smaller than light

---

[1] Inclusion of the electromagnetic stress in the energy balance in this manner has been questioned recently as unphysical [9]. Our thesis is, because Maxwellian stress applies in linear and angular momentum balances, for consistency it should apply in the energy balance an well, on the same footing as contact stresses of pressure and viscosity.

speed c, $(v/c)^2$ may be neglected in the equations. As a result, the coefficient γ of the transformation (Lorentz transformation) can be set to unit and hence[2]

$$E^* = E + v \times B, \qquad H^* = H - v \times D \qquad (38)$$

Applying Reynolds' transport theorem of Eq. (10) to the left side of Eq. (32), rewriting the first and third terms of the right side with Cauchy's fundamental theorem, transforming them into volume integral form together with the other surface integral forms, and then noting that the volume of integration is arbitrary yields

$$\rho \frac{du}{dt} + \left(\rho \frac{dv}{dt} - \nabla \cdot \tilde{T} - \rho b \right) \cdot v + \left(\rho I \frac{d\Omega}{dt} - \nabla \cdot \Lambda - \rho I \right) \cdot \Omega$$
$$= \tilde{T}^T : \nabla v + \lambda^T : \nabla \Omega - \nabla \cdot (E^* \times H^*) - \nabla \cdot q + \rho r \qquad (39)$$

where the identity $\nabla \cdot \left(\tilde{T} \cdot v \right) = \left(\nabla \cdot \tilde{T} \right) \cdot v + \tilde{T}^T : \nabla v$ is employed with superscript T denoting transpose. The Poynting term can be expressed in terms of variables of the stationary coordinate system by the use of Eq. (38). This yields the following representation:

$$-\nabla \cdot (E^* \times H^*) = -\nabla \cdot (E \times H) - [E \cdot \{\nabla \times (v \times D)\} + H \cdot \{\nabla \times (v \times B)\}]$$
$$+ [(\nabla \times E) \cdot (v \times D) + (\nabla \times H) \cdot (v \times B)]$$
$$+ \nabla \cdot [(v \times B) \times (v \times D)] \qquad (40)$$

Using the Maxwell relationships of Eqs (1) and (2) the third term of the right side of Eq. (40) transforms as follows.

$$[(\nabla \times E) \cdot (v \times D) + (\nabla \times H) \cdot (v \times B)]$$
$$= -\left(\frac{\partial g}{\partial t} \right) \cdot v + j^* \cdot (v \times B) \qquad (41)$$

where

$$j^* = j - \rho_e v \qquad (42)$$
$$g = D \times B \qquad (43)$$

j^* is electric current density as observed in a moving coordinate system for the case $\gamma = 1$, and g is electromagnetic energy density[3]. The second term on the right hand side of Eq. (42) is introduced permissably as its scalar product with $(v \times B)$ in Eq. (41) vanishes. The last term in Eq. (40) transforms as follows by vector identity and the definition of g:

$$\nabla \cdot [(v \times B) \times (v \times D)] = v \cdot (B \times D) \nabla \cdot v + v \cdot \nabla (v \cdot B \times D)$$
$$= -(v \cdot g) \nabla \cdot v - v \cdot \nabla (v \cdot g)$$
$$= -(v \cdot g) \nabla \cdot v - v \cdot (v \cdot \nabla g) - v \cdot (g \cdot \nabla v)$$
$$= -[\nabla \cdot (vg)] \cdot v - (vg : \nabla v) \qquad (44)$$

---
[2] A Galilean derivation of these relationships and Eq.(42) is given in Appendix B
[3] This choice of g agrees with the Minowski definition but differs from the Abraham definition which is $\epsilon_0 \mu_0 E \times H$. See references in [7]

By substituting Eqs. (44) and (41) into Eq. (40) and substituting the result into Eq. (39), taking account of the linear momentum balance of Eq. (14) and the internal angular momentum balance of Eq. (24), and the definition of $\tilde{\mathbf{T}}$ from Eq. (15)the balance of energy relationship is expressed as follows:

$$\rho \frac{du}{dt} = \mathbf{T}^T : \nabla \mathbf{v} + \mathbf{\Lambda}^T : \nabla \mathbf{\Omega} - \mathbf{T} : \boldsymbol{\epsilon} \cdot \mathbf{\Omega} - \nabla \cdot (\mathbf{E} \times \mathbf{H})$$
$$- [\mathbf{E} \cdot \{\nabla \times (\mathbf{v} \times \mathbf{D})\} + \mathbf{H} \cdot \{\nabla \times (\mathbf{v} \times \mathbf{B})\}]$$
$$+ \mathbf{j}^* \cdot (\mathbf{v} \times \mathbf{B}) - \nabla \cdot \mathbf{q} + \rho r \qquad (45)$$

Unknown parameters at this stage are stress tensor $\mathbf{T}$, couple stress tensor $\mathbf{\Lambda}$, polarizations $\mathbf{M}$ and $\mathbf{P}$, electric current density vector $\mathbf{j}^*$ and heat flux density vector $\mathbf{q}$.

Eq. (45) for internal energy is not yet in the form that is needed as the Poynting term in the equation requires further transformation. From the mathematical identity for a vector triple product we have

$$\nabla \times (\mathbf{v} \times \mathbf{B}) = (\mathbf{B} \cdot \nabla) \mathbf{v} - \mathbf{B} (\nabla \cdot \mathbf{v}) - (\mathbf{v} \cdot \nabla) \mathbf{B} \qquad (46)$$

and

$$\nabla \times (\mathbf{v} \times \mathbf{D}) = (\mathbf{D} \cdot \nabla) \mathbf{v} - \mathbf{D} (\nabla \cdot \mathbf{v}) - (\mathbf{v} \cdot \nabla) \mathbf{D} + \rho_e \mathbf{v} \qquad (47)$$

where use is made of Eqs. (3) and (4). Next the left sides of Eqs. (1) and (2) are rewritten in the material time derivative form by adding $(\mathbf{v} \cdot \nabla)\mathbf{B}$ and $(\mathbf{v} \cdot \nabla)\mathbf{D}$ to both sides of the equations, respectively. Then using Eqs. (46) and (47) the following equations are obtained.

$$\frac{d\mathbf{B}}{dt} = -\nabla \times \mathbf{E} + (\mathbf{B} \cdot \nabla) \mathbf{v} - \mathbf{B} (\nabla \cdot \mathbf{v}) - \nabla \times (\mathbf{v} \times \mathbf{B}) \qquad (48)$$

$$\frac{d\mathbf{D}}{dt} = -\mathbf{j} + \nabla \times \mathbf{H} + \rho_e \mathbf{v} + (\mathbf{D} \cdot \nabla) \mathbf{v} - \mathbf{D} (\nabla \cdot \mathbf{v}) - \nabla \times (\mathbf{v} \times \mathbf{D}) \qquad (49)$$

Scalar multiplication of both sides of (48) and (49) by $\mathbf{H}$ and $\mathbf{E}$ respectively, then adding the equations and rearranging, yields the desired expression for the divergence of the Poynting vector.

$$\nabla \cdot (\mathbf{E} \times \mathbf{H}) = -\mathbf{j}^* \cdot \mathbf{E} - \mathbf{E} \cdot \frac{d\mathbf{D}}{dt} - \mathbf{H} \cdot \frac{d\mathbf{B}}{dt}$$
$$+ \{\mathbf{E}\mathbf{D} + \mathbf{H}\mathbf{B} - (\mathbf{D} \cdot \mathbf{E} + \mathbf{B} \cdot \mathbf{H})\mathbf{I}\} : \nabla \mathbf{v}$$
$$- [\mathbf{E} \cdot \{\nabla \times (\mathbf{v} \times \mathbf{D})\} + \mathbf{H} \cdot \{\nabla \times (\mathbf{v} \times \mathbf{B})\}] \qquad (50)$$

Substituting this expression into Eq. (45) yields the energy balance in the following form.

$$\rho \frac{du}{dt} = \mathbf{T}^T : \nabla \mathbf{v} + \mathbf{\Lambda}^T : \nabla \mathbf{\Omega} - \mathbf{T} : \boldsymbol{\epsilon} \cdot \mathbf{\Omega}$$
$$- [\mathbf{E}\mathbf{D} + \mathbf{H}\mathbf{B} - (\mathbf{D} \cdot \mathbf{E} + \mathbf{B} \cdot \mathbf{H})\mathbf{I}] : \nabla \mathbf{v}$$
$$+ \mathbf{E} \cdot \frac{d\mathbf{D}}{dt} + \mathbf{H} \cdot \frac{d\mathbf{B}}{dt} + \mathbf{j}^* \cdot \mathbf{E}^* - \nabla \cdot \mathbf{q} + \rho r \qquad (51)$$

where, again
$$\boldsymbol{E}^* = \boldsymbol{E} + \boldsymbol{v} \times \boldsymbol{B} \tag{52}$$

Substituting the definitions of **D** and **B** from Eqs. (26,27) into the electromagnetic work terms of Eq. (51) after minor algebraic transformation yields the energy balance in final form.

$$\begin{aligned}
\rho \frac{du_m}{dt} =\ & \boldsymbol{\mathsf{T}}^T : \nabla \boldsymbol{v} + \boldsymbol{\Lambda}^T : \nabla \boldsymbol{\Omega} - \boldsymbol{T} : \boldsymbol{\epsilon} \cdot \boldsymbol{\Omega} \\
& - [\boldsymbol{E}\boldsymbol{D} + \boldsymbol{H}\boldsymbol{B} - (\boldsymbol{D}\cdot\boldsymbol{E} + \boldsymbol{B}\cdot\boldsymbol{H})\boldsymbol{\mathsf{I}}] : \nabla \boldsymbol{v} \\
& + \boldsymbol{E}\cdot\frac{d\boldsymbol{P}}{dt} + \boldsymbol{H}\cdot\frac{d\boldsymbol{M}}{dt} + \boldsymbol{j}^*\cdot\boldsymbol{E}^* - \nabla\cdot\boldsymbol{q} \\
& + \rho R - \left(\frac{\mu_0 H^2}{2} + \frac{\epsilon_0 E^2}{2}\right)(\nabla\cdot\boldsymbol{v})
\end{aligned} \tag{53}$$

where
$$u_m \equiv u - \frac{\epsilon_0 E^2}{2\rho} - \frac{\mu_0 H^2}{2\rho} \tag{54}$$

The u_m of Eq.(54) represents internal energy of the medium exclusive of field energy associated with the space the medium occupies.

8 Determination of Entropy Production Rate

Dividing both sides of the Gibbs equation (31) by ρdt and substituting from Eq. (9) gives

$$\begin{aligned}
\rho\frac{du^{(e)}}{dt} =\ & \rho T \frac{ds}{dt} - \left(p - \frac{\epsilon_0 E^{(e)2}}{2} - \frac{\mu_0 H^{(e)2}}{2} - \boldsymbol{P}\cdot\boldsymbol{E}^{(e)} - \boldsymbol{M}\cdot\boldsymbol{H}^{(e)}\right)\nabla\cdot\boldsymbol{v} \\
& + \boldsymbol{E}^{(e)}\cdot\frac{d\boldsymbol{D}^{(e)}}{dt} + \boldsymbol{H}^{(e)}\cdot\frac{d\boldsymbol{B}^{(e)}}{dt}
\end{aligned} \tag{55}$$

$\boldsymbol{H}^{(e)}$ termed the "effective field" in [5] and $\boldsymbol{E}^{(e)}$ are the equilibrium values of magnetic and electric field, respectively, that would produce the magnetization $\boldsymbol{M}$ or electric polarization $\boldsymbol{P}$ actually present in a given fluid element. Accordingly, substitution was made of $\boldsymbol{P} = \boldsymbol{P}^{(e)}$ and $\boldsymbol{M} = \boldsymbol{M}^{(e)}$.

Substituting for $\boldsymbol{D}^{(e)}$ and $\boldsymbol{B}^{(e)}$ in the electromagnetic work terms of Eq. (55) from Eq. (26,27) with (e) superscripted permits the equation to be expressed in final form as:

$$\begin{aligned}
\rho\frac{du_m^{(e)}}{dt} =\ & \rho T\frac{ds}{dt} - \left(p - \boldsymbol{P}\cdot\boldsymbol{E}^{(e)} - \boldsymbol{M}\cdot\boldsymbol{H}^{(e)}\right)\nabla\cdot\boldsymbol{v} \\
& + \boldsymbol{E}^{(e)}\cdot\frac{d\boldsymbol{P}}{dt} + \boldsymbol{H}^{(e)}\cdot\frac{d\boldsymbol{M}}{dt}
\end{aligned} \tag{56}$$

where $u_m^{(e)}$ is the equilibrium value of u_m.

$$u_m^{(e)} \equiv u^{(e)} - \frac{\epsilon_0 E^{(e)2}}{2\rho} - \frac{\mu_0 H^{(e)2}}{2\rho} \tag{57}$$

$u_m^{(e)}$ expresses the internal energy of the equilibrium medium exclusive of the field energy associated with the space occupied by the medium. The internal energy $u_m = u_m(t)$ appearing in the energy balance of Eq. (53) can be regarded as that of the medium at equilibrium though the fields $\boldsymbol{E}$ and $\boldsymbol{H}$ present are not the equilibrium fields[4]. *The assertion, held valid at all instants, is expressed as follows.*

$$u_m(t) = u_m^{(e)}(t) \tag{58}$$

and so

$$\frac{\mathrm{d}u_m(t)}{\mathrm{d}t} = \frac{\mathrm{d}u_m^{(e)}(t)}{\mathrm{d}t} \tag{59}$$

From thermodynamics, entropy change between states must be evaluated by a reversible process that leads from the initial state to the final state. Thus, an equation for entropy change is obtained by substituting for $\mathrm{d}u_m/\mathrm{d}t$ and $\mathrm{d}u_m^{(e)}/\mathrm{d}t$ in Eq.(59) from Eqs. (53) and (56), respectively. The result is

$$\rho T \frac{\mathrm{d}s}{\mathrm{d}t} = \mathbf{T}_v^t : \nabla \boldsymbol{v} + \boldsymbol{\Lambda}^t : \nabla \boldsymbol{\Omega} - \nabla \cdot \boldsymbol{q} + \rho R - \mathbf{T} : \boldsymbol{\epsilon} \cdot \boldsymbol{\Omega}$$
$$+ [(\boldsymbol{D} \cdot \boldsymbol{E} + \boldsymbol{B} \cdot \boldsymbol{H})] \mathbf{I} : \nabla \boldsymbol{v} - \left[\boldsymbol{P} \cdot \boldsymbol{E}^{(e)} + \boldsymbol{M} \cdot \boldsymbol{H}^{(e)}\right] \mathbf{I} : \nabla \boldsymbol{v}$$
$$\left(\boldsymbol{H} - \boldsymbol{H}^{(e)}\right) \cdot \frac{\mathrm{d}\boldsymbol{M}}{\mathrm{d}t} + \left(\boldsymbol{E} - \boldsymbol{E}^{(e)}\right) \cdot \frac{\mathrm{d}\boldsymbol{P}}{\mathrm{d}t} + \boldsymbol{j}^* \cdot \boldsymbol{E}^*$$
$$- \left(\epsilon_0 E^2 + \mu_0 H^2\right) \nabla \cdot \boldsymbol{v} \tag{60}$$

where $\mathbf{T}_v$ is viscous stress

$$\mathbf{T}_v = \mathbf{T} + p\mathbf{I} - \mathbf{T}_m \tag{61}$$

and $\mathbf{T}_m$ is the Maxwell type stress vector for the polarizable fluid.

$$\mathbf{T}_m = -\left(\frac{\epsilon_0 E^2}{2} + \frac{\mu_0 H^2}{2}\right) \mathbf{I} + \boldsymbol{D}\boldsymbol{E} + \boldsymbol{B}\boldsymbol{H} \tag{62}$$

The fifth term on the right side of Eq. (60) transforms as follows:

$$\mathbf{T} : \boldsymbol{\epsilon} \cdot \boldsymbol{\Omega} = \mathbf{T}_v : \boldsymbol{\epsilon} \cdot \boldsymbol{\Omega} + (\boldsymbol{P} \times \boldsymbol{E} + \boldsymbol{M} \times \boldsymbol{H}) \cdot \boldsymbol{\Omega}$$
$$= \mathbf{T}_v : \boldsymbol{\epsilon} \cdot \boldsymbol{\Omega} - \boldsymbol{E} \cdot (\boldsymbol{P} \times \boldsymbol{\Omega}) - \boldsymbol{H} \cdot (\boldsymbol{M} \times \boldsymbol{\Omega})$$
$$= \mathbf{T}_v : \boldsymbol{\epsilon} \cdot \boldsymbol{\Omega} - \left(\boldsymbol{E} - \boldsymbol{E}^{(e)}\right) \cdot (\boldsymbol{P} \times \boldsymbol{\Omega})$$
$$- \left(\boldsymbol{H} - \boldsymbol{H}^{(e)}\right) \cdot (\boldsymbol{M} \times \boldsymbol{\Omega}) \tag{63}$$

where $\boldsymbol{E}^{(e)} \cdot (\boldsymbol{P} \times \boldsymbol{\Omega}) = \boldsymbol{H}^{(e)} \cdot (\boldsymbol{M} \times \boldsymbol{\Omega}) = 0$ because $\boldsymbol{E}^{(e)}$ is colinear with $\boldsymbol{P}$, and $\boldsymbol{H}^{(e)}$ with $\boldsymbol{M}$.

---

[4] In the treatment of [8] $\boldsymbol{E}$ and $\boldsymbol{H}$ are not distinguished from $\boldsymbol{E}^{(e)}$ and $\boldsymbol{H}^{(e)}$ nor are the internal energy differences of Eqs. (54) and (57) introduced with the result that an incorrect expression is obtained for entropy production rate that subsequently yields an incorrect equation of magnetization relaxation.

Also,
$$\nabla v = \frac{1}{2}\left[\nabla v + (\nabla v)^t\right] + \frac{1}{2}\left[\nabla v - (\nabla v)^t\right]$$
$$= D^{(v)} + \epsilon \cdot \omega \tag{64}$$

where $D^{(v)} = \frac{1}{2}\left[\nabla v + (\nabla v)^t\right]$ is the strain rate tensor and $\omega = \frac{1}{2}\nabla \times v$ is the fluid rate of rotation, i.e., one-half the fluid vorticity. Thus, the first term on the right side of Eq. (60) can be written as:

$$\mathbf{T}_v^t : \nabla v = \mathbf{T}_v^t : \left(D^{(v)} + \epsilon \cdot \omega\right) \tag{65}$$

The sixth term on the right hand side of Eq. (60) transforms as follows:

$$[(D \cdot E + B \cdot H)]\mathsf{I} : \nabla v - \left[P \cdot E^{(e)} + MH^{(e)}\right]\mathsf{I} : \nabla v$$
$$= \left[(\epsilon_0 E^2 + \mu_0 H^2) + P \cdot \left(E - E^{(e)}\right) + M \cdot \left(H - H^{(e)}\right)\right]\nabla \cdot v \tag{66}$$

where $\mathsf{I} : \nabla v = \nabla \cdot v$. Substituting the various transformed terms into Eq. (60) yields the entropy production equation in final form:

$$\rho T \frac{ds}{dt} = \mathbf{T}_v^t : \left[D^{(v)} + \epsilon \cdot (\omega - \Omega)\right] + \Lambda^t : \nabla \Omega + j^* \cdot E^*$$
$$\nabla \cdot q + \rho \mathrm{R} + \left(E - E^{(E)}\right) \cdot \left\{\frac{dP}{dt} + P(\nabla \cdot v) - \Omega \times P\right\}$$
$$+ \left(H - H^{(e)}\right) \cdot \left\{\frac{dM}{dt} + M(\nabla \cdot v) - \Omega \times M\right\} \tag{67}$$

9 Clausius-Duhem Inequality

The second law of thermodynamics states that the entropy increase dS in a system is larger than the entropy $dS^{(s)}$ supplied with the difference between the two compensated by positive internal entropy production $dS^{(i)}$. This relationship is expressed as follows:

$$dS^{(i)} = dS - dS^{(s)} \geq 0 \tag{68}$$

The inequality applied to a continuum can be written in the following form.

$$\left(\frac{d}{dt}\right)\int_{V_m} \rho s^{(i)} dV = \left(\frac{d}{dt}\right)\int_{V_m} \rho s dV - \left[-\int_{S_m} n \cdot \frac{q}{T} dS + \int_{V_m} \frac{\rho \mathrm{R}}{T} dV\right] \geq 0 \tag{69}$$

where q/T is local entropy flux. Applying the Reynolds transport theorem to the time derivative terms and Gauss' theorem to the surface integral term, noting the arbitrariness of the volume, and multiplying by T on both sides yields

$$\rho T \frac{ds^{(i)}}{dt} = \rho T \frac{ds}{dt} + T\nabla \cdot \frac{q}{T} - \rho \mathrm{R} \geq 0 \tag{70}$$

Expansion of the term containing the heat flux vector gives

$$T\nabla \cdot \frac{q}{T} = \nabla \cdot q + q \cdot \nabla \cdot \ln\left(\frac{1}{T}\right) \qquad (71)$$

Substituting this expression into the inequality of (70) together with the expression for entropy production rate from Eq. (67) yields

$$\rho T \frac{ds^{(i)}}{dt} = q \cdot \nabla \ln\left(\frac{1}{T}\right) + j^* \cdot E^* + \Phi \geq 0 \qquad (72)$$

which is the Clausius-Duhem inequality for this system, where Φ is the dissipation function:

$$\Phi = \mathbf{T}_v^t : \left\{ D^{(v)} + \epsilon \cdot (\omega - \Omega) \right\} + \Lambda^t : \nabla \Omega$$
$$+ \left(E - E^{(e)}\right) \cdot \left\{ \frac{dP}{dt} + P(\nabla \cdot v) - \Omega \times P \right\}$$
$$+ \left(H - H^{(e)}\right) \cdot \left\{ \frac{dM}{dt} + M(\nabla \cdot v) - \Omega \times M \right\} \geq 0 \qquad (73)$$

and where it is assumed that heat flow and current flow produce dissipation in the presence of mechanical dissipation and each other, and that no cross couplings exist. Accordingly, the following conditions must be met.

$$q \cdot \nabla \ln\left(\frac{1}{T}\right) \geq 0 \qquad (74)$$

and

$$j^* \cdot E^* \geq 0 \qquad (75)$$

10 Determination of Constitutive Equations

Constitutive equations of q, j^*, $\mathbf{T}_v$, Λ, M and P must be determined such that the conditions (72) and (73) are met. Linear relationships are developed in the following.

10.1 Constitutive Equations of q and j^*

The conducting magnetic fluids are assumed to be isotropic, hence constitutive equations are determined consistent with (74) and (75) that q is proportional to E^*, and j^* proportional to E^*. Assuming linear variations this yields:

$$q = -k\nabla T \qquad (76)$$

$$j^* = \sigma E^* \qquad (77)$$

where (76) is Fourier's first law, and (77) is Ohm's law. Transforming the quantities in (77) to the stationary coordinate system using Eqs. (42) and (52) leads to

$$\boldsymbol{j} = \sigma\left(\boldsymbol{E} + \boldsymbol{v} \times \boldsymbol{B}\right) + \rho_e \boldsymbol{v} \tag{78}$$

This implicity recovers the Lorentz force law and gives the relationship between electric field and electric current density in a conductor containing free charge of density ϱ_e moving with velocity $\boldsymbol{v}$ when a magnetic flux density $\boldsymbol{B}$ exists.

10.2 Constitutive Equations for Stress and Surface Couple Tensors

The following decomposition is employed in this section.

$$\mathbf{T} = \check{\mathbf{T}}^{(s)} + \mathbf{T}^{(a)} + \frac{1}{3}\mathbf{I} tr\mathbf{T} \tag{79}$$

where $\mathbf{T}$ is any dyadic, $tr\mathbf{T}$ is the trace of $\mathbf{T}$, $\mathbf{T}^{(a)}$ the antisymmetric part of $\mathbf{T}$, and $\mathbf{T} - (1/3)\mathbf{I} tr\mathbf{T}$. The following identity applies for the indicated product of two dyadics:

$$\mathbf{T} : \mathbf{V} = \frac{1}{9}\left(tr\mathbf{T}\right)\left(tr\mathbf{V}\right) + \mathbf{T}^{(a)} : \mathbf{V}^{(a)} + \check{\mathbf{T}}^{(s)} : \check{\mathbf{V}}^{(s)} \tag{80}$$

The first term on the right is a product of scalars. The second term on the right can alternatively be written as the scalar product of two axial vectors. Applying the identity to the first term on the right side of Eq. (73) gives:

$$\mathbf{T}_v^t \left\{ \mathbf{D}^{(v)} + \boldsymbol{\epsilon} \cdot (\boldsymbol{\omega} - \boldsymbol{\Omega}) \right\} = \frac{1}{9} tr\left(\mathbf{T}_v^t\right)\left(tr\mathbf{D}^{(v)}\right)$$
$$+ \left(\mathbf{T}_v^T\right)^{(a)} : \boldsymbol{\epsilon} \cdot (\boldsymbol{\omega} - \boldsymbol{\Omega}) + \left(\check{\mathbf{T}}_v^t\right)^{(s)} : \check{\mathbf{D}}^{(v)} \tag{81}$$

Thus, non-negative values of entropy production are assured with the following relationships.

$$\left(\check{\mathbf{T}}_v^t\right)^{(s)} = 2\eta \check{\mathbf{D}}^{(v)}, \tag{82}$$

$$\left(\mathbf{T}_v^t\right)^{(a)} = -2\zeta \boldsymbol{\epsilon} \cdot (\boldsymbol{\omega} - \boldsymbol{\Omega}), \tag{83}$$

$$tr\left(\mathbf{T}_v^t\right) = \gamma \left(tr\mathbf{D}^{(v)}\right) \tag{84}$$

where η, ζ, and γ are positive valued proportionality factors, and where it is recognized in (83) that the product of antisymmetric dyadics is negative. $\mathbf{T}_v$ decomposed in the manner of Eq. (79) reads

$$\mathbf{T}_v = \check{\mathbf{T}}_v^{(s)} + \mathbf{T}_v^{(a)} + \frac{1}{3}\mathbf{I} tr\mathbf{T}_v$$
$$= \left(\mathbf{T}_v^t\right)^{(s)} - \left(\mathbf{T}_v^T\right)^{(a)} + \frac{1}{3}\mathbf{I} tr\mathbf{T}_v^t \tag{85}$$

Substituting values from (82,83,84) now yields the following constituted form of $\mathbf{T}_v$.

$$\mathbf{T}_v = 2\eta \breve{\mathbf{D}}^{(v)} + 2\zeta \boldsymbol{\epsilon} \cdot (\boldsymbol{\omega} - \boldsymbol{\Omega}) + \frac{\gamma}{3} \mathbf{I} tr \mathbf{D}^{(v)}$$
$$= \eta \left[\nabla \boldsymbol{v} + (\nabla \boldsymbol{v})^t \right] + 2\zeta \boldsymbol{\epsilon} \cdot (\boldsymbol{\omega} - \boldsymbol{\Omega}) + \lambda (\nabla \cdot \boldsymbol{v}) \mathbf{I} \qquad (86)$$

where $\lambda = (\gamma - 2\eta)/3$. The η is shear coefficient of viscosity, ζ is known as the vortex viscosity, and λ is the bulk coefficient of viscosity.

The constitutive relationship for surface couple stress is obtained from the second term on the right side of Eq. (73) following the analogous treatment. With assumption that asymmetrical couple stress is absent the result is

$$\boldsymbol{\Lambda} + \eta' \left[\nabla \boldsymbol{\Omega} + (\nabla \boldsymbol{\Omega})^t \right] + \lambda' (\nabla \cdot \boldsymbol{\Omega}) \mathbf{I} \qquad (87)$$

where η' and λ' are the shear and bulk coefficients of spin viscosity, respectively.

10.3 Constitutive Equation for Magnetization

Inspection of the last term on the right side of Eq. (73) shows that the term is non-negative provided the bracketed sum of magnetic terms is proportional to $(\boldsymbol{H} - \boldsymbol{H}^{(e)})$. The same conclusion was reached by Felderhof and Kroh [7]. In the regime of initial susceptibility $\chi_m = (1/\mu_0)(\partial M/\partial H)_{H=0}$ the terms are proportional to $(\boldsymbol{M}_0 - \boldsymbol{M})$ where $\boldsymbol{M}_0$ is the value of magnetization that would be in equilibrium in magnetic field $\boldsymbol{H}$, given for dilute fluids from the Langevin superparamagnetic relationship [15], and $\boldsymbol{M}$, as previously defined, is the actual magnetization at a given instant of time. Then the constitutive equation for magnetization may be written as:

$$\frac{d\boldsymbol{M}}{dt} + \boldsymbol{M}(\nabla \cdot \boldsymbol{v}) = \boldsymbol{\Omega} \times \boldsymbol{M} + \frac{1}{\tau_m}(\boldsymbol{M}_0 - \boldsymbol{M}) \qquad (88)$$

where τ_m is the magnetic relaxation time constant. Eq. (88) is the Shliomis relaxation relationship [4], generalized for compressible fluid. The corresponding equation for relaxation of electric polarization is obtained from the second to last term on the right side of Eq. (73). The electric polarization in usual magnetic fluids is small such that the contribution to the dissipation function is negligible, hence this relaxation equation will not be written here. With the electric dipolar relaxation time τ_e very small and approaching zero, the difference $(\boldsymbol{P}_0 - \boldsymbol{P})$ must also be small to prevent the ratio from becoming singular. Then the constitutive equation for electric dipole moment reduces to

$$\boldsymbol{P} \cong \boldsymbol{P}_0 = \chi_e \epsilon_0 \boldsymbol{E} \qquad (89)$$

where χ_E is the electric susceptibility.

11 The Constituted Equation Set

Substituting the constitutive equations of q, j, $\mathbf{T}_v^T$ and Λ, i.e., Eqs. (76), (78), (86) and (87), and T_m into Eqs. (14), (24) and (72) yields definite expressions for the equations of motion, internal angular momentum and entropy production rate. The system of basic equations in differential form becomes:

$$\frac{\partial \rho}{\partial t} + \nabla \cdot (\rho \boldsymbol{v}) = 0 \tag{90}$$

$$\rho \left(\frac{\partial \boldsymbol{v}}{\partial t} + \boldsymbol{v} \cdot \nabla \boldsymbol{v} \right) = -\nabla p + (\eta + \lambda) \nabla (\nabla \cdot \boldsymbol{v}) + \eta \nabla^2 \boldsymbol{v}$$
$$+ \nabla \times [2\zeta (\boldsymbol{\omega} - \boldsymbol{\Omega})] + \rho_e \boldsymbol{E} + \boldsymbol{j} \times \boldsymbol{B} + (\boldsymbol{P} \cdot \nabla) \boldsymbol{E}$$
$$+ (\boldsymbol{M} \cdot \nabla) \boldsymbol{H} + \boldsymbol{P} \times (\nabla \times \boldsymbol{E}) + \boldsymbol{M} \times (\nabla \times \boldsymbol{H}) + \rho \boldsymbol{b} \tag{91}$$

$$I \frac{d\boldsymbol{\Omega}}{dt} = (\lambda' + \eta') \nabla (\nabla \cdot \boldsymbol{\Omega}) + \eta' \nabla^2 \boldsymbol{\Omega} - 4\zeta (\boldsymbol{\omega} - \boldsymbol{\Omega})$$
$$+ \boldsymbol{M} \times \boldsymbol{H} + \boldsymbol{P} \times \boldsymbol{E} + \rho \boldsymbol{l} \tag{92}$$

$$\rho T \frac{ds}{dt} = \frac{\eta}{2} \left[\nabla \boldsymbol{v} + (\nabla \boldsymbol{v})^t \right]^2 + 4\zeta (\boldsymbol{\omega} - \boldsymbol{\Omega})^2 + \lambda (\nabla \boldsymbol{v})^2$$
$$+ \eta \left[\nabla \boldsymbol{\Omega} + \nabla \boldsymbol{\Omega}^t \right] : \nabla \boldsymbol{\Omega} + \lambda' (tr \boldsymbol{\Omega})^2$$
$$+ k \nabla^2 T + \frac{\boldsymbol{j}^{*2}}{\sigma} + \rho R + \frac{1}{\chi_m \tau_m} (\boldsymbol{M}_0 - \boldsymbol{M})^2 \tag{93}$$

$$\frac{d\boldsymbol{M}}{dt} + \boldsymbol{M} (\nabla \cdot \boldsymbol{v}) = \boldsymbol{\Omega} \times \boldsymbol{M} + \frac{1}{\tau_m} (\boldsymbol{M}_0 - \boldsymbol{M}) \tag{94}$$

$$\boldsymbol{P} = \chi_e \epsilon_0 \boldsymbol{E} \tag{95}$$

$$\frac{\partial \boldsymbol{B}}{\partial t} = -\nabla \times \boldsymbol{E} \tag{96}$$

$$\frac{\partial \boldsymbol{D}}{\partial t} + \boldsymbol{j} = \nabla \times \boldsymbol{H} \tag{97}$$

$$\nabla \cdot \boldsymbol{B} = 0 \tag{98}$$

$$\nabla \cdot \boldsymbol{D} = \rho_e \tag{99}$$

$$\boldsymbol{j} = \sigma (\boldsymbol{E} + \boldsymbol{v} \times \boldsymbol{B}) + \rho_e \boldsymbol{v} \tag{100}$$

$$\boldsymbol{j}^* = \boldsymbol{j} - \rho_e \boldsymbol{v} \tag{101}$$

$$\boldsymbol{D} = \epsilon_0 \boldsymbol{E} + \boldsymbol{P} \qquad \boldsymbol{B} = \mu_0 \boldsymbol{H} + \boldsymbol{M} \tag{102}$$

Note in Eq.(91) that terms may be combined as follows.

$$\boldsymbol{P} \cdot \nabla \boldsymbol{E} + \boldsymbol{P} \times (\nabla \times \boldsymbol{E}) = (\nabla \boldsymbol{E}) \cdot \boldsymbol{P} \tag{103}$$

$$\boldsymbol{M} \cdot \nabla \boldsymbol{H} + \boldsymbol{M} \times (\nabla \times \boldsymbol{H}) = (\nabla \boldsymbol{H}) \cdot \boldsymbol{M} \qquad (104)$$

The term $\boldsymbol{M} \times (\nabla \times \boldsymbol{H})$ disappears for non-conductive fluid unless the $\boldsymbol{E}$ field is time-varying and the corresponding electrical term disappears unless $\boldsymbol{B}$ is time-varying, with both forces small, even in high frequency fields.

The last term in Eq. (93) is useful in computing radio frequency heating of ferrofluids [16].

For negligible angular acceleration, spin diffusion, electric polarization, and distant source of body couple, Eq. (92) simplifies to

$$\boldsymbol{\Omega} = \frac{1}{4\zeta} \boldsymbol{M} \times \boldsymbol{H} + \boldsymbol{\omega} \qquad (105)$$

Substituting for $\boldsymbol{\Omega}$ in Eq. (94) yields the relaxation equation in the commonly useful form for incompressible fluid as

$$\frac{d\boldsymbol{M}}{dt} = \boldsymbol{\omega} \times \boldsymbol{M} - \frac{1}{\tau_m}(\boldsymbol{M} - \boldsymbol{M}_0) - \frac{1}{4\zeta} \boldsymbol{M} \times (\boldsymbol{M} \times \boldsymbol{H}) \qquad (106)$$

where, for dilute colloids, $\zeta = 3\eta\phi/2$ For incompressible, isothermal, non-conductive magnetic fluid absent of free charge and internal heat release, having negligible electric polarization, using Eq. (105) reduces Eq. (91) to the following simpler form of the linear momentum balance having wide applicability to systems in which relaxation rate of the magnetic moment plays an important role.

$$\rho \left(\frac{\partial \boldsymbol{v}}{\partial t} + \boldsymbol{v} \cdot \nabla \boldsymbol{v} \right) = -\nabla p + +\eta \nabla^2 \boldsymbol{v} + \rho \boldsymbol{b} + \boldsymbol{M} \cdot \nabla \boldsymbol{H} + \frac{1}{2} \nabla \times (\boldsymbol{M} \times \boldsymbol{H}) \qquad (107)$$

This equation further reduces to the following quasi-equilibrium equation when relaxation rate is rapid such that $\boldsymbol{M}$ and $\boldsymbol{H}$ are sensibly collinear [1].

$$\rho \left(\frac{\partial \boldsymbol{v}}{\partial t} + \boldsymbol{v} \cdot \nabla \boldsymbol{v} \right) = -\nabla p + +\eta \nabla^2 \boldsymbol{v} + \rho \boldsymbol{b} + M \nabla H \qquad (108)$$

M and H are the magnitudes of $\boldsymbol{M}$ and $\boldsymbol{H}$, respectively.

12 Conclusion

This work derived a general set of basic equations for flow and response of conductive magnetic fluids with internal rotation based on integral balance equations and thermodynamics. The work is free of arbitrary definitions of electromagnetic energy density and stress. Employing a strict formulation of entropy production in which applied electromagnetic field is distinguished from its equilibrium counterpart, the usual constitutive relationships emerge from the analysis without empirical assumption, and include the magnetization relaxation relationship. The work gives support to the Minkowski form of the electromagnetic momentum density $\boldsymbol{D} \times \boldsymbol{B}$.

This study has treated the Maxwell type stress of a magnetic fluid on the same basis as ordinary contact stresses of pressure and viscous stress and produces the accepted form of the linear and angular momentum balances. In addition, the work shows that the product of Maxwell type stress and velocity is useful in formulating the integral energy balance.

As a recommendation for further work it is desirable that a broader treatment be developed such that the stress tensor in Eq. (62) is expressed in rest frame variables. Such a theory would be exact to first order in velocity throughout. For additional remarks, see Appendix C.

Acknowledgments

The author is appreciative of communications with H. Brenner, U. Felderhof, C. Rinaldi-Ramos, M. Shliomis, T. Tanahashi, and M. Zahn.

References

1. J. L. Neuringer and R. E. Rosensweig, Phys. Fluids **7** (12), 1927 (1964).
2. J. S. Dahler and L. E. Scriven, Nature **192**, 36 (1961); D. W. Condiff and J. S. Dahler, Phys. Fluids **7** (6), 842 (1964).
3. M. I. Shliomis, Soviet Phys. JETP (Engl. transl.) 34 (6), 1291 (1972).
4. M. I. Shliomis, Soviet Phys. Uspekhi (Engl. transl.) 17 (2), 153 (1974).
5. M. I. Shliomis, Phys. Rev. E, Comment on "Magnetoviscosity and relaxation in ferrofluids" Phys.Rev.E.**64** (2001).
6. S. R. deGroot and P. Mazur,*Non-equilibrium thermodynamics*, Dover, Mineola,NY(1984).
7. B. U. Felderhof and H. J. Kroh, Jl. Chem. Phys. **110** (15), 7403 (1999).
8. K. Shizawa and T. Tanahashi, Bulletin of JSME, **29** (250), 1171 (1986); **29** (255), 2878 (1986).
9. C. Rinaldi and H. Brenner, "Body vs surface forces in continuum mechanics: Is the Maxwell stress tensor physically objective?", Phys. Rev E **65** (3); 036615 (2002).
10. J. A. Stratton, *Electromagnetic theory*, (McGraw-Hill, New York, 1941).
11. R. E. Rosensweig, Chapter 13 in G. Astarita, *Thermodynamics:An advanced textbook for Chemical Engineers*, Plenum Press, New York (1989).
12. B. T. Chu, Phys. Fluids 2 (5), 473 (1959).
13. I. E. Tarapov, Magnit. Gidrodin. **1**, 3 (1973).
14. E. Blums, Yu. A. Mikhailov, and R. Ozols, *Heat and mass transfer in MHD flows*, World Scientific, Singapore (1987).
15. R. E. Rosensweig, *Ferrohydrodynamics*, Cambridge University Press, New York (1985); reprinted with slight corrections by Dover, Mineola, New York (1997).
16. R. E. Rosensweig, Magnetohydrodynamics, **36** (4), 300 (2000).

Symbols

Latin

b	Body force per unit mass	(N kg^{-1})
B	Magnetic induction, tesla	(kg s^{-2} A^{-1})
D	Displacement vector	(m^{-2} s A)
$\mathbf{D}^{(v)}$	Strain rate tensor	(s^{-1})
E	Electric field vector in stationary coordinates, volts/meter	(m kg s^{-3} A^{-1})
E^*	Electric field vector in the moving medium, volt/meter	(m kg s^{-3} A^{-1})
g	Electromagnetic momentum vector	(N s m^{-3})
H	Magnetic field vector in stationary coordinates	(m^{-1} A)
H^*	Magnetic field vector in the moving medium	(m^{-1} A)
I	Unit tensor	(dimensionless)
I	Moment of inertia per unit volume	(kg m^{-1})
j	Current density in stationary coordinates	(A m^{-2})
j^*	Current density in the moving medium	(A m^{-2})
l	Body couple vector per unit mass transmitted from afar	(N m kg^{-1})
M	Magnetization, tesla	(kg s^{-2} A^{-1})
M_0	Equilibrium magnetization in field H	(kg s^{-2} A^{-1})
n	Unit outward facing normal	(dimensionless)
P	Electric polarization vector	(m^{-2} s A)
P_0	Equilibrium electric polarization vector in field E	(m^{-2} s A)
p	Pressure-like variable defined in Eq. (30)	(N m^{-2})
p_0	Pressure in absence of electromagnetic fields	(N m^{-2})
$p^{(e)}$	Pressure in presence of electromagnetic fields	(N m^{-2})
q	Heat flux vector	(J m^{-2} s^{-1})
r	Position vector	(m)
R	Internal heat release rate per unit mass	(J kg^{-1} s^{-1})
s	Entropy per unit mass	(J K^{-1} kg^{-1})
$\mathbf{s}$	Spin vector, i.e., angular momentum per unit mass	(m^2 s^{-1})
S	Total entropy of a system	(J K^{-1})
S_m	Surface area enclosing volume V_m	(m^2)
t	Time	(s)
t_n	Stress vector	(N m^{-2})
T	Temperature	(K)
T	Stress tensor less electromagnetic momentum flux	(N m^{-2})
$\tilde{\mathsf{T}}$	Stress tensor sum of pressure, viscous and electromagnetic contributions	(N m^{-2})
u	Internal energy per unit mass including field energy	(J kg^{-1})
u_m	Internal energy per unit mass excluding field energy	(J kg^{-1})
$u^{(e)}$	Value of u in equilibrated system	(J kg^{-1})
$u_m^{(e)}$	Value of u_m in equilibrated system	(J kg^{-1})
V_m	Volume containing mass	(m^3)

Greek

γ	Lorentz transformation coefficient	(dimensionless)
ϵ_0	Permittivity of free space, 8.854 x 10^{-12} farad/meter	(m^3 kg^{-1} s^4 A^2)
ϵ	Polyadic alternator	(dimensionless)
ζ	Vortex vorticity	(N s m^{-2})
η	Shear coefficient of viscosity	(N s m^{-2})
η'	Shear coefficient of spin viscosity	(N s)
λ	Bulk coefficient of viscosity	(N s m^{-2})
λ	Bulk coefficient of spin viscosity	(N s)
λ_n	Couple stress vector	(N m^{-1})
Λ	Couple stress tensor	(N m^{-1})
μ_0	Permeability of free space, 4π x 10^{-7} henry/meter	(m kg s^{-2} A^{-2})
ϱ	Mass density	(kg m^{-3})
ϱ_e	Charge density, coulomb/$meter^3$	(A s m^{-3})
σ	Electrical conductivity, siemen	(m^{-3} kg^{-1} s^3 A^2)
τ_m	Relaxation time constant	(s)
υ	Specific volume	(m^3 kg^{-1})
Φ	Dissipation function	(N s m^{-2})
χ_e	Electrical susceptibility	(dimensionless)
χ_m	Magnetic susceptibility	(dimensionless)
$\boldsymbol{\omega}$	Fluid rotational rate vector	(radian s^{-1})
$\boldsymbol{\Omega}$	Particle rotational rate vector	(radian s^{-1})

Appendix A: Notation

Vectors are set in bold Roman such as $\boldsymbol{v}$ or $\boldsymbol{B}$, or lower case bold Greek such as $\boldsymbol{\lambda}$.

Dyadics are indicated in bold sans serif such as **T** or bold upper case Greek such as $\boldsymbol{\Lambda}$.

T corresponds to $\hat{\boldsymbol{e}}_i\hat{\boldsymbol{e}}_j T_{ij}$ in indicial notation employing the Einstein summation convention; $\hat{\boldsymbol{e}}_i$ and $\hat{\boldsymbol{e}}_j$ are a unit vectors. T_{ij} is defined as the j th component of force acting on the surface having normal orientated in the i th direction.

$\mathbf{T}^T$ is the transpose of **T** and corresponds to $\hat{\boldsymbol{e}}_i\hat{\boldsymbol{e}}_j T_{ji}$.

The operator ∇ corresponds to $\hat{\boldsymbol{e}}_i \partial/\partial x_i$.

The nesting convention is followed as shown by these examples:

$$\boldsymbol{ab} \cdot \boldsymbol{cd} = (\boldsymbol{b} \cdot \boldsymbol{c})\,\boldsymbol{ad}$$
$$\boldsymbol{ab} : \boldsymbol{cd} = (\boldsymbol{b} \cdot \boldsymbol{c})(\boldsymbol{a} \cdot \boldsymbol{d})$$

$\nabla \cdot \mathbf{T}$ corresponds by the nesting convention to $\hat{\boldsymbol{e}}_i \partial T_{ij}/\partial x_i$ and not $\hat{\boldsymbol{e}}_i \partial T_{ij}/\partial x_j$.

I is the unit dyadic corresponding to $\hat{\boldsymbol{e}}_i\hat{\boldsymbol{e}}_j \delta_{ij}$ where δ_{ij} is the Kronecker delta function.

$$\delta_{ij} = \begin{cases} 0 & (i \neq j) \\ 1 & (i = j) \end{cases}$$

$\boldsymbol{\epsilon}$ is the alternating polyadic corresponding to $\hat{e}_i \hat{e}_j \hat{e}_k \epsilon_{ijk}$ where

$$\epsilon_{ijk} = \begin{cases} 1 & (ijk = 123, 231, or\, 312) \\ 0 & (i = j, i = k, or\, j = k) \\ -1 & (ijk = 132, 213, or\, 321) \end{cases}$$

From these definitions, $\boldsymbol{\epsilon} : \boldsymbol{ab} = -\boldsymbol{a} \times \boldsymbol{b}$ and $\nabla \cdot (\mathbf{T} \cdot \boldsymbol{v}) = \mathbf{T}^T : \nabla \boldsymbol{v} + \boldsymbol{v} \cdot (\nabla \cdot \mathbf{T})$.

Appendix B: Galilean Transformation of E, H and j

The transformation of these field vectors is carried out in the spirit of the text with integral statement of the basic law taken as the point of departure.

Faraday's Law and the E Field

Faraday's law applicable to moving media relates induced voltage around an arbitrarily chosen closed circuit to the time rate of change of induction enclosed by the circuit. Choosing the circuit to coincide with a moving material line of fixed identity, the relationship may be written as:

$$\oint \boldsymbol{E}^* \cdot \mathrm{d}\boldsymbol{l} = \frac{\mathrm{d}\Phi_f}{\mathrm{d}t} \tag{109}$$

where $\boldsymbol{E}^*$ is the electric field in the frame (rest frame) of an observer moving with the local velocity, $\mathrm{d}\boldsymbol{l}$ is the differential of path length, and Φ is the number of flux linkages defined by

$$\Phi_f = \int \boldsymbol{B} \cdot \boldsymbol{n} \mathrm{d}S \tag{110}$$

where $\boldsymbol{n}$ is a unit vector normal to differential area of S, the enclosed surface. The Reynolds' transport theorem[5] for differentiating a surface integral gives

$$\frac{\mathrm{d}}{\mathrm{d}t} \int \boldsymbol{B} \cdot \boldsymbol{n} \mathrm{d}S = \int \left[\frac{\partial \boldsymbol{B}}{\partial t} + (\nabla \cdot \boldsymbol{B}) \boldsymbol{v} + \nabla \times (\boldsymbol{B} \times \boldsymbol{v}) \right] \cdot \boldsymbol{n} \mathrm{d}S \tag{111}$$

Applying Stokes's theorem to the left side of (109) converts the line integral to a surface integral.

$$\oint \boldsymbol{E}^* \cdot \mathrm{d}\boldsymbol{l} = -\int (\nabla \times \boldsymbol{E}^*) \cdot \boldsymbol{n} \mathrm{d}S \tag{112}$$

---

[5] R. Aris, Vectors, tensors and the basic equations of fluid mechanics. Dover, Mineola, NY (1989). An alternate derivation is given in W. K. H. Panofsky and M Phillips, Classical electricity and Magnetism, Second Edition, Addison-Wesley, Reading, Massachusetts (1962, pages 160–300).

Combining the expressions, recognizing that the surface of integration is arbitrary, and noting from the Maxwell law that $\nabla \cdot \boldsymbol{B} = 0$, yields the following differential relationship upon rearrangement.

$$\nabla \times (\boldsymbol{E}^* - \boldsymbol{v} \times \boldsymbol{B}) = -\frac{\partial \boldsymbol{B}}{\partial t} \tag{113}$$

As a tenet of relativity, including Galilean relativity, the form of a basic relationship of physics is invariant to the choice of inertial (constant velocity) reference frame. Accordingly, comparing (113) with the form of the Faraday law expressed in terms of laboratory coordinates, namely

$$\nabla \times \boldsymbol{E} = -\frac{\partial \boldsymbol{B}}{\partial t} \tag{114}$$

shows that the relationship between the electric field vectors is given by

$$\boldsymbol{E}^* = \boldsymbol{E} + \boldsymbol{v} \times \boldsymbol{B} \tag{115}$$

This relationship is given in Eq.(38) of the text. The observer in the moving frame senses an additional electric field due to motion through the induction field.

Ampere's Law and the $\boldsymbol{H}$ and $\boldsymbol{j}$ Fields

These relationships can be derived from a treatment of Maxwell's extension of Ampere's law. In its general integral form applicable to a medium in motion,

$$\oint \boldsymbol{H}^* \cdot \mathrm{d}\boldsymbol{l} = \int \boldsymbol{j}^* \cdot \boldsymbol{n} \mathrm{d}S + \frac{\mathrm{d}}{\mathrm{d}t} \int \boldsymbol{D} \cdot \boldsymbol{n} \mathrm{d}S \tag{116}$$

Again the asterisk indicates the rest frame. From the Reynolds' transport theorem, as introduced above,

$$\frac{\mathrm{d}}{\mathrm{d}t} \int \boldsymbol{D} \cdot \mathrm{d}S = \int \left[\frac{\partial \boldsymbol{D}}{\partial t} + (\nabla \cdot \boldsymbol{D}) \boldsymbol{v} + \nabla \times (\boldsymbol{D} \times \boldsymbol{v}) \right] \cdot \boldsymbol{n} \mathrm{d}S \tag{117}$$

Applying Stokes's theorem to the left side of (116) converts the line integral to a surface integral.

$$\oint \boldsymbol{H}^* \cdot \mathrm{d}\boldsymbol{l} = -\int (\nabla \times \boldsymbol{H}^*) \cdot \boldsymbol{n} \mathrm{d}S \tag{118}$$

Combining the expressions, recognizing that the surface of integration is arbitrary, and noting from the Maxwell law that $\nabla \cdot \boldsymbol{D} = \rho_e$, yields the following differential relationship upon rearrangement.

$$\nabla \times (\boldsymbol{H}^* + \boldsymbol{v} \times \boldsymbol{D}) = (\boldsymbol{j}^* + \boldsymbol{v}\rho_e) + \frac{\partial \boldsymbol{D}}{\partial t} \tag{119}$$

This may be compared with the form of Ampere's extended equation written in laboratory coordinates which is

$$\nabla \times \boldsymbol{H} = \boldsymbol{j} + \frac{\partial \boldsymbol{D}}{\partial t} \tag{120}$$

Comparison of the two foregoing expressions yields the transformation relationships for magnetic field and electrical current density. These are:

$$\boldsymbol{H}^* = \boldsymbol{H} - \boldsymbol{v} \times \boldsymbol{D} \tag{121}$$

$$\boldsymbol{j}^* = \boldsymbol{j} - \rho_e \boldsymbol{v} \tag{122}$$

Motion relative to charges alters the perception of electric current density as shown by (122). Motion relative to the displacement field from (121) produces a component of magnetic field in the observer's frame.

Appendix C:
Comment on Electromagnetic Body Couple Density

From the exact expression for the rest frame stress tensor $\mathbf{T}_m^*$ known from analysis of motionless fluid [15] and having the form given by Eq.(62), the associated body couple density is given by $\boldsymbol{vec}\mathbf{T}_m^* = \boldsymbol{M}^* \times \boldsymbol{H}^* + \boldsymbol{P}^* \times \boldsymbol{E}^*$. Expansion of the semi-relativistic expressions [10] relating field vectors between frames gives to first order in velocity,

$$\boldsymbol{M}^* = \boldsymbol{M} + \mu_0 \boldsymbol{v} \times \boldsymbol{P} \tag{123}$$
$$\boldsymbol{H}^* = \boldsymbol{H} - \boldsymbol{v} \times \boldsymbol{D} \tag{124}$$
$$\boldsymbol{P}^* = \boldsymbol{P} - \epsilon_0 \boldsymbol{v} \times \boldsymbol{M} \tag{125}$$
$$\boldsymbol{E}^* = \boldsymbol{E} + \boldsymbol{v} \times \boldsymbol{B} \tag{126}$$

Thus,

$$\begin{aligned}\boldsymbol{M}^* \times \boldsymbol{H}^* + \boldsymbol{P}^* \times \boldsymbol{E}^* &= \boldsymbol{M} \times \boldsymbol{H} + \boldsymbol{P} \times \boldsymbol{E} - \boldsymbol{B} \times (\boldsymbol{v} \times \boldsymbol{D}) + \boldsymbol{D} \times (\boldsymbol{v} \times \boldsymbol{B}) \\ &+ \frac{\boldsymbol{H}}{c^2} \times (\boldsymbol{v} \times \boldsymbol{E}) - \frac{\boldsymbol{E}}{c^2} \times (\boldsymbol{v} \times \boldsymbol{H}) \\ &\cong \boldsymbol{M} \times \boldsymbol{H} + \boldsymbol{P} \times \boldsymbol{E} + \boldsymbol{v} \times \boldsymbol{g} \end{aligned} \tag{127}$$

where the second equality results as a vector identity following neglect of the terms reciprocal to c^2, and $\boldsymbol{g} = \boldsymbol{D} \times \boldsymbol{B}$ is the electromagnetic momentum density. In comparison to (127), the body couple corresponding to Eq. (62) in the text is absent the $\boldsymbol{v} \times \boldsymbol{g}$ term. This limits the text result to quasi-static flows of low velocity. Newton's equations are galilei-invariant but electromagnetism automatically brings in Lorentz-invariance, hence it should be expected that only a completely relativistic formulation could be totally satisfactory.

Ferrohydrodynamics: Retrospective and Issues

Mark I. Shliomis

Department of Mechanical Engineering, Ben-Gurion University of the Negev,
P.O.B. 653, Beer-Sheva 84105, Israel

Abstract. Two basic sets of hydrodynamic equations for magnetic colloids (so-called *ferrofluids*) are reviewed. Starting from the *quasistationary* ferrohydrodynamics, we then give a particular attention to an expanded model founded on the concept of *internal rotation*. A specific relation between magnetic and rotational degrees of freedom of suspended grains provides a coupling of the fluid magnetization with the fluid dynamics. Hence a complete set of constitutive equations consists of the equation of ferrofluid motion, the Maxwell equations, and the *magnetization equation*. There are three kinds of the latter. Two of them were derived phenomenologically as a generalization of the Debye relaxation equation in case of *spinning* magnetic grains, while one of them was derived microscopically from the Fokker-Planck equation. Testing the magnetization equations, we compare their predictions about the dependence of the *rotational viscosity* on the magnetic field and the shear rate.

1 Quasistationary Ferrofluid Dynamics

Physics and hydrodynamics of ferrofluids has begun from the basic work [1] and the following long series of papers by Rosensweig and co-workers, included later on in his monograph [2]. Those papers and the book laid a serious scientific foundation for further research and gave an impetus to great variety of ferrofluid applications in industry, technology, and medicine. A set of equations describing ferrofluid dynamics ("ferrohydrodynamics" – the term of Rosensweig) first proposed in [1] consists of the equation of ferrofluid motion

$$\rho \frac{d\boldsymbol{v}}{dt} = -\boldsymbol{\nabla} p + \eta \nabla^2 \boldsymbol{v} + M\boldsymbol{\nabla} H, \qquad \frac{d}{dt} = \frac{\partial}{\partial t} + (\boldsymbol{v} \cdot \boldsymbol{\nabla}), \qquad (1)$$

the *magnetic state* equation $M = M(H,T)$, for which it is natural to employ [3,4] the Langevin formula

$$M = nmL(\xi), \qquad \xi = mH/k_\text{B}T, \qquad L(\xi) = \coth \xi - \xi^{-1}, \qquad (2)$$

(here $m = M_\text{d}V$ being the magnetic moment of a single subdomain magnetic particle, $V = \pi d^3/6$ the particle volume, n their number density, and M_d stands for the domain magnetization of dispersed ferromagnetic material) and the equations

$$\text{div}\,\boldsymbol{v} = 0, \qquad \text{rot}\,\boldsymbol{H} = 0, \qquad \text{div}\,\boldsymbol{B} = 0, \qquad (\boldsymbol{B} = \boldsymbol{H} + 4\pi\boldsymbol{M}) \qquad (3)$$

indicating that the ferrofluid is considered to be incompressible and nonconducting. The volume density of magnetic forces in (1), $\boldsymbol{F} = M\boldsymbol{\nabla} H$, is calculated as divergence of the *stress tensor of magnetic field* in ferrofluid, $F_i = \partial \sigma_{ik}^{\mathrm{mag}}/\partial x_k$, where

$$\sigma_{ik}^{\mathrm{mag}} = \tfrac{1}{4\pi} H_i B_k - \frac{\boldsymbol{H}}{8\pi} \cdot \left[\boldsymbol{B} - \rho \left(\frac{\partial \boldsymbol{B}}{\partial \rho}\right)_{\boldsymbol{H},T}\right] \delta_{ik}. \tag{4a}$$

According to the assumption (2), ferrofluid magnetization is nonsensitive to the velocity field and relaxes *instantaneously* to the equilibrium (Langevin) value. Then in the low-field limit, when $\boldsymbol{B} = \mu \boldsymbol{H}$ with a constant $\mu = 1 + 4\pi\chi$, equation (4a) takes the well-known form [5]

$$\sigma_{ik}^{\mathrm{mag}} = \frac{\mu H_i H_k}{4\pi} - \frac{H^2}{8\pi}\left[\mu - \rho\left(\frac{\partial \mu}{\partial \rho}\right)_T\right]\delta_{ik}. \tag{4b}$$

For a non-magnetic fluid ($\mu = 1$) this expression is reduced to the Maxwell stress tensor of magnetic field

$$\sigma_{ik}^{\mathrm{mag}} = \tfrac{1}{4\pi}\left(H_i H_k - \tfrac{1}{2} H^2 \delta_{ik}\right).$$

Equation (4a) may be rewritten as

$$\sigma_{ik}^{\mathrm{mag}} = \tfrac{1}{4\pi}(H_i B_k - \tfrac{1}{2}H^2 \delta_{ik}) - \frac{\boldsymbol{H}}{2}\cdot\left[\boldsymbol{M} - \rho\left(\frac{\partial \boldsymbol{M}}{\partial \rho}\right)_{\boldsymbol{H},T}\right]\delta_{ik}. \tag{4c}$$

Its divergence consists of two terms,

$$\boldsymbol{F} = (\boldsymbol{M}\cdot\boldsymbol{\nabla})\boldsymbol{H} - \boldsymbol{\nabla}\left\{\frac{\boldsymbol{H}}{2}\cdot\left[\boldsymbol{M} - \rho\left(\frac{\partial \boldsymbol{M}}{\partial \rho}\right)_{\boldsymbol{H},T}\right]\right\}, \tag{5}$$

where the first is called sometimes the *Kelvin force*, and the second represents the *magnetostrictive* force. Being the *potential* force, the latter is always included into the pressure gradient in (1), i.e., it is equilibrated automatically by the hydrostatic or hydrodynamic pressure. Further, since the equilibrium magnetization is collinear with the local magnetic field, $\boldsymbol{M} = (M/H)\boldsymbol{H}$, the Kelvin force takes the form $(\boldsymbol{M}\cdot\boldsymbol{\nabla})\boldsymbol{H} = (M/H)(\boldsymbol{H}\cdot\boldsymbol{\nabla})\boldsymbol{H}$. Hence, using an identity

$$(\boldsymbol{H}\cdot\boldsymbol{\nabla})\boldsymbol{H} = \tfrac{1}{2}\boldsymbol{\nabla} H^2 - \boldsymbol{H}\times\mathrm{rot}\,\boldsymbol{H}$$

and taking into account the condition (3) of the absence of free electrical currents, $\mathrm{rot}\,\boldsymbol{H} = 0$, we arrive at the pointed above expression $\boldsymbol{F} = M\boldsymbol{\nabla} H$.

With allowance for (2) and the equality

$$L(\xi) = \frac{\mathrm{d}}{\mathrm{d}\xi} \ln \frac{\sinh \xi}{\xi},$$

the Kelvin force can be written in the case of *isothermal* fluids, $T = \mathrm{const}$, as

$$\boldsymbol{F} = nk_{\mathrm{B}}T\boldsymbol{\nabla} \ln \frac{\sinh \xi}{\xi}. \tag{6}$$

If magnetic grains are distributed uniformly all over the ferrofluid sample, $n = $ const, the last expression takes the form

$$\boldsymbol{F} = \boldsymbol{\nabla}\left(nk_\mathrm{B}T\ln\frac{\sinh\xi}{\xi}\right), \qquad (7)$$

i.e., the magnetic force proves to be *potential*. It makes an evidence of Bernoulli theorem for non-vortical ferrofluid flow [2]

$$p + \rho g z + \tfrac{1}{2}\rho v^2 - nk_\mathrm{B}T\ln\frac{\sinh\xi}{\xi} = \text{const} \qquad (8)$$

(g is the gravity acceleration and z is the vertical coordinate). This relationship is to high extent useful in qualitative investigations of equilibriums and flows of ferrofluids under nonuniform magnetic fields. Note, however, that a nonuniformity of the field induces an inhomogeneity of the particle distribution which in its turn leads to an alteration in magnetic force. Indeed, as seen from (6), each a particle is acted upon by the force

$$\boldsymbol{F}/n = -\boldsymbol{\nabla}U, \qquad U = -k_\mathrm{B}T\ln\frac{\sinh\xi}{\xi}, \qquad (9)$$

where U is the potential. The equilibrium (Boltzmann's or "barometric") particle distribution is given by the formula $n = n_0\exp(-U/k_\mathrm{B}T)$. Thus we find, using (9),

$$n = n_0\frac{\sinh\xi}{\xi}, \qquad n_0\int_\mathcal{V}\frac{\sinh\xi(\boldsymbol{r})}{\xi(\boldsymbol{r})}\mathrm{d}^3\boldsymbol{r} = \bar{n}\mathcal{V}. \qquad (10)$$

The constant n_0 is determined from the indicated normalization condition expressing conservation of the total particle number in the ferrofluid volume $\mathcal{V}$; $\bar{n}$ is the mean density of the particle number. Substituting $n(\xi)$ from (10) into (6) or (9) gives the resulting *self-consistent* magnetic force density

$$\boldsymbol{F} = n_0 k_\mathrm{B}T\frac{\sinh\xi}{\xi}\boldsymbol{\nabla}\ln\frac{\sinh\xi}{\xi} \equiv n_0 k_\mathrm{B}T\boldsymbol{\nabla}\frac{\sinh\xi}{\xi} = \boldsymbol{\nabla}(nk_\mathrm{B}T). \qquad (11)$$

When the fluid is at rest, this force is equilibrated by the pressure gradient, $\boldsymbol{F} = \boldsymbol{\nabla}p$ [see (1)], which emerges automatically. Thus, one can say of *magnetic pressure* $p_m = nk_\mathrm{B}T$ created by the thermal agitation of the magnetic grains in an inert liquid. As it should be, p_m is similar to the pressure of an ideal gas of molecules.

One should not, however, be in hurry with the replacement

$$\bar{n}k_\mathrm{B}T\ln\frac{\sinh\xi}{\xi} \Longrightarrow n_0 k_\mathrm{B}T\frac{\sinh\xi}{\xi} \qquad (12)$$

in the Bernoulli theorem (8), because an equilibrium particle concentration $n(\xi)$ sets in for a very long time. The equilibrium is settled owing to diffusion and *magnetophoresis* which proceed extremely slow due to smallness of the particle diffusion coefficient D. Actually, according to the Einstein formula, $D = k_\mathrm{B}T/3\pi\eta d$, it is inversely proportional to the particle diameter ($d \simeq 10$ nm) and hence appears to be *two orders* as less than for molecules. Therefore, concentration inhomogeneities never arise in a flowing ferrofluid since they simply have no time to be formed. As a consequence, there exist two scenarios of the onset of convection in magnetized ferrofluids [6]. Namely, if an applied temperature difference increases faster than the limit imposed by particle diffusion, the ferrofluid behaves as a pure (i.e., single-component) fluid. In this case, only *stationary* instability occur upper the threshold of convection [7]. In the opposite case, when the concentration gradient (induced by the temperature gradient due to magnetophoresis and the Soret effect) is built up undisturbed by convection, the theory [6,8] predicts *oscillatory* instability in a certain region of parameters. Thus, the replacement (12) should be carried out, strictly speaking, only for a genuine equilibrium, i.e., when we deal with the *ferrohydrostatics*.

When ferrofluid moves in a magnetic field, the pseudovector $\boldsymbol{M}$ can depends, in principle, on the two available pseudovectors: the field $\boldsymbol{H}$ and the fluid vorticity $\boldsymbol{\Omega} = \frac{1}{2}\mathrm{rot}\,\boldsymbol{v}$. Meanwhile the discussed hydrodynamic model [1] assumes that the magnetization is independent on a ferrofluid flow, whereas the flow does depend on the magnetization – see (1). This discrimination has yet another side: instead of the magnetization *dynamics* equation – the one for $\mathrm{d}\boldsymbol{M}/\mathrm{d}t$ – the theory proposes an *equilibrium* expression (2). According to the latter, $\boldsymbol{M}$ is collinear with the field $\boldsymbol{H}$ at any moment and determined by its instantaneous value. This assumption means that the magnetization *relaxation time* is considered to be zero. Hence it is natural to regard the ferrohydrodynamics [1] as a *quasistationary* theory. As it has been shown by the author [4], this theory has a definite field of applicability.

The point is that, when magnetic field is shifted in direction, the particle magnetic moment $\boldsymbol{m}$ – and the ferrofluid magnetization $\boldsymbol{M} = n\langle\boldsymbol{m}\rangle$ as well – can regain equilibrium by two processes [3,4]. In *Brownian* relaxation the vector $\boldsymbol{m}$ rotates in unison with the particle itself. The rotation is resisted by viscous torque due to surrounding carrier liquid and characterized by Brownian time of rotational diffusion $\tau_\mathrm{B} = 3\eta V/k_\mathrm{B}T$. In the *Néel* mechanism of relaxation, $\boldsymbol{m}$ rotates inside the particle (with the reference time τ_N), whereas the particle itself does not rotate. If the field $\boldsymbol{H}$ is turned off at the instant $t = 0$, the magnetization $\boldsymbol{M}(0)$ would decay to zero with the reduced relaxation time [4,9]

$$\tau = \frac{\tau_\mathrm{N}\tau_\mathrm{B}}{\tau_\mathrm{N} + \tau_\mathrm{B}}. \tag{13}$$

In a real ferrofluid, any relation between τ_N and τ_B is possible. Since, however, τ_N increases exponentially with the increase of the particle volume V while τ_B is simply proportional to V, the equality $\tau_\mathrm{N} = \tau_\mathrm{B}$ is fulfilled at a certain particle diameter d_S (the so-called *Shliomis' diameter*). For $d < d_\mathrm{S}$ it holds the condition $\tau_\mathrm{N} \ll \tau_\mathrm{B}$, i.e., according to (13), $\tau \simeq \tau_\mathrm{N}$: ferrofluid magnetization relaxes owing to *internal* diffusion of the particle magnetic moments. If, conversely, $d > d_\mathrm{S}$, then $\tau_\mathrm{N} \gg \tau_\mathrm{B}$ so the Néel process is *frozen* and magnetization relaxation proceeds via Brownian rotary diffusion of the particles in the liquid matrix, $\tau \simeq \tau_\mathrm{B}$. The larger the particle size the better it holds the condition of "freezing up" of the particle magnetic moment into the particle body.

The quasistationary ferrohydrodynamics is valid just for colloids of Néel particles. Indeed, the coupling between magnetic and mechanical degrees of freedom of such particles breaks down (hence M does not depends on Ω), and the magnetization relaxes to the direction of H almost immediately – for $d \simeq (0.3-0.5)d_S$ there is $\tau \simeq \tau_N \sim 10^{-9}$ s (!) – as the model assumes.

The opposite case of suspensions of *rigid* magnetic dipoles (i.e., the grains with "freezing in" magnetic moments, $\tau_N \gg \tau_B$) is considered below. In this model with the infinitely strong coupling and the finite relaxation time $\tau \simeq \tau_B \sim 10^{-4} - 10^{-5}$ s, the vector M is not obliged to be collinear with H. Therefore, besides the magnetic force density $(M \cdot \nabla)H$, ferrofluid can undergo a *magnetic torque* of the density $M \times H$. The model of rigid dipoles is widely used in the theory of magnetic fluids. Despite its simplicity, it allows one to explain a wide complicated tangle of magnetic and hydrodynamic phenomena emerging in ferrofluids under the field. Below we derive and discussed the basic equations for the model.

2 Ferrohydrodynamics: Allowance for Internal Rotation

The main peculiarity of ferrofluids is a specific relation between the magnetic and rotational degrees of freedom of suspended magnetic grains of which the fluids are composed. Therefore the concept of *internal rotation* first applied to ferrofluids in [3] has proved to be very fruitful. The model [3] takes into account that the volume density of the angular momentum of ferrofluids consists of both the visible ("orbital") and the internal ("spin") parts. The former, $L = \rho(r \times v)$, is associated with the translational motion of magnetic grains and molecules of the solvent. The latter, S, is caused by the rotation of the grains themselves and should be treated as an independent variable along with the fluid velocity v, density ρ, and pressure p. However, an appropriate thermodynamic coordinate is the difference $S - I\Omega$ where Ω is the local angular velocity of the fluid and I means the volume density of the particles moment of inertia. For a suspension of spherical grains $I = \rho_s \phi d^2/10$ where ϕ is the volume fraction of dispersed solid phase, ρ_s is the material density of the solids, and d is the mean particle diameter. It is convenient to set $S = I\omega_p$ where ω_p is the macroscopic (i.e., averaged over physically small volume) angular velocity of the particles. It is worth noting that rotation of the particles induces micro-flow-fields in the surrounding carrier liquid that in turn contain an additional angular momentum. This contribution may be included into S by means of a re-normalization of the moment of inertia I. In fact, an effective ("hydrodynamic") diameter d_h of a magnetic grain has to be a little larger than its true diameter d due to such an attached micro-vortex.

Any deviation of ω_p from Ω gives rise to dissipation processes due to redistribution of angular momentum between L and S forms. (The angular momentum conservation law refers, sure, to the total angular momentum $L+S$). These processes contribute the stress tensor σ_{ik}. For an ordinary (nonmagnetic) suspension the tensor has been derived by the methods of irreversible thermodynamics in [10]:

$$\sigma_{ik} = -p\delta_{ik} + \eta\left(\frac{\partial v_i}{\partial x_k} + \frac{\partial v_k}{\partial x_i}\right) + \frac{1}{2\tau_s}(S_{ik} - I\Omega_{ik}), \qquad (14)$$

where

$$S_{ik} = \epsilon_{ikl}S_l, \qquad \Omega_{ik} = \tfrac{1}{2}(\partial v_k/\partial x_i - \partial v_i/\partial x_k) = \epsilon_{ikl}\Omega_l,$$

and ϵ_{ikl} stands for antisymmetric unit tensor. Apart from the viscosity η, (14) contains once more kinetic coefficient: the spin relaxation time $\tau_s = I/6\eta\phi = \rho_s d^2/60\eta$. For $d = 10\,\text{nm}$ and $\eta = 10^{-2}\,\text{P}$ this formula gives $\tau_s \sim 10^{-11}\,\text{s}\,(!)$. Thus, the difference $\boldsymbol{\omega}_\text{p} - \boldsymbol{\Omega}$ instantly decays whereupon the hydrodynamic description is reduced to the common set of hydrodynamic equations. Ferrofluids, however, give us an opportunity to maintain this difference by an extraneous magnetic torque which acts directly upon the particles rotation:

$$6\eta\phi(\boldsymbol{\omega}_\text{p} - \boldsymbol{\Omega}) = \boldsymbol{M} \times \boldsymbol{H}. \qquad (15)$$

Eliminating the last term in (14) with the aid of the torque balance equation (15) and including in σ_{ik} the stress tensor of magnetic field σ_{ik}^{mag} from (4c), one gets [3,4]

$$\sigma_{ik} = -p\delta_{ik} + \eta\left(\frac{\partial v_i}{\partial x_k} + \frac{\partial v_k}{\partial x_i}\right) + \tfrac{1}{2}(M_i H_k - M_k H_i) + \tfrac{1}{4\pi}(H_i B_k - \tfrac{1}{2}H^2\delta_{ik}); \quad (16\,\text{a})$$

the magnetostrictive diagonal part of σ_{ik}^{mag} has been included here in the isotropic tensor of the pressure. Substituting σ_{ik} from (16 a) into the *momentum conservation law* $\rho\,\text{d}v_i/\text{d}t = \partial\sigma_{ik}/\partial x_k$ yields the equation of ferrofluid motion

$$\rho\frac{\text{d}\boldsymbol{v}}{\text{d}t} = -\boldsymbol{\nabla} p + \eta\nabla^2\boldsymbol{v} + (\boldsymbol{M}\cdot\boldsymbol{\nabla})\boldsymbol{H} + \tfrac{1}{2}\,\text{rot}\,(\boldsymbol{M}\times\boldsymbol{H}). \qquad (17)$$

At the calculation of the divergency of the stress tensor (16 a) we have used equations (3). Note that under the assumption of *collinearity* of $\boldsymbol{M}$ with $\boldsymbol{H}$ the last term in (17) vanishes and the Kelvin force density $(\boldsymbol{M}\cdot\boldsymbol{\nabla})\boldsymbol{H}$ takes the form $M\boldsymbol{\nabla}H$, so that (17) is reduced to (1).

The *angular momentum conservation law* reads $\text{d}S_{ik}/\text{d}t = \sigma_{ki} - \sigma_{ik}$. This equation governs the particle rotation. Meanwhile an extreme smallness of the corresponding relaxation time τ_s has allowed us to neglect the inertia term $\text{d}\boldsymbol{S}/\text{d}t$ in comparison with the relaxation term $(\boldsymbol{S} - I\boldsymbol{\Omega})/\tau_s$. As the result, the equation of particle rotation is reduced to the torque-balance equation (15) which expresses *symmetry* of the stress tensor (16): $\sigma_{ki} = \sigma_{ik}$. Indeed, on substitution $B_k = H_k + 4\pi M_k$ in (16 a) we are convinced of the symmetry:

$$\sigma_{ik} = -(p + \tfrac{1}{8\pi}H^2)\delta_{ik} + \eta\left(\frac{\partial v_i}{\partial x_k} + \frac{\partial v_k}{\partial x_i}\right) + \tfrac{1}{2}(M_i H_k + M_k H_i) + \tfrac{1}{4\pi}H_i H_k. \quad (16\,\text{b})$$

The system of equations (17) and (3) is still not complete since it does not determine the ferrofluid magnetization. The latter influences the fluid motion and depends itself on the motion as well. There are two basic ways to derive the missing magnetization equation. Both the ways have been proposed by the author with co-workers [3,11,12] and discussed in [2,4,9,13,14].

3 Phenomenological Magnetization Equation I

Originally the magnetization equation has been derived phenomenologically [3] as a modification of the Debye relaxation equation [15]. To get a generalized equation, one should introduce a local reference frame Σ', in which the suspended magnetic grains are quiescent *on the average*, i.e., $\boldsymbol{\omega}'_\text{p} = 0$. It is natural to assume that the magnetization relaxation is described in the system by the simplest Debye-like equation

$$\frac{\text{d}'\boldsymbol{M}}{\text{d}t} = -\frac{1}{\tau_\text{B}}(\boldsymbol{M} - \boldsymbol{M}_0), \tag{18}$$

where the *equilibrium* magnetization $\boldsymbol{M}_0$ is described by the Langevin formula (2)

$$\boldsymbol{M}_0 = nmL(\xi)\boldsymbol{\xi}/\xi, \qquad \boldsymbol{\xi} = m\boldsymbol{H}/k_\text{B}T, \tag{19}$$

and τ_B is the above determined Brownian time of rotational particle diffusion since reorientation of rigid magnetic dipoles is possible only when turning the particles themselves. Equation (18) assumes that any deviation of $\boldsymbol{M}$ – either in direction or magnitude – from its equilibrium value $\boldsymbol{M}_0$ decays according to the simple exponential law $(\boldsymbol{M} - \boldsymbol{M}_0) \sim \exp(-t/\tau_\text{B})$. The frame of reference Σ' rotates with respect to the fixed ("laboratory") system Σ with the angular velocity $\boldsymbol{\omega}_\text{p}$. The rates of change of any vector $\boldsymbol{A}$ in systems Σ and Σ' are related by the kinematic expression

$$\frac{\text{d}\boldsymbol{A}}{\text{d}t} = \boldsymbol{\omega}_\text{p} \times \boldsymbol{A} + \frac{\text{d}'\boldsymbol{A}}{\text{d}t}. \tag{20}$$

Substituting here $\boldsymbol{A} = \boldsymbol{M}$, $\boldsymbol{\omega}_\text{p}$ from (2), and $\text{d}'\boldsymbol{M}/\text{d}t$ from (7), we obtain the equation sought [3]:

$$\frac{\text{d}\boldsymbol{M}}{\text{d}t} = \boldsymbol{\Omega} \times \boldsymbol{M} - \frac{1}{\tau_\text{B}}(\boldsymbol{M} - \boldsymbol{M}_0) - \frac{1}{6\eta\phi}\boldsymbol{M} \times (\boldsymbol{M} \times \boldsymbol{H}). \tag{21}$$

If we assume – following [1] – instant relaxation of the magnetization, $\tau_\text{B} \to 0$, then (21) is reduced to $\boldsymbol{M} = \boldsymbol{M}_0$ in complete agreement with the quasistationary theory, see (2). Multiplying (21) by $\boldsymbol{M}$ we find

$$\text{d}M^2/\text{d}t = (2/\tau_\text{B})(M^2 - \boldsymbol{M} \cdot \boldsymbol{M}_0),$$

whence it follows that the last (relaxation) term in (21) describes a process of approach of the vector $\boldsymbol{M}$ to its equilibrium orientation without change of its length. As the result, the relaxation rates of the longitudinal and transverse (to the field) components of magnetization appears to be different. Let us calculate them. In the linear approximation in $\boldsymbol{M} - \boldsymbol{M}_0$ one can split the nonequilibrium part of magnetization into components parallel and perpendicular to the field:

$$\boldsymbol{M} - \boldsymbol{M}_0 = \frac{\boldsymbol{H}\left[\boldsymbol{H} \cdot (\boldsymbol{M} - \boldsymbol{M}_0)\right]}{H^2} + \frac{\boldsymbol{H} \times (\boldsymbol{M} \times \boldsymbol{H})}{H^2}. \tag{22}$$

Substituting this expression into (21) gives the linear magnetization equation

$$\frac{d\boldsymbol{M}}{dt} = \boldsymbol{\Omega} \times \boldsymbol{M} - \frac{\boldsymbol{H}\,[\boldsymbol{H}\cdot(\boldsymbol{M}-\boldsymbol{M}_0)]}{H^2 \tau_\parallel} - \frac{\boldsymbol{H}\times(\boldsymbol{M}\times\boldsymbol{H})}{H^2 \tau_\perp}, \qquad (23)$$

where $\tau_\parallel = \tau_B$ and

$$\frac{1}{\tau_\perp} = \frac{1}{\tau_B} + \frac{nmL(\xi)H}{6\eta\phi} = \frac{1}{\tau_B}\left[1 + \frac{1}{2}\xi L(\xi)\right],$$

so that

$$\tau_\perp = \frac{2\tau_B}{2 + \xi L(\xi)}. \qquad (24)$$

Let us revert to the Debye–like equation (18) and demonstrate that this *postulated* by myself [3] relaxation equation can be easily *derived* [12] from irreversible thermodynamics (IT). With this purpose it is convenient to take advantage of the method of IT proposed by Landau and first applied by him just to the description of relaxation of the order parameter in a nonequilibrium system [16]. An equilibrium value of the parameter ($\boldsymbol{M}_0$ in our case) corresponds to the minimum of an appropriate thermodynamic potential Φ (usually the Gibbs or Helmholtz free energy) depending on the magnetization $\boldsymbol{M}$ and other thermodynamic variables. Thus, at the equilibrium $\partial\Phi/\partial\boldsymbol{M} = 0$. Out of equilibrium this condition is not satisfied, so the relaxation process occurs: $\boldsymbol{M}$ changes in time approaching $\boldsymbol{M}_0$. For small deviations from equilibrium, the derivative $\partial\Phi/\partial\boldsymbol{M}$ and the relaxation rate $d\boldsymbol{M}/dt$ are small. The relation between the two derivatives in the Landau theory is reduced to simple proportionality:

$$\frac{d\boldsymbol{M}}{dt} = -\gamma \frac{\partial\Phi}{\partial\boldsymbol{M}} \qquad (25)$$

with a constant coefficient $\gamma > 0$. Hence we have

$$\frac{d\Phi}{dt} = \frac{\partial\Phi}{\partial\boldsymbol{M}} \cdot \frac{d\boldsymbol{M}}{dt} = -\gamma\left(\frac{\partial\Phi}{\partial\boldsymbol{M}}\right)^2 < 0 \qquad (26)$$

as it *should be*: when a system moves to equilibrium, its free energy *decreases*. In the case of a weakly nonequilibrium state of the system, one can substitute in (25) and (26) the expansion

$$\frac{\partial\Phi}{\partial\boldsymbol{M}} = \left(\frac{\partial\Phi}{\partial\boldsymbol{M}}\right)_0 + \left(\frac{\partial^2\Phi}{\partial\boldsymbol{M}^2}\right)_0 (\boldsymbol{M}-\boldsymbol{M}_0) + \ldots,$$

where subscript 0 marks the *point of equilibrium*. As the first derivative in this point is equal to zero and the second one is positive, equation (25) turns into (18) with $\tau_B^{-1} = \gamma(\partial^2\Phi/\partial\boldsymbol{M}^2)_0$ and (26) takes the form

$$\frac{d\Phi}{dt} = -\frac{(\boldsymbol{M}-\boldsymbol{M}_0)^2}{\gamma \tau_B^2}. \qquad (27)$$

Thus, (18) and hence (21) are well corroborated by the method of IT. Notice that some different thermodynamic method based on a strict formulation of entropy production leads [17] to the same phenomenological magnetization equation: Eq. (21) coincides with Eq. (106) from [17]. Equations (3), (17) and (21) constitute the complete set of conventional ferrohydrodynamic equations.

4 Magnetization Equation Derived Microscopically

Aforecited phenomenological method allows to obtain *linear* relaxation term in kinetic equation (18) and corresponding *quadratic* term for the rate of free energy diminution (27). It is clear that such terms are valid only for small departures from equilibrium. Indeed, equation (21) allows to describe well the *rotational viscosity* (see Sect. 6) for arbitrary intensity ξ of a stationary magnetic field but only sufficiently small values of $\Omega\tau_B$. The applicability of (21) in this case was corroborated by real [18-20] and numerical [21] experiments as well as by numerical [22] and analytical [23] solutions of the Fokker–Planck equation. [It is well worth noting that the restriction $\Omega\tau_B \ll 1$ is not so much to the point. Owing to small values of τ_B for ferrofluids based on low-viscous carrier liquids (with $\eta_0 \sim 10^{-2}$ P) the above inequality is satisfied at all reasonable fluid vorticities. As far as we know, violations of this condition took place only in experiments [24-26] with glycerine-based ferrofluids ($\eta_0 \sim 10$ P)]. Meanwhile, under consideration of strongly nonequilibrium situations characterized by large values of dimensionless fluid vorticity $\Omega\tau_B$ and especially frequency $\omega\tau_B$ of alternating magnetic field $H \propto \cos\omega t$, one should employ a more precise magnetization equation. Such a *macroscopic* equation has been derived [11] from the kinetic Fokker–Planck equation which provides the *microscopic* description of particle diffusion in colloids.

The Fokker–Planck equation for a ferrofluid moving in magnetic field $\boldsymbol{H}$ has the form [9,11]

$$2\tau_B \frac{\partial W}{\partial t} = \hat{\boldsymbol{R}} \cdot (\hat{\boldsymbol{R}} - 2\tau_B\, \boldsymbol{\Omega} - \boldsymbol{e} \times \boldsymbol{\xi}\,)W\,, \qquad (28)$$

where $\boldsymbol{e} = \boldsymbol{m}/m$ is the unit vector along the particle magnetic moment, $\boldsymbol{\xi} = m\boldsymbol{H}/k_B T$, and $\hat{\boldsymbol{R}} = \boldsymbol{e} \times \partial/\partial\boldsymbol{e}$ is the infinitesimal rotation operator. Equation (28) determines the orientational distribution function $W(\boldsymbol{e},t)$ of particles magnetic moments. The macroscopic magnetization is determined by the relation $\boldsymbol{M}(t) = nm\langle\boldsymbol{e}\rangle$ where angular brackets denote statistical averaging with the distribution function. Multiplying (28) by $\boldsymbol{e}$ and integrate over the angles, we arrive at the equation

$$\tau_B \frac{\mathrm{d}\langle\boldsymbol{e}\rangle}{\mathrm{d}t} = \tau_B \boldsymbol{\Omega} \times \langle\boldsymbol{e}\rangle - \langle\boldsymbol{e}\rangle - \tfrac{1}{2}\langle\boldsymbol{e}\times(\boldsymbol{e}\times\boldsymbol{\xi})\rangle\,, \qquad (29)$$

which however is not closed. Indeed, along with the first moment of the distribution function, $\langle\boldsymbol{e}\rangle$, equation (29) contains the second moment (the last term in the equation). It is easy to make sure that the equation for the second moment includes the third one, and so on, thus there is the infinite chain of cross-linked equations. Ideally, however, one would like to have only one equation since only the first moment – magnetization – has a clear physical meaning. An original scheme of closure of the first-moment equation (29), titled the *effective field method* (EFM), has been proposed in [11]. Let us explain the fruitful physical idea.

In equilibrium ($\Omega = 0$) under a constant magnetic field the stationary solution of (28) is the Gibbs distribution

$$W_0(e) = \frac{\xi}{4\pi \sinh \xi} \exp(\boldsymbol{\xi} \cdot \boldsymbol{e}). \tag{30}$$

An averaging of the vector $\boldsymbol{e}$ with function (30) gives expression (19) for the equilibrium magnetization. Note that only in true equilibrium the magnetization is one or another function of the field. In a nonequilibrium state there is *no connection* between $\boldsymbol{M}$ and $\boldsymbol{H}$: any arbitrary magnetization may be created – in principle – even in the absence of the field. Nevertheless, one may consider any value of $\boldsymbol{M}$ as an equilibrium magnetization in a certain – specially prepared – magnetic field. This *effective field* $\boldsymbol{H}_e$ is related to the *nonequilibrium* magnetization by the *equilibrium* relation:

$$\boldsymbol{M} = nmL(\zeta)\,\boldsymbol{\zeta}/\zeta \qquad \boldsymbol{\zeta} = m\boldsymbol{H}_e/k_B T. \tag{31}$$

During the equilibrium settling process, the magnetization (31) relaxes to its equilibrium value (19) as the effective field $\boldsymbol{H}_e$ (or $\boldsymbol{\zeta}$) approaches the true field $\boldsymbol{H}$ (or $\boldsymbol{\xi}$). Comparing (19) and (31) we see that the latter is obtained by averaging of $\boldsymbol{e}$ with the distribution function

$$W_e(e) = \frac{\zeta}{4\pi \sinh \zeta} \exp(\boldsymbol{\zeta} \cdot \boldsymbol{e}), \tag{32}$$

which differs from the Gibbs distribution (30) by replacement of the true dimensionless field $\boldsymbol{\xi}$ by the effective field $\boldsymbol{\zeta}$. Carrying out the averaging in (29) with the function (32), we find the sought equation [11]

$$\frac{d\boldsymbol{M}}{dt} = \boldsymbol{\Omega} \times \boldsymbol{M} - \left[1 - \frac{(\boldsymbol{\xi} \cdot \boldsymbol{\zeta})}{\zeta^2}\right]\frac{\boldsymbol{M}}{\tau_B} - \frac{1}{L(\zeta)}\left[\frac{1}{L(\zeta)} - \frac{1}{\zeta}\right]\frac{\boldsymbol{M} \times (\boldsymbol{M} \times \boldsymbol{H})}{6\eta\phi}. \tag{33}$$

This equation together with (31) determines the dependence $\boldsymbol{M}(t;\boldsymbol{H},\boldsymbol{\Omega})$ in an implicit form, where the dimensionless effective field $\boldsymbol{\zeta}$ is the parameter. In the case of small departures from equilibrium, the effective field might be represented as a sum of the true field and some small correction: $\boldsymbol{\zeta} = \boldsymbol{\xi} + \boldsymbol{\nu}$. Then from (19) and (31) in the linear approximation in $\boldsymbol{\nu}$ we get

$$\boldsymbol{M} - \boldsymbol{M}_0 = nm\left[\frac{dL(\xi)}{d\xi}\boldsymbol{\nu}_\| + \frac{L(\xi)}{\xi}\boldsymbol{\nu}_\perp\right], \tag{34}$$

where the components

$$\boldsymbol{\nu}_\| = \boldsymbol{\xi}(\boldsymbol{\nu} \cdot \boldsymbol{\xi})/\xi^2, \qquad \boldsymbol{\nu}_\perp = \boldsymbol{\xi} \times (\boldsymbol{\nu} \times \boldsymbol{\xi})/\xi^2$$

are parallel and perpendicular to the true field, respectively. Employing the relation (34), one can reduce (33) to the linear magnetization equation (23), where relaxation times of the longitudinal and transverse components of magnetization are

$$\tau_\| = \frac{\mathrm{d}\ln L(\xi)}{\mathrm{d}\ln\xi}\tau_\mathrm{B}\,, \qquad \tau_\perp = \frac{2L(\xi)}{\xi - L(\xi)}\tau_\mathrm{B}\,. \qquad (35)$$

It is well-established that equation (33) derived by EFM describes magnetization processes in real magnetic colloids very well. Predictions of the theory are well corroborated by experiments on positive [27] and negative [28,29] rotational viscosities, and on magneto-vortical birefringence [24] in ferrofluids. Solutions of (33) agree perfectly with the results of numerical integration of the non-stationary Fokker-Planck equation [13], and what is more, with the numerical simulation of Brownian dynamics performed by Cebers [21] on the basis of Langevin equation for rotational motion of magnetic grains in the presence of magnetic field. Phenomenological equation (21) is certainly far simpler for analysis than (33). However, the latter guarantees the correct description of magnetization even if its deviation from equilibrium value is large (e.g., at $\Omega\tau \gg 1$), that is when (21) leads to erroneous results. Interestingly, a new phenomenological magnetization equation derived recently [12] from irreversible thermodynamics (see below) is free partially from the above mentioned shortcoming of (21).

5   Phenomenological Magnetization Equation II

We have shown in Sect. 3 how does Debye equation (18) originate from the potential $\Phi(M)$. Instead of the magnetization, one can choose as an independent variable the effective field and introduce the potential $\tilde{\Phi}(H_\mathrm{e})$. Then instead of (25) we obtain in similar fashion

$$\frac{\mathrm{d}H_\mathrm{e}}{\mathrm{d}t} = -\tilde{\gamma}\frac{\partial\tilde{\Phi}}{\partial H_\mathrm{e}}\,, \qquad \tilde{\gamma} > 0.$$

Acting further as in Sect. 3, we arrive at the equation (cf. (18))

$$\frac{\mathrm{d}'H_\mathrm{e}}{\mathrm{d}t} = -\frac{1}{\tau_\mathrm{B}}\left(H_\mathrm{e} - H\right), \qquad (36)$$

where we set $\tilde{\gamma}^{-1} = (\partial^2\tilde{\Phi}/\partial H_\mathrm{e}^2)\tau_\mathrm{B}$. Under this choice, (36) turns into (18) in the low-field limit, when the true magnetization and its equilibrium value take the form $M = \chi H_\mathrm{e}$ and $M_0 = \chi H$, respectively; here $\chi = nm^2/3k_BT$ stands for the initial magnetic susceptibility.

Equation (36) is written out in a rotating coordinate system Σ'. Reverting to the immobile system Σ by the general formula (20) and eliminating ω_p with the aid of torque-balance equation (15), we obtain [12]

$$\frac{\mathrm{d}H_\mathrm{e}}{\mathrm{d}t} = \Omega \times H_\mathrm{e} - \frac{1}{\tau_\mathrm{B}}\left(H_\mathrm{e} - H\right) - \frac{1}{6\eta\phi}H_\mathrm{e}\times(M\times H). \qquad (37)$$

This equation (marked below as Sh'01) coincides with the previous phenomenological equation (21) (marked as Sh'72) in the limit of low magnetic field. However, due to nonlinearity of the Langevin magnetization law, equations (21) and (37) predict very different magnetization values for large enough magnitudes of ξ. Note that as shown above, both these equations are in full agreement with fundamental principles and hence both they are correct from phenomenological point of view: irreversible thermodynamics gives preference neither to (21) nor to (37).

In the case of small deviations from equilibrium, equation (37) can be linearized with respect to $\boldsymbol{H}_e - \boldsymbol{H}$ and $\boldsymbol{M} - \boldsymbol{M}_0$. Substituting $\boldsymbol{\nu} = \boldsymbol{\zeta} - \boldsymbol{\xi} = (m/k_B T)(\boldsymbol{H}_e - \boldsymbol{H})$ into (34) gives

$$(\boldsymbol{M} - \boldsymbol{M}_0)_\| = 3\chi \frac{dL(\xi)}{d\xi}(\boldsymbol{H}_e - \boldsymbol{H})_\|, \qquad \boldsymbol{M}_\perp = 3\chi \frac{L(\xi)}{\xi}(\boldsymbol{H}_e)_\perp .$$

Employing these relationships and the equality

$$\frac{d\boldsymbol{M}}{dt} = \frac{d\boldsymbol{M}}{d\boldsymbol{H}_e} \cdot \frac{d\boldsymbol{H}_e}{dt} = 3\chi \left[\frac{L(\xi)}{\xi} \frac{d(\boldsymbol{H}_e)_\|}{dt} + \frac{L(\xi)}{\xi} \frac{d(\boldsymbol{H}_e)_\perp}{dt} \right]$$

we revert to (23) with $\tau_\| = \tau_B$ and $\tau_\perp$ from (24). Thus, in weakly nonequilibrium situations the old (21) and the new (37) phenomenological magnetization equations are reduced to each other at any field strength.

6 Testing Magnetization Equations

6.1 Rotational Viscosity in a Stationary Field

As a checking on applicability of the discussed above magnetization equations we choose their predictions about the *rotational* or *spin viscosity* η_r of ferrofluids. Below we shall compare η_r obtained from (21) and (37) with its almost exact value resulting from the EFM equation (33).

The Einstein formula for viscosity of suspension, $\eta = \eta_0(1 + 2.5\phi)$, was obtained without taking into account the rotational motion of suspended particles relative to carrier liquid. If however ferrofluid is subject to the field $\boldsymbol{H}$, the latter impedes free particle rotation – see (15). But any deviation of the particles angular velocity $\boldsymbol{\omega}_p$ from the angular velocity of the fluid $\boldsymbol{\Omega}$ leads to an additional dissipation which is just manifested in rotational viscosity η_r [3]. Let us calculate it.

The boundary wall streamlined by a ferrofluid is acted (on unit area) by the force $f_i = [\sigma_{ik}]n_k$ where [ ] denotes difference evaluated across the fluid–solid interface and $\boldsymbol{n}$ is the normal to the interface. The friction (tangential) force exerted on the wall is $f_\tau = [\sigma_{\tau n}]$. By using the electrodynamic boundary conditions $[H_\tau] = 0$ and $[B_n] = 0$, we get from (16 a)

$$f_\tau = \eta(\partial v_\tau/\partial x_n) + \tfrac{1}{2}(M_\tau H_n - M_n H_\tau) . \tag{38}$$

For the Poiseuille or Couette flow, $\boldsymbol{v} = (0, v(x), 0)$, in a transversal magnetic field, $\boldsymbol{H} = (H, 0, 0)$, the magnetization has two components: $\boldsymbol{M} = (M_x, M_y, 0)$, so (38) may be written in the form $f_\tau = 2(\eta + \eta_r)\Omega$, where $2\Omega = \partial v/\partial x$ and rotational viscosity is defined as

$$\eta_r = M_y H/4\Omega. \tag{39}$$

Thus, the additional viscosity is expressed through the off-axis component of magnetization M_y. For small $\Omega \tau_B$, when all three cited above magnetization equations are reduced to (23), this component is also small, $M_y \propto \Omega \tau_B$:

$$\boldsymbol{M}_\perp = \tau_\perp \boldsymbol{\Omega} \times \boldsymbol{M}_0, \quad \text{i.e.,} \quad M_y = \tau_\perp M_0 \Omega. \tag{40}$$

Eliminating M_y from (39)-(40), we find

$$\eta_r = \tfrac{1}{4} \tau_\perp M_0 H. \tag{41}$$

Substituting here $M_0 = nmL(\xi)$ and $\tau_\perp$ from (24), we arrive at the formula [3]

$$\eta_r(\xi) = \frac{3}{2}\eta\phi \frac{\xi L(\xi)}{2 + \xi L(\xi)} = \frac{3}{2}\eta\phi \frac{\xi - \tanh\xi}{\xi + \tanh\xi}. \tag{42}$$

So, both phenomenological equations, (21) and (37), predict the same dependence of rotational viscosity on magnetic field strength. In the absence of the field an individual particle "rolls" freely along corresponding shear surface with angular velocity $\boldsymbol{\omega}_p$ equal to $\boldsymbol{\Omega}$, so that $\eta_r(0) = 0$. Conversely, $\eta_r(\xi)$ attains its limiting value $\eta_r(\infty) = \tfrac{3}{2}\eta\phi$ (the saturation) when *rolling* of the particle is replaced by *slipping*: the field of sufficiently large intensity guarantees constancy of the particle's orientation, not allowing it to twist with the fluid. Note that the saturation value of η_r does *not depend* on a concrete form of the magnetization equation but follows directly from the equation of fluid motion (17). Actually, in the limit under consideration $\boldsymbol{\omega}_p = 0$, so that (15) takes the form $\boldsymbol{M} \times \boldsymbol{H} = -6\eta\phi\boldsymbol{\Omega}$. Substituting this torque in (17) and grouping there the second and fourth terms,

$$\eta \nabla^2 \boldsymbol{v} + \tfrac{1}{2} \text{rot}\, (\boldsymbol{M} \times H) = \left(\eta + \tfrac{3}{2}\eta\phi\right)\nabla^2 \boldsymbol{v},$$

we immediately arrive at $\eta_r(\infty) = \tfrac{3}{2}\eta\phi$ since the quantity added here to the ordinary viscosity should be regarded as the rotational one.

The EFM-equation (33) yields a result somewhat different from (42). Substituting of $\tau_\perp$ from (35) into (41) gives

$$\eta_r(\xi) = \frac{3}{2}\eta\phi \frac{\xi L^2(\xi)}{\xi - L(\xi)}. \tag{43}$$

Figure 1 shows that though at first sight functions (42) and (43) do not appear alike, they agree closely in the entire range of their argument. Both of them approach the saturation value $\eta_r(\infty) = \tfrac{3}{2}\eta\phi$ at $\xi \gg 1$. In the figure we

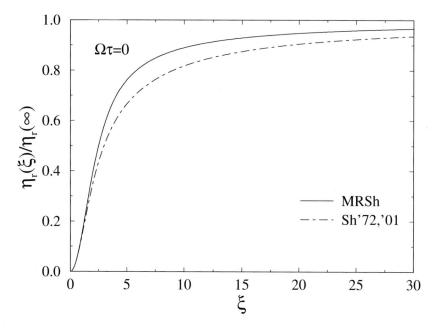

Fig. 1. Reduced rotational viscosity in the Newtonian limit $\Omega\tau_B \to 0$ versus the dimensionless field strength, as calculated from (42) (Sh'72 [3] and Sh'01 [12]) and (43) (MRSh [11]).

plot the reduced rotational viscosity $\eta_r(\xi)/\eta_r(\infty)$ as a function of ξ. The upper curve calculated by (43) of the EFM [11] represents a very good approximation. Actually, as shown in [9,22,23], it hardly differs from the direct solution of the Fokker-Planck equation (28) in linear approximation in $\Omega\tau_B$. The phenomenological equations, (21) and (37), also provide a quite satisfactory description of the rotational viscosity. Both they result in the lower curve in Fig. 1 that is described by the *Shliomis' formula* (42). This function agrees with (43) in the low- and high-field limits and deviates from it, at most, on 15 % in the entire range of the argument ξ. As seen from (39), η_r does not depend on the flow vorticity till M_y is proportional to Ω, what takes place only if $\Omega\tau_B \ll 1$. For finite values of $\Omega\tau_B$ the viscosity (39) does depend on Ω. As a result, the function $\sigma_{\tau n}(\Omega)$ deviates from the linear one, i.e., a ferrofluid acquires rheological properties.

Proceeding to the calculation of the non-Newtonian viscosity on the basis of (37), it is convenient to pass from the fields $\boldsymbol{H}$ and $\boldsymbol{H}_e$ to their nondimensional values $\boldsymbol{\xi}$ and $\boldsymbol{\zeta}$:

$$\frac{d\boldsymbol{\zeta}}{dt} = \boldsymbol{\Omega} \times \boldsymbol{\zeta} - \frac{1}{\tau_B}(\boldsymbol{\zeta} - \boldsymbol{\xi}) - \frac{L(\zeta)}{2\tau_B \zeta} \boldsymbol{\zeta} \times (\boldsymbol{\zeta} \times \boldsymbol{\xi}). \tag{44}$$

At the stated above arrangement of the applied magnetic field with respect to the fluid flow, the last equation admits a steady solution in which the effective field $\boldsymbol{\zeta}$ tracks the true field $\boldsymbol{\xi}$ with lag angle α, i.e., $\boldsymbol{\zeta} = (\zeta\cos\alpha, \zeta\sin\alpha, 0)$. The dependence of ζ and α upon ξ and $\Omega\tau_B$ is given by

$$\sqrt{\xi^2 - \zeta^2} = \frac{2\Omega\tau_B\,\zeta}{2 + \zeta L(\zeta)}, \qquad \cos\alpha = \frac{\zeta}{\xi}. \tag{45}$$

Substituting $M_y = nmL(\zeta)\sin\alpha$ in (39) and using (45), we obtain

$$\eta_r = \frac{3}{2}\eta\phi\,\frac{\zeta L(\zeta)}{2 + \zeta L(\zeta)}. \qquad \text{(Sh'01)} \tag{46}$$

By the same way we find from (33)

$$\sqrt{\xi^2 - \zeta^2} = \frac{2\Omega\tau_B\,\zeta L(\zeta)}{\zeta - L(\zeta)}, \qquad \cos\alpha = \frac{\zeta}{\xi}, \tag{47}$$

that results in

$$\eta_r = \frac{3}{2}\eta\phi\,\frac{\zeta L^2(\zeta)}{\zeta - L(\zeta)}. \qquad \text{(MRSh)} \tag{48}$$

The solution of equation (21) can be presented in a similar form. Let us introduce a new variable $\boldsymbol{\zeta}$ instead of $\boldsymbol{M}$ by the relation $\boldsymbol{M} = M_0(\boldsymbol{\zeta}/\xi)$ where $M_0 = nmL(\xi)$. It is worth noting that $\boldsymbol{\zeta}$ is no more an effective field unlike ζ in preceding relationships (45)–(48). By substituting the components $M_x = M_0(\zeta/\xi)\cos\alpha$ and $M_y = M_0(\zeta/\xi)\sin\alpha$ in (21), we get $\cos\alpha = M/M_0 = \zeta/\xi$ and

$$\sqrt{\xi^2 - \zeta^2} = \frac{2\Omega\tau_B\,\xi\zeta}{2\xi + \zeta^2 L(\xi)}. \tag{49}$$

This expression together with the definition (39) yield

$$\eta_r = \frac{3}{2}\eta\phi\,\frac{\zeta^2 L(\xi)}{2\xi + \zeta^2 L(\xi)}. \qquad \text{(Sh'72)} \tag{50}$$

In the limit $\Omega\tau_B \ll 1$, one can neglect the value $\Omega\tau_B$ in (45), (47) and (49), after what all three of these relationships are reduced to $\zeta = \xi$. Eliminating now ζ from (13) and (17), we see that, as it should be, expressions (50) and (46) obtained from the old and the new phenomenological magnetization equations turn into (42), while the EFT formula (48) is transformed into (43). When, however, the ferrofluid is subjected to a sufficiently large shear rate, $\Omega\tau_B \geq 1$, the flow induces – along with the Brownian motion – a quotient *demagnetization* since under the viscous shear the magnetic grains tend to be rotated out of alignment with the magnetic field. Formally, this effect originates from decreasing the parameter ζ determined by (45), (47) and (49). According to these equations, $\zeta = \xi$ when $\Omega\tau_B = 0$ but the more there is of $\Omega\tau$ at constant ξ, the less there is of ζ. The reduction of the magnetization leads in turn to some decrease in the rotational viscosity. This decrease, imperceptible in practice up to $\Omega\tau_B \simeq 1$, then becomes very significant. Figures 2–4 illustrates the dependence of the viscosity increase on the magnetic-field strength for three values of the product $\Omega\tau_B$. Interestingly, under the finite shear rate the viscosities given by (46) and (50) do not coincide

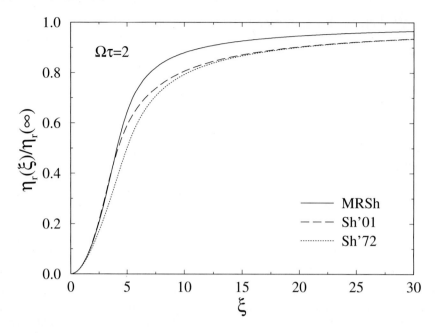

Fig. 2. Dependence of the rotational viscosity on the field for the dimensionless shear rate $\Omega\tau_B = 2$, as calculated from (47)–(48) MRSh, (45)–(46) Sh'01, and (49)–(50) Sh'72.

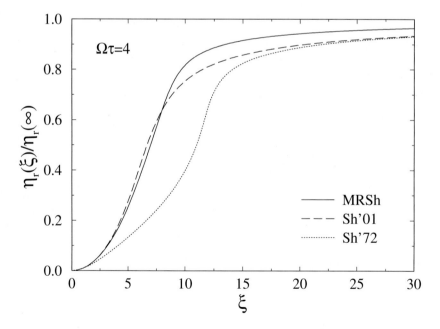

Fig. 3. Same as Fig. 2, but for $\Omega\tau_B = 4$.

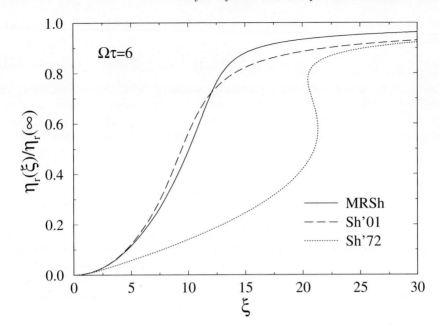

Fig. 4. Same as Figs. 2 and 3, but for $\Omega\tau_B = 6$.

with each other any more. As seen from the plot, the higher the shear the more discrepancy between viscosity values predicted by the new and the old phenomenological equations. At high shear in a high field, the old equation [(21), Sh'72] predicts a *hysteresis* of viscosity (see Fig. 4), which however is corroborated neither by direct calculations of [13,20] nor by the solution (47)–(48) of the EFM equation (33). The new equation [(37), Sh'01] also does not predict such a hysteresis, but it provides us with a quite satisfactory viscosity description in a wide region of parameters ξ and $\Omega\tau_B$. Indeed, in this entire region the solutions of (33) and (37) agree closely, as shown in Figs. 2–4. Thus, in the case of a *stationary* magnetic field, equation (37) can be recommended for an employment on the same level with (33). It is worth nothing that all the above calculations, carried out for a shear flow, apply equally to a rigid rotation of a ferrofluid with an angular velocity Ω in a constant transversal magnetic field, $H \perp \Omega$, and to a quiescent ferrofluid subjected to a uniform rotating field $H = (H \cos \Omega t, H \sin \Omega t, 0)$ as well.

6.2 "Negative Viscosity" under an Alternating Magnetic Field

For a stationary field and small $\Omega\tau_B$, equation (23) yields the magnetic torque

$$M \times H = -\tau_\perp M_0 H \Omega. \tag{51}$$

On comparing this expression with the torque-balance equation (15),

$$M \times H = 6\eta\phi(\Omega - \omega_p), \tag{52}$$

and employing the definition (41), $\eta_r = \frac{1}{4}\tau_\perp M_0 H$, we find a clear and physically transparent relationship

$$\eta_r = \frac{3}{2}\eta\phi\frac{\Omega - \omega_p}{\Omega}. \tag{53}$$

Note that being obtained for a stationary magnetic field, (53) is valid also for oscillatory field, since (51) and (52) remain to be true after averaging over the period of the field variation as well. According to (53), $\eta_r > 0$ if magnetic grains rotate slowly than the fluid ($\omega_p < \Omega$) and then decelerate the flow. Such a situation takes place in a stationary (see Sect. 6.1) or slow time-varying magnetic field. The field prevents free particle rotation with the fluid angular velocity Ω and thereby forces the surrounding liquid to flow past the particles. It leads to an additional dissipation of the kinetic energy of the fluid and is manifested in the rotational viscosity $\eta_r > 0$.

An *alternating*, linearly polarized magnetic field $\boldsymbol{H} = (H_0\cos\omega t, 0, 0)$ induces rotational swings of the grains but does not single out any preferred direction of their rotation. Therefore, an averaging over a physically small element of ferrofluid volume results in spin of zero: $\omega_p = 0$. However, any flow with a vorticity not zero is sufficient to break the degeneracy of the rotation direction and leads to a nonzero macroscopic spin rate of the magnetic grains. When the field frequency is high enough, the grains rotate faster than the fluid ($\omega_p > \Omega$) and then they *spin up* the flow. The acceleration happens, naturally, at the expense of ac-field energy and manifests itself in a reduction of the total viscosity, i.e., in a *negative* rotational viscosity: $\eta_r < 0$ [see (53)].

This "negative viscosity" effect has been predicted by Shliomis and Morozov [30] and verified experimentally by Bacri's [28] and Rehberg's [29] groups. The first prediction has been made on the base of Debye-like equation (21); afterwards the results [30] were recalculated [28] by using a more precise EFM equation (33). However, in the case of a low field amplitude, $\xi_0 \equiv mH_0/k_BT \ll 1$, and a low shear rate, $\Omega\tau_B \ll 1$, both these equations admit the same simple analytical solution [30]

$$\eta_r = \frac{1}{8}\eta\phi\xi_0^2\frac{1 - \omega^2\tau_B^2}{(1 + \omega^2\tau_B^2)^2}. \tag{54}$$

As seen, at $\omega\tau_B = 1$ the rotational viscosity changes its sign, passing from the domain of positive ($\omega\tau_B < 1$) to that of negative ($\omega\tau_B > 1$) values. In the latter domain, it attains the minimum at $\omega\tau_B = \sqrt{3}$ and tends to zero at $\omega\tau_B \to \infty$ since in this limit ferrofluid does not have enough time for re-magnetization: the grains cease to feel the magnetic field.

To solve (33) for finite amplitudes of oscillating magnetic field, it is convenient to present this equation in the dimensionless form measuring the time in units of τ_B:

$$\frac{d}{dt}\left[L(\zeta)\frac{\boldsymbol{\zeta}}{\zeta}\right] = \tau_B\boldsymbol{\Omega}\times\left[L(\zeta)\frac{\boldsymbol{\zeta}}{\zeta}\right] - \frac{L(\zeta)}{\zeta}(\boldsymbol{\zeta} - \boldsymbol{\xi}) - \frac{1}{2\zeta^2}\left[1 - \frac{3L(\zeta)}{\zeta}\right]\boldsymbol{\zeta}\times(\boldsymbol{\zeta}\times\boldsymbol{\xi}). \tag{55}$$

The problem contains usually a small parameter $\Omega\tau_B$. In this case (55) may be solved by the theory of perturbations. In zero approximation in the parameter, the effective field is evidently parallel to the true field, while its magnitude is determined by the equation

$$\frac{d\zeta}{dt} = -\left[\frac{d\ln L(\zeta)}{d\ln \zeta}\right]^{-1}(\zeta - \zeta_0 \cos\omega t). \tag{56}$$

In the linear approximation in $\Omega\tau_B$, (55) yields

$$\boldsymbol{M} = M^{(0)}\boldsymbol{h} + M^{(1)}\tau_B\boldsymbol{\Omega}\times\boldsymbol{h}, \quad M^{(0)} = nmL(\zeta), \quad M^{(1)} = M^{(0)}\Psi(\zeta),$$

where $\boldsymbol{h} = \boldsymbol{H}/H$ is the unit vector along the field and function $\Psi(\zeta)$ is satisfied to the linear equation

$$\frac{d\Psi}{dt} = 1 - \frac{\zeta_0}{2}\left[\frac{1}{L(\zeta)} - \frac{1}{\zeta}\right]\Psi\cos\omega t. \tag{57}$$

The magnetic torque $\boldsymbol{M}\times\boldsymbol{H}$ can be averaged over the period of field variation $2\pi/\omega$ because we are interested in $\omega \sim \tau_B^{-1}$ while τ_B is always much less than the characteristic hydrodynamic time $\sim \rho l^2/\eta$, where l is the reference spatial scale of the fluid motion. For the mean magnetic torque one gets

$$\overline{\boldsymbol{M}\times\boldsymbol{H}} = -6\eta\phi\, g(\xi_0, \omega\tau_B)\boldsymbol{\Omega}, \quad g = \tfrac{1}{2}\xi_0\,\overline{L(\zeta)\Psi(\zeta)\cos\omega t}. \tag{58}$$

Averaging now the equation of fluid motion (17) over the time and substituting into (17) the mean torque (58), one may group the second and fourth terms in this equation:

$$\eta\nabla^2\boldsymbol{v} - 3\eta\phi g\,\mathrm{rot}\,\boldsymbol{\Omega} = \eta\left(1 + \tfrac{3}{2}\phi g\right)\nabla^2\boldsymbol{v}.$$

Thus the rotational viscosity is $\eta_r = \tfrac{3}{2}\eta\phi g$. In accordance with (52) and (58), the angular velocity of the particles may be written as $\boldsymbol{\omega_p} = (1-g)\boldsymbol{\Omega}$, so that η_r can be presented in the form (53). For arbitrary ξ_0 and $\omega\tau_B$ the problem (56)–(57) was solved numerically [28]. Results of the computations are displayed in the *map of viscosity*, Fig. 5, where a set of isolines of reduced viscosity $g(\xi_0, \omega\tau_B) = \eta_r(\xi_0, \omega\tau_B)/\eta_r(\infty, 0)$ is presented in the plane $(\xi_0, \omega\tau_B)$. The isoline $g = 0$ parts the plane into two regions: $g > 0$ for $\omega\tau_B \leq 1$ and $g < 0$ for $\omega\tau_B \geq 1$. A similar map of viscosity, but calculated from the phenomenological magnetization equation (21), Sh'72, is shown in Fig. 6. A strong similarity between a plot of experimentally obtained isolines and that of Fig. 5 has been observed in [28]. As for the map in Fig. 6, the experimental data [28] agree with it on the whole, but there are some discrepancies in details. Namely, in the experiment [28] as well as in Fig. 5 the neutral isoline, $g = 0$, bends to the right, while in Fig. 6 it bends slightly to the left. Still more there is the difference between the isolines of Fig. 5 and those as calculated from the new phenomenological equation (37), Sh'01. Generally, this equation is very appropriate for description of ferrofluid magnetization in a stationary magnetic field (see Sect. 6.1), whereas in an alternating field it is working satisfactory only in the low-frequency limit $\omega\tau_B \ll 1$. [The only exception to the rule represents itself the case of *rotating* magnetic field of the kind of $\boldsymbol{H} = (H_0\cos\omega t, H_0\sin\omega t, 0)$.

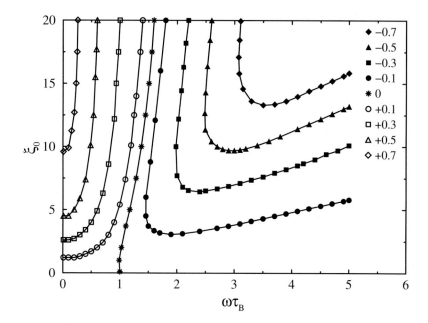

Fig. 5. Isolines of reduced viscosity $\eta_r(\xi_0, \omega\tau_B)/\eta_r(\infty, 0) = g$ in the plane $(\xi_0, \omega\tau_B)$ for $-0.7 \le g \le +0.7$, as calculated from EFM equation (33).

Such a field is reduced evidently to the stationary field $\boldsymbol{H} = (H_0, 0, 0)$ acting upon the fluid rotating as a whole with the angular velocity $\boldsymbol{\Omega} = (0, 0, \omega)$. Thus, as (37) is valid for any shear rate $\Omega\tau_B$ in a stationary field, it is also valid for any frequency of the field rotation $\omega\tau_B$].

A difference between discussed magnetization equations is also manifested at the relaxation from an equilibrium magnetization in a quiescent ferrofluid after the field is suddenly switched off. Then the fluid remains at rest, $\Omega = 0$, so $\boldsymbol{M}$ and $\boldsymbol{H}_e$ are always parallel to $\boldsymbol{H}$. Hence equations (21), (33) and (37) are reduced to

$$\mathrm{d}M/\mathrm{d}t = -(M - M_0)/\tau_B, \tag{59 a}$$

$$\mathrm{d}M/\mathrm{d}t = -(1 - H/H_e)M/\tau_B, \tag{59 b}$$

and

$$\mathrm{d}H_e/\mathrm{d}t = -(H_e - H)/\tau_B, \tag{59 c}$$

respectively. In Fig. 7 we plot the decay of reduced magnetization $M(t)/M_0$ according to (59) with $M_0 = nmL(\xi)$ for some initial field magnitudes $\xi = mH/k_BT$. As the true field H is switched off at the moment $t = 0$, equations (59 a) and (59 b) coincide with each other at $t > 0$, when their solution reads

$$M(t)/M_0 = \exp(-t/\tau_B), \tag{60}$$

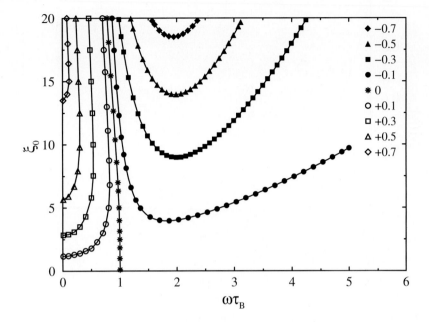

Fig. 6. Same as Fig. 5, but calculated from Debye-like equation (21).

i.e., it does not depend on ξ. Analogously, (59 c) has the solution $H_e(t) = H \exp(-t/\tau_B)$, so that we find

$$M(t)/M_0 = L(\xi e^{-t/\tau_B})/L(\xi). \tag{61}$$

The last decay predicted by the new magnetization equation (37), Sh'01, is exponential only in the limit $\xi \ll 1$, while the old phenomenological equation (21) and – what is much more important – the EFM equation (33) predicts exponential decay of the magnetization for any values of ξ. This difference in relaxation behavior side by side with the difference in the ferrofluid viscosity can be of relevance for testing the magnetization equations and the interpretation of corresponding experiments. At the description of nonstationary situations (like that as presented in Fig. 7) one should, however, *a priori* give preference to (21) and (33) before (37) since all previous predictions of (33) always were realized.

6.3 Response to Rotating Magnetic Field

When the field rotates in the (x, y)-plane, ferrofluid magnetization also rotates in this plane with the same speed as the field but lags behind in phase by a certain angle due to the finite relaxation time:

$$\boldsymbol{H} = (H_0 \cos \omega t, H_0 \sin \omega t, 0), \quad \boldsymbol{M} = [M \cos(\omega t - \alpha), M \sin(\omega t - \alpha), 0]. \tag{62}$$

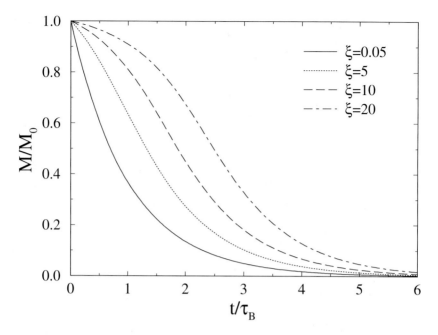

Fig. 7. Time dependence of the reduced magnetization $M(t)/M_0$ after the field ξ is switched off, as described by (61). The lowest curve also represents the solution (60) for any field.

Here M and α should be determined by one of three discussed above magnetization equations. For weakly nonequilibrium situations, when all of them are reduced to (23), we find

$$M = M_0 \cos\alpha, \quad \tan\alpha = (\omega - \Omega)\tau_\perp, \qquad (63)$$

where $M_0 = nmL(\xi)$ and $\tau_\perp$ is defined by (24) or (35), depending on the model we use.

Substituting from (62) into (16 b) gives for a nontrivial (non-diagonal) component of the stress tensor

$$\sigma_{xy} = \eta\left(\frac{\partial v_x}{\partial y} + \frac{\partial v_y}{\partial x}\right) + \tfrac{1}{8\pi}H_0^2 \sin 2\omega t + \tfrac{1}{2}M_0 H_0 \cos\alpha \sin(2\omega t - \alpha). \qquad (64)$$

On averaging over the period of the field variation both the magnetic terms in (64) disappear, so the rotating field does not induce any flow inside *the bulk* of the fluid. Interestingly, the magnetic grains do rotate with the angular velocity

$$\omega_p = \frac{M_0 H_0}{6\eta\phi} \cdot \frac{\omega\tau_\perp}{1 + \omega^2\tau_\perp^2},$$

while the fluid remains quiescent [31]. The point is, each a grain forces to spin the near-by mass of the viscous liquid thereby becoming a center of microscopic vortex, the size of which does not exceed the mean distance between the grains $\sim \phi^{-1/3} d$. But such a motion is not hydrodynamic yet, because averaging of the *microwhirls* over physically small fluid volume does not result in *macroscopic vorticity* $\boldsymbol{\Omega} = \frac{1}{2}\mathrm{rot}\,\boldsymbol{v}$. Indeed, microwhirls created by the neighboring grains *compensate* each other like elementary electric currents of neighboring molecules in Ampere's model of ferromagnetism do that.

The uniformity of both the internal moment of momentum $\boldsymbol{S} = I\boldsymbol{\omega}_\mathrm{p}$ and the magnetization $\boldsymbol{M}$ breaks off at the ferrofluid border. A jump-like increase in magnetization on the boundary surface (outside the fluid $\boldsymbol{M} = 0$) results in magnetic tangential surface stress. The latter just provide the coupling of rotating field and the fluid motion.

Let us consider ferrofluid confined within the layer by horizontal planes $x = 0, l$. Under the field (62) rotating in the vertical plane (x, y), the fluid can come into the motion along axis y, $\boldsymbol{v} = [0, v(x), 0]$, so that $\boldsymbol{\Omega} = (0, 0, \Omega)$ where $\Omega = \frac{1}{2}(\mathrm{d}v/\mathrm{d}x)$. The tangential force density (38) acting the upper surface is

$$f_y = [\sigma_{yx}] = \eta \frac{\mathrm{d}v}{\mathrm{d}x} - \tfrac{1}{2}(M_x H_y - M_y H_x). \tag{65}$$

Substituting from (62) and (63) into (65) yields

$$f_y = \eta \frac{\mathrm{d}v}{\mathrm{d}x} - \tfrac{1}{2} M H_0 \sin\alpha = 2\eta\Omega - \tfrac{1}{2} M_0 H_0 \frac{(\omega - \Omega)\tau_\perp}{1 + (\omega - \Omega)^2 \tau_\perp^2}. \tag{66}$$

As seen from (66), if the surface $x = l$ is free, $f_y = 0$, the rotating field generates the flow with the constant vorticity

$$\Omega \approx \frac{M_0 H_0}{4\eta} \cdot \frac{\omega \tau_\perp}{1 + \omega^2 \tau_\perp^2}, \tag{67}$$

so the plane Couette flow – with the linear velocity profile $v(x) = 2\Omega x$ – arises. [Note that we have neglected in the right-hand side of (67) the value Ω in comparison with ω since $\Omega/\omega \leq \frac{1}{4}\phi\xi^2$ at $\xi \ll 1$ and $\Omega/\omega \leq \frac{3}{2}\phi$ at $\xi \gg 1$]. If, conversely, the upper surface is fixed, $\Omega = 0$, to hold it one needs to apply the force in the opposite direction:

$$f_y = -\tfrac{1}{2} M_0 H_0 \frac{\omega \tau_\perp}{1 + \omega^2 \tau_\perp^2}. \tag{68}$$

Let the field (62) rotates in horizontal plane around a long $(l \gg R)$ vertical cylinder of the radius R completely filled with ferrofluid. Then the lateral walls of the cylinder are acted upon by the tangential force density

$$f_\varphi = [\sigma_{\varphi r}] = \eta\left(\frac{\mathrm{d}v}{\mathrm{d}r} - \frac{v}{r}\right) + \tfrac{1}{2} M H_0 \sin\alpha, \tag{69}$$

cf. (65), where $v(r)$ is the azimuthal component of the fluid velocity and H_0 stands for the field amplitude *inside* the fluid. Restricting ourselves by the case of low enough field strength, $\xi < 1$, we obtain from (63) and (69)

$$f_\varphi = \eta\left(\frac{\mathrm{d}v}{\mathrm{d}r} - \frac{v}{r}\right) + \tfrac{1}{2}\chi H_0^2 \frac{(\omega - \Omega)\tau_\mathrm{B}}{1 + (\omega - \Omega)^2 \tau_\mathrm{B}^2}, \quad \Omega = \tfrac{1}{2}\left(\frac{\mathrm{d}v}{\mathrm{d}r} + \frac{v}{r}\right). \tag{70}$$

This relationship allows us to calculate the *magnetic torque* $\mathcal{T} = 2\pi R^2 l f_\varphi$ acting on the *motionless* cylinder ($v = dv/dr = 0$):

$$\mathcal{T} = \chi H_0^2 \mathcal{V} \frac{\omega \tau_B}{1 + \omega^2 \tau_B^2}, \tag{71 a}$$

where $\mathcal{V} = \pi R^2 l$ is the volume of cylinder. Satisfying the boundary conditions of continuity of H_φ and B_r on the fluid border $r = R$, one can express the torque via the intensity $\mathcal{H}_0$ of *external* magnetic field [32]:

$$\mathcal{T} = \frac{\chi \mathcal{H}_0^2 \mathcal{V} \omega \tau_B}{\sqrt{(1 + \omega^2 \tau_B^2)[(1 + 2\pi\chi)^2 + \omega^2 \tau_B^2]}}. \tag{71 b}$$

This expression agrees well with experimental data [33,34]. To interpret (71 b), it should be noted that the *stationary* fields $\boldsymbol{H}$ and $\boldsymbol{\mathcal{H}}$ are coupled by the relation

$$\boldsymbol{H} = \frac{\boldsymbol{\mathcal{H}}}{1 + 2\pi\chi}$$

where 2π is the *demagnetization factor* of cylinder in the direction transverse to its axis of symmetry (axis z). Therefore, at the low frequency of the field rotation, $\omega\tau_B \ll 1$, the magnetic torque (71 b) may be presented in its usual form [5]

$$\boldsymbol{\mathcal{T}} = \boldsymbol{\mathcal{M}} \times \boldsymbol{\mathcal{H}}, \tag{72}$$

where we have introduced the total magnetic moment of the cylinder

$$\boldsymbol{\mathcal{M}} = M\mathcal{V}, \qquad M = \frac{\chi \mathcal{H}}{1 + 2\pi\chi}$$

and have taken into account that the small angle between $\boldsymbol{\mathcal{M}}$ and $\boldsymbol{\mathcal{H}}$ is equal to $\omega\tau_B$.

Revert to Eq. (70). If the cylinder is not fixed, the equation $f_\varphi = 0$ determines the velocity of free fluid rotation $v(r) = \omega r$, i.e., $\Omega = \omega$: the fluid rotates *as a whole* about the vertical axis. Finally, if the cylinder is fixed, but not completely filled with ferrofluid, the latter comes into motion due to magnetic tangential stresses on the free fluid surface. The moving surface involves an adjacent fluid layer and forms the circular hydrodynamic flow first observed in [34]. This phenomenon known as *rotational effect* has long held fascination for investigators. Many theories of the spin-up motion were devoted to the effect, but only recently it was clarified by experiments and theory of Rosensweig et al. [35] and Pshenichnikov et al. [32,33,36].

7 Conclusion

Thus, there exist two basic models of the ferrofluid dynamics. One of the two represents a quasistationary theory (Sect. 1), which provides an irreproachable description of any ferrohydrostatic problems. It is also widely employed at the study of thermal convection since in non-isothermal situations the magnetic force is not a potential one and then it is not reduced to the simple re-normalization of the pressure.

The general theory (Sects. 2–5) rejects assumptions of instant relaxation of the magnetization M and its collinearity with the magnetic field H, and makes allowance for the particle rotation with respect to the surrounding liquid. The result is that the product $M \times H$ can take some finite value and so constitutes a body couple. The model consists of hydrodynamic and Maxwell equations plus a magnetization equation. In the capacity of the latter, one should use macroscopic equation (33) derived microscopically from the Fokker–Planck equation by the *effective field method*. Equation (33) guarantees a correct quantitative description of magnetization processes even if deviations of magnetization from its equilibrium value are sufficiently large. However, in the case of a weakly nonequilibrium situation, one should give preference to phenomenological equations (21) or (37): They describe such a state sufficiently well, being at the same time far simpler for analysis than (33). A deviation from the state of equilibrium may be often characterized quantitatively by the value of the dimensionless shear rate $\Omega\tau_B$ or/and the field frequency $\omega\tau_B$. As shown in Sect. 6, equation (37) can be recommended for high shear but only stationary or slow time-varying magnetic field, $\omega\tau_B \ll 1$. On the contrary, equation (21) is valid even for $\omega\tau_B \sim 1$ but only at low shear. Finally, in a low field, $\xi \ll 1$, all of three discussed above magnetization equations coincide with each other.

Acknowledgements

I thank Ronald Rosensweig for his comments on the manuscript and Alexei Krekhov for providing Figs. 5 and 6. This work was supported by the Alexander von Humboldt Foundation by dint of the Meitner–Humboldt research award, and the Israel Science Foundation under Grant No. 336/00.

References

1. J.L. Neuringer, R.E. Rosensweig: Phys. Fluids **7**, 1927 (1964)
2. R.E. Rosensweig: *Ferrohydrodynamics* (Cambridge University Press, Cambridge 1985)
3. M.I. Shliomis: Sov. Phys. JETP **34**, 1291 (1972)
4. M.I. Shliomis: Sov. Phys. Usp. **17**, 153 (1974)
5. L.D. Landau, E.M. Lifshitz: *Electrodynamics of Continuous Media*, second ed. (Pergamon Press, New York 1984)
6. M.I. Shliomis: 'Convective Instability of Magnetized Ferrofluids: Influence of Magnetophoresis and Soret Effect'. In: *Thermal nonequilibrium phenomena in fluid mixtures*, ed. by W. Köhler, S. Wiegand (LNP, Springer, 2001)
7. B.A. Finlayson: J. Fluid Mech. **40**, 753 (1970)
8. M.I. Shliomis, B.L. Smorodin: J. Magn. Magn. Mater. (2002) (in press).
9. Yu.L. Raikher, M.I. Shliomis: Adv. Chem. Phys. **87**, 595 (1994)
10. M.I. Shliomis: Sov. Phys. JETP **24**, 173 (1967)

11. M.A. Martsenyuk, Yu.L. Raikher, M.I. Shliomis: Sov. Phys. JETP **38**, 413 (1974)
12. M.I. Shliomis, Phys. Rev. E **64**, 063501 (2001)
13. M.I. Shliomis, T.P. Lyubimova, D.V. Lyubimov: Chem. Eng. Comm. **67**, 275 (1988)
14. M.I. Shliomis, Phys. Rev. E **64**, 060501(R) (2001)
15. P. Debye: *Polar Molecules* (Dover, New York 1929)
16. L.D. Landau, I.M. Khalatnikov: Dokl. Akad. Nauk SSSR **96**, 469 (1954)
17. R.E. Rosensweig: this issue, p. 63.
18. J.P. McTague: J. Chem. Phys. **51**, 133 (1969)
19. E.N. Mozgovoi, E.Ya. Blum, A.O. Tsebers: Magnetohydrodinamics **9**, 52 (1973)
20. O. Ambacher, S. Odenbach, K. Stierstadt: Z. Phys. B–Condensed Matter **86**, 29 (1992)
21. A.O. Cebers: Magnetohydrodynamics **20**, 343 (1984); **21**, 357 (1985); E. Blums, A. Cebers, M. Maiorov: *Magnetic Fluids* (W. de Gruyter, Berlin 1997)
22. A.C. Levi, R.F. Hobson, F.R. McCourt: Canad. J. Phys. **51**, 180 (1973)
23. B. U. Felderhof: Magnetohydrodinamics **36**, 396 (2000)
24. B.M. Heegaard, J.-C. Bacri, R. Perzynski, M.I. Shliomis: Europhys. Lett. **34**, 299 (1996)
25. F. Gazeau, B.M. Heegaard, J.-C. Bacri, A. Cebers, R. Perzynski: Europhys. Lett. **35**, 609 (1996)
26. F. Gazeau, C. Baravian, J.-C. Bacri, R. Perzynski, M.I. Shliomis: Phys. Rev. E **56**, 614 (1997)
27. J.P. Embs, H.W. Müller, C. Wagner, K. Knorr, M. Lücke: Phys. Rev. E **61**, R2196 (2000)
28. J.-C. Bacri, R. Perzynski, M.I. Shliomis, G.I. Burde: Phys. Rev. Lett. **75**, 2128 (1995)
29. A. Zeuner, R. Richter, I. Rehberg: Phys. Rev. E **58**, 6287 (1998)
30. M.I. Shliomis, K.I. Morozov: Phys. Fluids **6**, 2855 (1994)
31. V.M. Zaĭtzev, M.I. Shliomis: J. Appl. Mech. Tech. Phys. **10**, 696 (1969)
32. A.F. Pshenichnikov, A.V. Lebedev: Magnetohydrodinamics **36**, 317 (2000)
33. A.V. Lebedev, A.F. Pshenichnikov: J. Magn. Magn. Mater. **122**, 227 (1993)
34. R. Moskowitz, R.E. Rosensweig: Appl. Phys. Lett. **11** (1967)
35. R.E. Rosensweig, J. Popplewell, R.J. Johnston: J. Magn. Magn. Mater. **85**, 171 (1990)
36. A.F. Pshenichnikov, A.V. Lebedev, M.I. Shliomis: Magnetohydrodinamics **36**, 339 (2000)

Supplementary Glossary

M magnetization
H magnetic field
H_e effective magnetic field
B magnetic induction: $B = H + 4\pi M$
F magnetic force (volume density)
v ferrofluid velocity
Ω flow vorticity: $\Omega = \frac{1}{2}\operatorname{rot} v$
S internal angular momentum (volume density): $S = I\omega_p$
ω_p angular velocity of magnetic particles
I moment of inertia of particles in a unit volume
p pressure
σ_{ik} stress tensor
T temperature
k_B Boltzmann's constant
ρ mass density
ϕ concentration of magnetic grains (volume fraction)
n number density of magnetic grains
m magnetic moment of a single particle: $m = M_d V$
V volume of a single particle
d particle diameter
D particle diffusion coefficient
M_d domain magnetization of the particle material
W orientational distribution function of particle magnetic moments
e unit vector along particle magnetic moment: $e = m/m$
h unit vector along magnetic field: $h = H/H$
ξ dimensionless magnetic field: $\xi = mH/k_BT$
ζ dimensionless effective magnetic field: $\zeta = mH_e/k_BT$
$L(\xi)$ Langevin function: $L(\xi) = \coth\xi - \xi^{-1}$
η shear viscosity
η_r rotational viscosity
χ initial magnetic susceptibility
Φ thermodynamic potential
t time
τ_B Brownian diffusion time
τ_N Néel diffusion time
$\tau_\perp$ transverse magnetization relaxation time
$\tau_\parallel$ longitudinal magnetization relaxation time
ω magnetic field frequency

Ferrofluid Dynamics

Hanns Walter Müller[1] and Mario Liu[2]

[1] Max-Planck Institut für Polymerforschung, Ackermannweg 10, D-55128 Mainz, Germany
[2] Institut für Theoretische Physik, Universität Tübingen, D-72676 Tübingen, Germany

Abstract. It is demonstrated how the complete structure of Newtonian ferrofluid dynamics, including magneto-dissipative effects, is derived from general principles. Ferrofluids are taken as homogeneous magnetizable fluids with a magnetic relaxation time sufficiently large to compare with other hydrodynamic time scales. The derivation makes no reference to the angular momentum of the ferromagnetic grains. The results are independent of most microscopic details, such as the form or shape of the particles or whether magneto-relaxation takes place via the intrinsic Néel process or by Brownian rotational diffusion. Both the Debye theory of Shliomis and his effective-field approach are shown to be special cases of the new set of equations.

1 Introduction and Motivation

Ferrofluids, or magnetic fluids, are colloidal suspensions of nano-sized ferromagnetic particles stably dispersed in a carrier liquid. When exposed to an external magnetic field, they behave paramagnetically, with susceptibilities χ unusually large for liquids. This property opens up a wide range for interesting and promising future applications [1,2]. Equally important, ferrofluid systems are also attractive under a theoretical point of view. Combining general fluid mechanics with electromagnetism causes ferrofluids to display many unexpected behavior, severely testing and with time perfecting our understanding of the hydrodynamics of polarizable media [1,3–5].

The fundamental theory for ferrofluid hydrodynamics has been worked out by Rosensweig and is well described in the first seven chapters of his text book [1]. Thereby he follows the so called *quasi-equilibrium* approach, which is based on the assumption that the local magnetization $\mathbf{M}(\mathbf{r},t)$ is in steadfast equilibrium with the local magnetic field, i.e. $\mathbf{M}(\mathbf{r},t) = \mathbf{M}^{\mathrm{eq}}[\mathbf{H}(\mathbf{r},t)]$. This theory covers a wide range of interesting and relevant effects, such as the deformation of a ferrofluid drop in a homogeneous magnetic field or the instability of a ferrofluid-air interface, when it is exposed to a magnetic field perpendicular to the surface (Rosensweig instability).

However, very soon it became evident that the *quasi-equilibrium* approximation does not suffice to account for all ferrohydrodynamic behavior, even in stationary flow configurations at static applied fields. Among the more remarkable flow phenomena which go beyond the *quasi-equilibrium* approach is the magneto-viscous effect: A ferrofluid tube flow in the presence of a static magnetic field experiences an extra dissipation, which manifests macroscopically as an enhanced effective shear viscosity [6,3]. Even more spectacular is the acceleration of the flow in response to a high-frequency AC-field [7–9] (also denoted as "negative viscosity", though only a negative viscosity increment was observed). That way, the oscillating magnetic field pumps energy into the rotating motion of the ferromagnetic grains resulting in an acceleration of the flow.

The occurrence of the magneto-viscous effects is intimately related to the fact that the local magnetization $\mathbf{M}(\mathbf{r},t)$ deviates from its equilibrium value $\mathbf{M}^{eq}[\mathbf{H}(\mathbf{r},t)]$, where $\mathbf{H}(\mathbf{r},t)$ is the local magnetic field. Significant increments $\delta\mathbf{M} = (\mathbf{M}-\mathbf{M}^{eq})$ are expected to appear whenever the magnetic relaxation time τ compares to the other relevant hydrodynamic time scales. Phenomena related to the finiteness of τ are commonly denoted as magneto-dissipative effects.

As outlined in Rosensweig's textbook, the microscopic mechanism responsible for the magneto-relaxation is either due to particle rotation against the viscosity of the liquid carrier (Brownian rotational diffusion) or by re-orientation of the magnetic moments relative to the crystallographic orientation of the ferromagnetic grains (Néel relaxation). Which of these mechanisms predominates depends on the specific anisotropy energy of the employed ferromagnetic material, the size of the suspended grains, and the viscosity of the carrier liquid. Since real ferrofluid suspensions usually exhibit a broader particle size distribution, it is in general a combination of both processes which determines the effective magnetic relaxation time for a given ferrofluid species.

Assuming that the Brownian mechanism is the principal source of dissipation, the intuitive picture of particles rotating against the viscous carrier lead Shliomis [3] to his theory for magneto-dissipative ferrohydrodynamics. To that end he included both the magnetization $\mathbf{M}$ and the mechanical angular momentum density $\mathbf{S}$ of the grains as additional thermodynamic variables. After eliminating the latter, an extra momentum flux remains, which enters the stress tensor in the form

$$\Delta\Pi_{ij} = \tfrac{1}{2}\varepsilon_{ijk}(\mathbf{H}\times\mathbf{M})_k, \qquad (1)$$

This term exactly compensates the antisymmetric part of Maxwell's stress $H_i B_j$, if $\mathbf{H}$ and $\mathbf{M}$ are non-parallel. Clearly, being treated as a separate independent variable, the magnetization requires an extra evolution equation. According to Shliomis there are two versions of this equations, of which the first is a phenomenological relaxation equation with a Debye-like relaxation term in the form $\delta\mathbf{M}/\tau$. In the ferrofluid literature this approach is frequently referred to as the *Debye theory*. In combination with Eq (1), many magneto-dissipative phenomena, especially the elevated shear viscosity, were successfully explained. The second variant of his relaxation equation for $\mathbf{M}$ is more elaborate as it is derived from a microscopic, statistical investigation of the rotary diffusion of magnetic particles. Since the problem was solved with the assistance of the effective field method, this second variant is commonly denoted as the *effective-field theory*, or *EFT*. The latter is rather more complicated and unwieldy than Debye as it provides an evolution equation for a quantity called the "effective magnetic field" from which the magnetization is to be determined in a subsequent step.

The *EFT* was found to explain the "negative viscosity" experiment much more convincingly than the Debye-like approach [8].

Comparing both theories, and emphasizing that the *EFT* is the more rigorous and accurate one, Shliomis concluded that it is valid for all experimentally relevant situations [8]. The *Debye theory*, on the other hand, he considers to be adequate only in the limit of small deviations from the magnetization equilibrium [8], $\delta M \ll M^{\text{eq}}$, implying the hydrodynamic low frequency limit $\omega\tau \ll 1$, where $1/\omega$ is the characteristic time scale of the experiment.

We do think that these assessments, referring to both (i) the general validity of the *EFT* and (ii) the limited validity of the phenomeonological *Debye approach* are in need of a clarification:

First, owing to their microscopic input, *EFT* is in it essence a *microscopic theory*, with necessarily rather specific inputs. In the present case, ferrofluids are considered as suspensions of noninteracting, spherical, equal sized Brownian rigid dipoles. In the framework of these restricting simplifications, *EFT* is rigorously valid. But one always has to be aware of the limitations and deficiencies purchased with the above idealizing assumptions. Regarding to the fact that real ferrofluids are suspension of interacting, non-spherical, poly-dispersed particles, whose magnetic relaxation usually involves both Brownian and Néel processes, *EFT* cannot be expected to be a sufficient approach under all circumstances.

Second, the potential of a proper *macroscopic* theory is much larger than for any *microscopic* approach, because it is constructed on the sole base of general principles, without any specific microscopic information. For ferrofluids such a macroscopic theory is very similar to Shliomis' *Debye theory* but it will not suffer from the above constraint to the low frequency regime. In the following we shall denote such a modified approach as the *rectified Debye theory*. Demonstrating how such a macroscopic theory – we shall call it *ferrofluid dynamics* (*FFD*) – can be derived from the concepts of non-equilibrium thermodynamics is the purpose of this article.

Generally speaking, any macroscopic theory consists of two separate ingredients. First the structure of the equations, which is solely based on conservation laws and symmetries, and second the material-specific parameters such as susceptibilities and transport coefficients. The aim of the present lecture is to show how the general structure of a hydrodynamics for ferrofluids is derived. No attempt will be undertaken to provide values for the material-dependent coefficients. Following the standard approach in macroscopic physics, those quantities (such as viscosities or susceptibilities) are usually measured. So we leave them to be determined by a series of suitable experiments. An alternative way is to calculate the coefficients from an appropriate microscopic model. Such a model is for example *EFT* but one must not forget that it is valid in its specified range of validity. In particular it does not necessarily yield – and indeed it does not, as we shall see – the most general structure of equations, which are compatible with symmetries and conservation laws.

2 Outline of the Strategy

In this section we give a layout of our strategy for deriving the structure of the ferrofluid dynamics. We are guided by the general observation that hydrodynamic theories are a very successful tool to account for low frequency large wave number phenomena in condensed matter physics. For isotropic non-magnetic fluids the relevant variables are the conserved quantities: densities of energy, mass and momentum. For magnetizable fluids it is obvious to supplement the set of variables by the magnetic field $\mathbf{H}$ (or the magnetic induction $\mathbf{B}$). That way we recover the thermodynamic variables of the *quasi-equilibrium theory*. Rosensweig assumes that the magnetization relaxation is instantaneous on the time scale of the other hydrodynamic processes of interest. In other words, $\mathbf{M}$ is in steadfast equilibrium with the magnetic field, $\mathbf{M}(\mathbf{r},t) = \mathbf{M}^{\mathrm{eq}}(\mathbf{H}(\mathbf{r},t))$. Thus magneto-dissipative effects, which necessarily imply $\delta \mathbf{M} \neq 0$, are disregarded.

It is usually believed that the interpretation of magneto-dissipative effects requires to incorporate the magnetization as an extra thermodynamic degree of freedom with its own separate evolution equation. But this is incorrect: Recall that linear electrodynamics is well able to account for dissipative effects by introducing an imaginary part of the electrical susceptibility. Focusing to the more general case of simultaneous electro- and magneto-dynamic processes, the appropriate description has been worked out in Liu's *hydrodynamic Maxwell theory* (*HMT*) [10]. Meanwhile this theory has been demonstrated to be well able to explain magneto-dissipative phenomena such as the field enhanced shear viscosity or the fluids spin up in a rotating field [11]. But like any hydrodynamic theory, the *HMT* is valid for small deviations from equilibrium. i.e. it is restricted to the case $\delta M \ll M^{\mathrm{eq}}$ and the low frequency limit $\omega \tau \ll 1$. In ferrofluids τ is typically of order $10^{-4} - 10^{-3}$ s and thus the above low frequency constraint is rather severe and easily violated (for instance by the "negative viscosity" experiment). When this happens, it is justified to include the magnetization as an independent variable to render the theory applicable also to the case $\omega \tau \simeq 1$ and $\delta M/M^{\mathrm{eq}} \simeq 1$.

In the following section we shall derive the equation of motion for the magnetization. This includes also the necessary modifications in all the other equations, which are related to the fact that $\mathbf{M}$ is turning independent. We shall denote this approach as *ferrofluid dynamics* (*FFD*). The method is standard non-equilibrium thermodynamics, with the sole input of conservation laws and symmetries. Besides the assumption that the magneto-relaxation can be covered by a single relaxation time τ there is no further material specific input. Consequently our approach is fairly general, it remains valid even if the magnetic particles interact appreciably with each other, if the particles are of non-spherical shape, or if they are of different size. Moreover, the theory holds irrespective of the microscopic relaxation mechanism, whether it is governed by the Brownian or the Néel mechanism. Furthermore, since no reference is made to the angular momentum $\mathbf{S}$ of the grains, the result is valid both for suspensions and for homogeneous magnetizable continua. Note however, that the present approach does not cover non-Newtonian rheological effects.

3 Derivation of the Equations

In the present section we derive the structure of ferrofluid dynamics. As outlined above, the granularity of the suspension is coarse-grained, and the ferrofluid is treated as a magnetizable continuum build up of two homogeneous constituents. The variables are the conserved quantities, the electromagnetic field, and the magnetization as the only one being non-hydrodynamic. The concentration field is taken into account by the mass density of the ferromagnetic material ρ_c. This is appropriate since the magnetophoresis and Soret effect, which are fairly pronounced in ferrofluids, may build up perceptible concentration gradients.

The above arguments imply that the thermodynamic energy density u is a function of the entropy density s, total density ρ, concentration ρ_c, magnetic field $\mathbf{B}$, magnetization $\mathbf{M}$, and the momentum density $\mathbf{g} = \rho\mathbf{v}$,

$$\mathrm{d}u = T\mathrm{d}s + \mu \mathrm{d}\rho + \mu_c \mathrm{d}\rho_c + \mathbf{v}\cdot\mathrm{d}\mathbf{g} + \mathbf{H}\cdot\mathrm{d}\mathbf{B} + \mathbf{h}\cdot\mathrm{d}\mathbf{M}. \tag{2}$$

Eq. (2) is to be understood as the definition for the conjugate variables such as temperature T, chemical and relative chemical potentials μ and μ_c, velocity field $\mathbf{v}$ etc.. In particular, the quantity $\mathbf{h}$ is associated to the magnetization. With $\mathbf{M} \equiv \mathbf{B} - \mathbf{H}$, or $\partial H_i/\partial M_j = -\delta_{ij}$ for given B, together with the thermodynamic Maxwell relation, $\partial H_i/\partial M_j = \partial h_j/\partial B_i$, we obtain

$$\mathbf{h} = \mathbf{B}^{\mathrm{eq}}(\mathbf{M}, s, \rho_c, \rho) - \mathbf{B} = \mathbf{H}^{\mathrm{eq}} - \mathbf{H}. \tag{3}$$

Eq.(3) results from the requirement that u has to be minimal with respect to $\mathbf{M}$ at $\mathbf{M} = \mathbf{M}^{\mathrm{eq}}(\mathbf{B})$, or equivalently $\mathbf{h} \equiv \partial u/\partial \mathbf{M} = 0$. So $\mathbf{B}^{\mathrm{eq}}(\mathbf{M})$ is the inverse function of the equilibrium magnetization curve $\mathbf{M}^{\mathrm{eq}}(\mathbf{B})$. Subtracting $\mathbf{M}$ from both $\mathbf{B}^{\mathrm{eq}}$ and $\mathbf{B}$, we may also write $\mathbf{h} = \mathbf{H}^{\mathrm{eq}} - \mathbf{H}$, where again $\mathbf{H}^{\mathrm{eq}}(\mathbf{M})$ is the inverse function of $\mathbf{M}^{\mathrm{eq}}(\mathbf{H})$. Note that the function $\mathbf{H}^{\mathrm{eq}}(\mathbf{M})$ is frequently referred to in the ferrofluid literature as the "effective field".

The conserved variables satisfy continuity equations,

$$\dot{\rho} + \nabla \cdot (\rho\mathbf{v}) = 0, \quad \dot{\rho}_c + \nabla \cdot (\rho_c\mathbf{v} - \mathbf{j}^D) = 0, \tag{4}$$

$$\dot{u} + \nabla \cdot \mathbf{Q} = 0, \quad \dot{g}_i + \nabla_j(\Pi_{ij} - \Pi_{ij}^D) = 0; \tag{5}$$

the equations of motion for s and $\mathbf{M}$ are

$$\dot{s} + \nabla \cdot (s\mathbf{v} - \mathbf{f}^D) = R/T, \tag{6}$$

$$\dot{\mathbf{M}} + (\mathbf{v}\cdot\nabla)\mathbf{M} + \mathbf{M} \times \boldsymbol{\Omega} = \mathbf{X}^D, \tag{7}$$

where R is the entropy production and $\boldsymbol{\Omega} = \nabla \times \mathbf{v}/2$ the vorticity.

Assuming that no external electric field is applied, the appearance of an electric field is due solely to electromagnetic induction. Taking the ferrofluid to be dielectrically neutral (i.e. $\mathbf{D} = \mathbf{E}$) the electric contributions to the equations of motion are smaller by a factor $(v/c)^2$ than their magnetic counterparts (c is the speed of light and v a typical velocity). Accordingly, we shall set it to zero.

(See [10,12] for the cases where an external electric field is applied.) As a result, we may use the Maxwell equations in the static approximation

$$\nabla \cdot \mathbf{B} = 0, \quad \nabla \times \mathbf{H} = 0. \tag{8}$$

The fluxes in Eqs (4-7) still need to be derived – although for some of them their convective contributions such as $\rho_c \mathbf{v}$ or $(\mathbf{v} \cdot \nabla)\mathbf{M} + \mathbf{M} \times \mathbf{\Omega}$ have already been made explicit. To derive the unknown flux contributions we employ the so-called standard procedure of hydrodynamics: Take the temporal derivative of Eq (2), substitute $\dot{u}$, $T\dot{s}$, $\mu\dot{\rho}$... using the above equations of motion, and most importantly, require that the resultant equation to hold identically (cf [10,13] and references therein). This yields the energy flux $\mathbf{Q}$, the momentum flux Π_{ij}, and the entropy production R as

$$\Pi_{ij} = \Pi_{ji} = [A + H_k B_k - u]\delta_{ij} + g_i v_j - H_i B_j + \tfrac{1}{2}(h_j M_i - h_i M_j), \tag{9}$$
$$Q_i = Av_i - Tf_i^D - \mu_c j_i^D - v_j \Pi_{ji}^D + \tfrac{1}{2}[\mathbf{v} \times (\mathbf{h} \times \mathbf{M})]_i \tag{10}$$
$$R = \mathbf{f}^D \cdot \nabla T + \mathbf{j}^D \cdot \nabla \mu_c - \mathbf{X}^D \cdot \mathbf{h} + \Pi_{ij}^D v_{ij}, \tag{11}$$

where $A \equiv Ts + \mu\rho + \mu_c \rho_c + \mathbf{v} \cdot \mathbf{g}$, $v_{ij} \equiv \tfrac{1}{2}(\nabla_i v_j + \nabla_j v_i)$. To make the set of equations closed and complete one still has to determine the dissipative fluxes $\mathbf{f}^D$, $\mathbf{j}^D$, $\mathbf{X}^D$, Π_{ij}^D. The form of the entropy production R as given in Eq. (11) implies that they are linear combinations of the forces ∇T, $\nabla \mu_c$, $-\mathbf{h}$, v_{ij}^0, v_{kk}, such that R is always positive. (We take $v_{ij}^0 \equiv v_{ij} - \tfrac{1}{3}v_{kk}\delta_{ij}$.) What now follows is the construction of the fluxes on the basis of symmetry considerations and specific assumptions, the second of which are subject to experimental verifications or microscopic scrutiny.

3.1 Weak Field Limit

If the applied magnetic field is weak, the system can be considered to be approximately isotropic. In this case we have the usual diagonal relations for the diffusive entropy and concentration currents, viscous stresses and especially the magnetic relaxation [14,2],

$$\mathbf{f}^D = \kappa \nabla T + \xi_1 \nabla \mu_c, \quad \mathbf{j}^D = \xi \nabla \mu_c + \xi_1 \nabla T, \tag{12}$$
$$\Pi_{ij}^D = 2\eta_1 v_{ij}^0 + \eta_2 v_{kk} \delta_{ij}, \quad \mathbf{X}^D = -\zeta \mathbf{h}, \tag{13}$$

The transport coefficients κ, ξ, $\kappa\xi - \xi_1^2$, η_1, η_2 and ζ are positive functions of thermodynamic variables. In particular they also depend on the magnitude of the magnetization M. As discussed in the introductory section, their actual values need to be determined either experimentally or on the basis of an appropriate microscopic model. This is beyond the scope of the present investigation.

In compliance with results given by Blums et al. [2] and a work by Felderhof and Groh [4], we observe that the magneto-relaxation term $\mathbf{X}^D$ is proportional to $\mathbf{h} = \mathbf{H}^{eq} - \mathbf{H}$. This is in contrast to the Debye-like increment $\delta\mathbf{M} = \mathbf{M} - \mathbf{M}^{eq}$ suggested by Shliomis [3]. We therefore denote $\mathbf{X}^D$ according to Eq. (13) as the *rectified* relaxation term. Although $\delta\mathbf{M}$ and $\mathbf{h}$ are in general not linearly related they do in two special cases: (i) For small deviations from local equilibrium, where $\delta\mathbf{M}$ and $\mathbf{H}^{\text{eq}} - \mathbf{H}$ are proportional to each other and (ii) if the applied field is sufficiently weak for the linear constitutive relation $\mathbf{M}^{\text{eq}} = \chi\mathbf{H}$ to apply.

We now turn to compare our stress tensor Π_{ij} with the traditional formulation. We start with the observation that the last term in (9) is equivalent to the magneto-dissipative element as given in Eq.(1). This term accounts for magneto-dissipation when the vectors $\mathbf{H}$ and $\mathbf{M}$ are twisted relative to each other. This happens for instance in the McTague experiment [6], where the vorticity of the flow deflects the magnetization vector out of the equilibrium direction.

We emphasize that the condition for the appearance of magneto-dissipative effects, $\mathbf{M} \neq \mathbf{M}^{\text{eq}}$, does not necessarily require the two vectors $\mathbf{H}$ and $\mathbf{M}$ to point in different directions. For instance, magneto-dissipation is also expected to occur if $\mathbf{M}$ and $\mathbf{H}$ oscillate *parallel* to each other but with a *temporal* phase lag. Note that in this situation Eq.(1) is inoperative, so the Shliomis theory does not yield a magneto-dissipative stress here. However, as will become clear in a moment, *FFD* does. To facilitate further comparison with the conventional notation of the stress tensor we rewrite the diagonal contribution to [Eq. (9)] in terms of the zero-field pressure $p_0(\rho, T)$. To that end we perform a Legendre transformation to the independent variables T, ρ, $\mathbf{v}$, H, and M (for simplicity we neglect the concentration dynamics by assuming $\rho_c = const$). The associated thermodynamic free energy $f = u - sT - \mathbf{v}\cdot\mathbf{g} - \mathbf{H}\cdot\mathbf{B}$ has the total differential

$$\mathrm{d}f = -s\mathrm{d}T + \mu\mathrm{d}\rho - \mathbf{g}\cdot\mathrm{d}\mathbf{v} - \mathbf{B}\cdot\mathrm{d}\mathbf{H} + \mathbf{h}\cdot\mathrm{d}\mathbf{M}. \tag{14}$$

Integrating out velocity and magnetic dependencies leads to

$$f(T,\rho,\mathbf{v},\mathbf{H},\mathbf{M}) = f_0(\rho,T) - \frac{1}{2}\rho v^2 - \frac{1}{2}H^2 - \mathbf{M}\cdot\mathbf{H} + \int_0^M H^{\text{eq}}(T,\rho,M')\mathrm{d}M', \tag{15}$$

where $f_0(T,\rho) = f(T,\rho,\mathbf{v} = \mathbf{H} = \mathbf{M} = 0)$. Using the thermodynamic relation for the zero-field pressure $p_0(T,\rho) = -f_0 + \rho\partial f_0/\partial\rho$ and the integral identity

$$\mathbf{H}^{\text{eq}}\cdot\mathbf{M} - \int_0^M (1-\rho\partial_\rho)H^{\text{eq}}(T,\rho,M')\mathrm{d}M' = \int_0^{H^{\text{eq}}}(1-\rho\partial_\rho)M^{\text{eq}}(T,\rho,H')\mathrm{d}H' \tag{16}$$

the diagonal contribution to the stress [square bracket of Eq (9)] transforms into

$$\delta_{ij}\left[p_0 + \tfrac{1}{2}H^2 + \int_0^{H^{\text{eq}}}(1-\rho\partial_\rho)M^{\text{eq}}(H')\mathrm{d}H' - \mathbf{h}\cdot\mathbf{M}\right]. \tag{17}$$

The last term $\mathbf{h}\cdot\mathbf{M}$ is missing from previous works. As outlined above it accounts for magneto-dissipation in a situation where the off-equilibrium increment $\mathbf{h}$ is *parallel* to the magnetization. For an illustration of what are the physical significances of the two different magneto-dissipative stresses, $(1/2)\varepsilon_{ijk}(\mathbf{H}\times\mathbf{M})_k \equiv (1/2)\varepsilon_{ijk}(\mathbf{M}\times\mathbf{h})_k$ and $\delta_{ij}(\mathbf{h}\cdot\mathbf{M})$ let us consider the force exerted by a ferrofluid on a rigid container wall. Multiplying the antisymmetric element with the surface normal vector leads to a *tangential* traction proportional to $\mathbf{H}\times\mathbf{M}$. On the other hand, the diagonal tensor element implies a magneto-dissipative *normal* force linear in $(\mathbf{h} \cdot \mathbf{M})$. This term may be probed by a measurement of the pressure drop across an interface between a ferrofluid and a non-magnetic medium. Since it is a dynamical effect, the interface must be exposed to a time-dependent magnetic field. The expected effect is maximized when the oscillation frequency ω approaches the inverse magnetic relaxation time $1/\tau$.

The reason magneto-dissipative normal stresses have not been discussed until now, may be due to the present focus on incompressible flow configurations. For that kind of flow problems any extra normal stress simply re-normalizes the pressure while leaving the velocity profile unchanged. If, however, normal forces are directly recorded, the associated magneto-dissipative contribution is likely to be measurable. An alternative way to probe the implications of the extra normal stress is the study of compressible flow configurations such as sound. We expect that the coupling between density oscillations and the magnetization will contribute appreciably to the attenuation.

3.2 Strong Magnetic Fields

In the weak field limit discussed in the previous subsection we have seen that the evolution equation for $\mathbf{M}$ as well as the stress tensor Π_{ij} deviate from the expressions given by the previous approach. In the presence of a strong magnetic field, the system can no longer be considered to be isotropic. So, further modifications arise since the resulting uniaxial symmetry of the system leads to a proliferation of the transport coefficients. This is where experimental input becomes imperative. For instance, each of the coefficients of Eq (12) turns into three, as in $\kappa \to \kappa \delta_{ij} + \kappa_\| M_i M_j + \kappa_\times \epsilon_{ijk} M_k$. Similarly, the two viscosities turn into 7. It is not very useful to present all these complications here because the set of isotropic coefficients needs yet to be measured. Strictly speaking, if the directions of $\mathbf{M}$ and $\mathbf{H}$ do not coincide (as is usually the if magneto-dissipative effects come into effect), the system becomes biaxial. This leads to an extra complication, which however will be ignored here.

In the following we focus on the complete uniaxial form of $\mathbf{X}^D$, since the magnetization dynamics belongs to the best studied aspects of ferrofluid physics. Following the linear construction scheme for deriving the contributions to the dissipative flux $\mathbf{X}^D$ leads us to

$$\begin{aligned}X_i^D = &-(\zeta \delta_{ij} + \zeta_\| M_i M_j + \zeta_\times \epsilon_{ijk} M_k) h_j \\ &+\lambda_1 M_i v_{kk} + \lambda_2 M_j v_{ij}^0 + \lambda_3 M_i M_j M_k v_{jk}^0 \\ &+\lambda_4 \epsilon_{ikj} M_k M_\ell v_{j\ell}^0.\end{aligned} \quad (18)$$

Then the Onsager symmetry relations enforce the following counter terms in the stress tensor

$$\begin{aligned}\Pi_{ij}^D = &\{\eta_2 v_{kk} + [\lambda_1 - \tfrac{1}{3}(\lambda_2 + M^2 \lambda_3)] M_k h_k\} \delta_{ij} \\ &+2\eta_1 v_{ij}^0 + \tfrac{1}{2}\lambda_2(M_i h_j + M_j h_i) + \lambda_3 M_i M_j M_k h_k \\ &+\tfrac{1}{2}\lambda_4 [M_j (\mathbf{M} \times \mathbf{h})_i + (i \leftrightarrow j)].\end{aligned} \quad (19)$$

Although these expressions appear complicated, one must realize that the uniaxial, and not the isotropic, case is the generic one: If we take M as small to arrive at the isotropic case, we must for consistency also neglect all term $\sim M^2$ in the Maxwell stress, which is considered too crude an approximation to be employed frequently.

The appearance of the parameters ζ, $\zeta_\|$ and $\zeta_\times$ in X_i^D implies different relaxation times for the respective magnetization components parallel to $\mathbf{h}$, parallel to $\mathbf{M}$ and perpendicular to both of these directions. The latter one is analogous to the Righi-Leduc effect in heat conduction or the Hall effect of an electrical conductor.

In the *Debye theory*, the *EFT*, or the isotropic case considered in the previous subsection, the only velocity gradient entering $\dot{\mathbf{M}}$ is the flow vorticity $\boldsymbol{\Omega}$. This is in contrast to Eq. (18), which suggests that a compressional flow v_{kk}, and even more importantly, an elongational flow v_{ij}^0 will do this too. The coefficients λ_i are material dependent and need to be measured for each ferrofluid. They are *reactive* transport coefficients (as opposed to dissipative ones), because they do not enter the expression for the entropy production (11). Nevertheless, as these coefficients appear as a product either with velocity gradients [Eq. (18)] or with the magneto-dissipative increment $\mathbf{h}$ [Eq. (19)], they can only be evaluated by an appropriate off-equilibrium experiment. It is noteworthy that a term similar to $v_{ij}^0 M_j$ is known to appear in the dynamics of nematic liquid crystals [15], where it gives rise to the "flow alignment" of the nematic director in response to a shear flow.

For a quantitative experimental test of the whether λ_2 for a given ferrofluid material is of significant size one has to study the magnetization dynamics in flow geometries which allow to control the relative strength between elongational and rotational flow contributions. This requirement is easy to fulfill by the flow between two rotating cylinders (Couette-Taylor setup). Allowing the two cylinders to rotate independently with distinct angular frequencies Ω_1 and Ω_2, enables to pass over from a rigid rotation ($v_{ij} = 0$), where $\Omega_1 = \Omega_2$, to a flow with a finite shear rate, where $v_{ij} \propto (\Omega_1 - \Omega_2) \neq 0$. Recording the magnetization vector within the sample (or equivalently the magnetic field outside of the sample) as a function of $\Omega_1 - \Omega_2$ allows to evaluate the transport coefficient λ_2.

4 Comparison with Existing Theories

In the introductory section we outlined that the macroscopic ferrofluid dynamics set up here provides the general macroscopic framework, any more specific (microscopic) approach has to fit in. In the present section the focus is on the relaxation equation for the magnetization. The modifications/extensions of the stress tensor as compared to previous theories have already been discussed in the last paragraph.

It will be shown here that both the *Debye theory* and *EFT* can be embedded into the formulae given by *FFD*, each with a specific choice of parameters.

4.1 The Debye Theory

In the first variant of his description, Shliomis [3] introduces a phenomenological equation for **M**, with a linear Debye-like relaxation term proportional to $\delta\mathbf{M} = \mathbf{M} - \mathbf{M}^{\mathrm{eq}}$. This equation results from what we consider to be a "mesoscopic" approach, because the derivation makes use of some specific microscopic input. The ferromagnetic grains are considered to be Brownian rigid dipoles. To account for their rotation the angular momentum density **S** is introduced as an extra dynamical variable. But not really, since in a later stage of the derivation, **S** is re-eliminated owing to the smallness of the particles' inertia. The phenomenological relaxation equation for **M** which remains after this adiabatic substitution takes on the following form

$$\frac{d\mathbf{M}}{dt} - \mathbf{\Omega} \times \mathbf{M} = \frac{1}{\tau_B}(\mathbf{M}^{\mathrm{eq}} - \mathbf{M}) - \frac{\mathbf{M} \times (\mathbf{M} \times \mathbf{H})}{6\eta_1 \varphi}. \tag{20}$$

Here $(d/dt) \equiv \partial_t + (\mathbf{v} \cdot \nabla)$, τ_B is the Brownian relaxation time, and φ the volume concentration. As pointed out above and by other authors [2,4], the proper relaxation term being rather more in line with the construction rules of non-equilibrium thermodynamics is proportional to **h** rather than $\delta\mathbf{M}$. This is what we have referred to as the *rectified* relaxation term. In the limit of small deviations from the magnetization equilibrium, we have a linear relationship between $\delta\mathbf{M}$ and **h**

$$\mathbf{M} - \mathbf{M}^{\mathrm{eq}} = \frac{M^{\mathrm{eq}}}{H}\mathbf{h} + \left(\frac{\partial M^{\mathrm{eq}}}{\partial H} - \frac{M^{\mathrm{eq}}}{H}\right)\frac{\mathbf{M} \cdot \mathbf{h}}{(M^{\mathrm{eq}})^2}\mathbf{M} + \mathcal{O}(h^2). \tag{21}$$

Replacing the leading order approximation of (21) into Eqs (7,18), we recover Shliomis' *Debye theory* (20) by the following choice of the transport coefficients

$$\lambda_1 = \lambda_2 = \lambda_3 = \lambda_4 = \zeta_\times = 0, \tag{22}$$

$$\zeta = \frac{1}{\tau_B}\frac{M^{\mathrm{eq}}}{H} + \frac{(M^{\mathrm{eq}})^2}{6\eta_1\varphi}, \tag{23}$$

$$\zeta_\parallel (M^{\mathrm{eq}})^2 = \frac{1}{\tau_B}\left(\frac{\partial M^{\mathrm{eq}}}{\partial H} - \frac{M^{\mathrm{eq}}}{H}\right) - \frac{(M^{\mathrm{eq}})^2}{6\eta_1\varphi}. \tag{24}$$

4.2 Effective-Field Theory

On the basis of a kinetic equation for rotary diffusion, Martsenyuk et al. [16] constructed the Fokker-Planck equation for the probability distribution of the particle's orientation. The authors relied on the idealizing assumptions that the ferrofluid is composed of (i) spherical (ii) mono-dispersed (iii) non-interacting (iv) rigid dipoles. Form the resultant infinite hierarchy of equations for the momenta of **M**, a separate equation for the magnetization is deduced by employing the method of the effective field. Thereby the magnetization

$$\mathbf{M} = M_s \mathcal{L}(\xi_e)\frac{\boldsymbol{\xi}_e}{\xi_e} \tag{25}$$

is taken to be a function of the dimensionless, effective field $\xi_e = (m\mathbf{H}^{\text{eq}})/(k_B T)$, with M_s denoting the saturation magnetization of the ferrofluid, m the magnetic moment of an individual particle, $\mathcal{L}(x) = \coth x - 1/x$ the Langevin function and k_B the Boltzmann constant. In terms of the actual non-dimensional magnetic field $\xi = (m\mathbf{H})/(k_B T)$, the effective field is governed by the ordinary differential equation

$$\frac{d}{dt}\left[\mathcal{L}_e \frac{\xi_e}{\xi_e}\right] = \mathbf{\Omega} \times \left[\mathcal{L}_e \frac{\xi_e}{\xi_e}\right] - \frac{1}{\tau_B}\frac{\mathcal{L}_e}{\xi_e}(\xi_e - \xi) - \qquad (26)$$
$$\frac{1}{2\tau_B \xi_e^2}\left(1 - \frac{3\mathcal{L}_e}{\xi_e}\right)\xi_e \times (\xi_e \times \xi),$$

where $\mathcal{L}_e = \mathcal{L}(\xi_e)$. This equation has to be solved for ξ_e for given ξ. Then in a second step, the magnetization is to be deduced from ξ_e via (25).

We point out that this somewhat unwieldy two-step procedure can be circumvented. Without approximation Eq. (26) can be recast in the following, rather more explicit form

$$2\tau_B\left\{\frac{d}{dt}\mathbf{M} - \mathbf{\Omega} \times \mathbf{M}\right\} = \qquad (27)$$
$$-\left[3\chi - \frac{M}{H^{\text{eq}}}\right]\mathbf{h} - 3\left[\frac{M}{H^{\text{eq}}} - \chi\right]\frac{\mathbf{M} \cdot \mathbf{h}}{M^2}\mathbf{M}.$$

Here $\chi = mM_s/(3k_B T)$ denotes the initial Langevin susceptibility. It is noteworthy to point out that the *EFT* approach yields a relaxation term, which is proportional to $\mathbf{h}$ rather than $\delta \mathbf{M}$, i.e. well in compliance with the *rectified Debye theory* outlined above. So it does not come as a surprise that Eq. (27) is recovered as a special case of the general structure of *FFD* as given by Eqs (7,18). The following particular parameter choice applies

$$\lambda_1 = \lambda_2 = \lambda_3 = \lambda_4 = \zeta_\times = 0, \qquad (28)$$

$$\zeta = \frac{1}{2\tau_B}\left[3\chi - \frac{M}{H^{\text{eq}}}\right], \qquad (29)$$

$$\zeta_\| = \frac{3}{2}\frac{1}{\tau_B}\frac{1}{M^2}\left[\frac{M}{H^{\text{eq}}} - \chi\right]. \qquad (30)$$

Clearly, owing to its microscopic input the *EFT* provides a dependence for the transport coefficients ζ and $\zeta_\|$ on the strength of the magnetization, a feature which cannot be accomplished by the macroscopic *FFD*. That is why *EFT* is a complementary approach rather than being competitive. On the other hand, the disappearance of many of the transport coefficients [see Eq. (28)], an immediate consequence of the idealizing assumptions, reflects the limitations of *EFT*. Real ferrofluids generally do not meet these approximations. So one either has to develop a more elaborate theory, or – following the standard approach in macroscopic physics – to determine the set of transport coefficients by a series of appropriate experiments.

5 Conclusion

The general structure of the hydrodynamic equations for ferrofluids is derived here. The gain in rigour is paid by a loss of specific information on the transport parameters. We did not provide the numerical values of the transport coefficients here, nor their dependence on the thermodynamic variables. This task is left open for a series of experiments. An alternative way – albeit less reliable and less quantitative – is to derive this information from a microscopic theory such as *EFT*. Besides giving numerical estimates for the transport coefficients their primary advantage is to provide physical insight into their dependencies on the thermodynamic variables.

In spite of the complete lack of microscopic specifics in the present derivation, the resultant theory does have some restrictions that we need to keep in mind. They arise due to the assumption that a unique characteristic time τ is sufficient to characterize the magnetic relaxation process. As a result, any microscopic features (such as poly-dispersity) that influence this time are to be handled with some care. For instance, a ferrofluid consisting of two populations, each with a distinct relaxation time, will not be well accounted for at higher frequencies, outside the hydrodynamic regime.

References

[*] hwm@mpip-mainz.mpg.de, liu@itp.uni-hannover.de
1. R.E. Rosensweig, *Ferrohydrodynamics*, (Cambridge University Press, Cambridge, 1985).
2. E. Blums, A. Cebers, M.M. Maiorov, *Magnetic Fluids*, (Walter de Gruyter, Berlin 1997).
3. M. I. Shliomis, Sov. Phys. JETP, **34**, 6 1291 (1972); Usp. Fiz. Nauk **112**, 427 (1974) [Sov. Phys. Usp. **17**, 153 (1974)]; M.I. Shliomis, J. Mag. Mag. Mat. **159** 236 (1996).
4. U. Felderhof, B. Kroh, J. Chem. Phys. **110**, 7403 (1999).
5. H. W. Müller, M. Liu, Phys. Rev. E **64**, 061405 (2001).
6. J. P. McTague, J. Chem. Phys. **51**, 133 (1969).
7. M.I. Shliomis and K. I. Morozov, Phys. Fluids **6**,2855 (1994).
8. J.-C. Bacri, R. Perzynski, M.I. Shliomis, G.I. Burde, Phys. Rev. Lett. **75**, 2128 (1995).
9. A. Zeuner, R. Richter, I. rehberg, Phys. Rev. E **58**, 6287 (1998).
10. Mario Liu, Phys. Rev. Lett. **70**, 3580 (1993); **74**, 4535 (1995); **80**, 2937, (1998).
11. S. Lissek, H.W. Müller, M. Liu, *Hydrodynamic Maxwell Theory for Ferrofluids*, submitted.
12. Y.M. Jiang and M. Liu, Phys. Rev. Lett. **77**, 1043 (1996); Phys. Rev. **E**, 6685 (1998).
13. K. Henjes and M. Liu, Ann. Phys. **223**, 243 (1992); M. Liu and K. Stierstadt, cond-mat/0010261.
14. L.D. Landau, E.M. Lifshitz, *Fluid Mechanics* (Pergamon, Oxford, 1987).
15. P. G. de Gennes, J. Prost, *The Theory of liquid crystals* Clarendon, Oxford (1983).
16. M.A. Martsenyuk, Y.L. Raikher, and M.I. Shliomis, Sov. Phys. JETP **38**, 413 (1974); Y. L. Raikher, M. I. Shliomis, in *Relaxation Phenomena in Condensed Matter*, ed. by W. Coffey, Advances in Chemical Physics Series **87**, Wiley (1994).

Heat and Mass Transfer Phenomena

Elmars Blums

Institute of Physics, University of Latvia, LV-2169, Salaspils, Latvia

Abstract. This section deals with main problems of the heat and mass transfer in magnetic colloids. The analysis is mainly based on the general model given in the Chapter written by R. E. Rosensweig. Hydrodynamic and thermal problems are simplified considering incompressible liquids and neglecting the effects of polarization and electric conductivity as well as ignoring some other secondary effects that usually can be neglected in ferrofluid experiments. Contrarily, the analysis of mass transfer accounts for new sedimentation phenomena and cross effects of interrelated heat and mass transfer. Since the description given by Rosensweig is of general theoretical nature, while the present work mainly focusses on experimental problems, the various equations needed will not be cited from Rosensweig's article but from the original literature since the transfer from the general model to the experimentally needed relations is often cumbersome and would exceed the frame of this presentation.

The following problems shortly are discussed: the equation of energy conservation accounting for new adiabatic effects and dissipation due to internal rotation, the thermomagnetic and magnetosolutal instabilities, the specifics of magnetic convection in fluids of non-uniform temperature and concentration, the magnetophoresis and thermodiffusion of colloidal particles under the effect of a magnetic field and some phenomena of combined thermal and Soret-driven convection.

1 Introduction

In main applications the magnetic fluids mostly are used to position the colloid at a certain part of devices by means of magnetic forces. Therefore, for a long time the most important research problems are the increase of fluid magnetization, the specific properties of free ferrofluid surfaces in the presence of an external field and the magnetoviscous and magnetorheological effects. Presently the research topics undergo serious changes. Particularly, there start to be popular non-potential bulk forces in fluids of a nonuniform magnetization. Possibility to induce an intensive convection by means of a magnetic field opens the promising design of new challenging applications (magnetically controlled thermosyphons for technological purposes, augment of heat transfer for cooling of high power electric transformers etc.). Besides, the temperature or the concentration gradients cause new thermo-and magnetophoretic transport processes, which can influence the long-time stability of ferrocolloids in many technical devices. In the present paper main problems of heat and mass transfer in the presence of a magnetic field shortly are discussed.

2 Energy Conversion

Many authors have analyzed the energy conservation equation in magnetizable media. From the second law of thermodynamics for the quasi-equilibrium model of ferrofluids it becomes [1]

$$\rho c_p \frac{dT}{dt} = T\frac{1}{\bar{V}}\frac{\partial \bar{V}}{\partial T}\frac{dp}{dt} - T\frac{\partial \mathbf{M}}{\partial T}\mu_0 \frac{d\mathbf{H}}{dt} + Q - div\mathbf{j}_q \qquad \left(\frac{d}{dt} = \frac{\partial}{\partial t} + \mathbf{v}\cdot\nabla\right) \quad (1)$$

The left part of equation (1) characterizes the accumulated thermal energy of the fluid (ρ is the fluid density and c_p is the specific heat capacity). The first two terms of the right side reflect the adiabatic effects of fluid compression and magnetization ($\bar{V}$ and $\mathbf{M}$ are the specific volume and the magnetization), the Q represents the internal heat source density, and the last term describes the conductive heat transfer. According to Fik's law the heat flux $\mathbf{j}_q$ is proportional to the temperature gradient:

$$\mathbf{j}_q = -\lambda \nabla T \quad (2)$$

(λ is the fluid thermal conductivity). The internal gradients of fluid temperature are created by the conductive heat input or output through the volume borders as well as by the internal heat source Q.

Due to a low compressibility of liquids the adiabatic rise of their temperature under the effect of an increasing pressure always is very small [1]. The thermal effect of fluid magnetization at room temperatures also is not strong. Only close to the Curie temperature, the adiabatic magnetization can cause a considerable change in the fluid temperature. This phenomenon at early stage of the magnetic fluid research was proposed to use for a thermomagnetic energy conversion [2]. The term Q in (1) represents the absorption of external radiation and the heat generation due to a viscous dissipation of the fluid kinetic energy. Magnetic colloids are not transparent to the radiation of optical and infrared frequencies; therefore, the internal effects of light absorption should be taken into account only in the event of thin fluid layers. If macroscopic volumes of liquids are considered, only the viscous dissipation of flow energy in Q usually should be taken into account. Apart from the conventional viscous term, the dissipation function for magnetic fluids contains an additional term, which reflects the effect of internal rotations. For Langevin-type magnetization $M = \varphi M_s L(\xi)$ of colloids ($\xi = \mu_0 mH/kT$, φ is the volume fraction of magnetic phase of a bulk saturation magnetization M_s), if they contain magnetically hard particles of a magnetic moment m, the heat source Q becomes [1]

$$Q = \frac{\eta}{2}\left(\frac{\partial v_i}{\partial x_k} + \frac{\partial v_k}{\partial x_i}\right)^2 + 6\varphi\eta\frac{\xi L(\xi)}{2+\xi L(\xi)}\left[\Omega^2 - \left(\frac{\mathbf{H}}{H}\cdot\mathbf{\Omega}\right)^2\right] \quad (3)$$

(η is the fluid dynamic viscosity). The thermal effect of internal rotations can be observed in the presence of a constant magnetic field only if it is oriented non-parallel to the flow vorticity $\mathbf{\Omega} = (1/2)\mathrm{rot}\mathbf{v}$. At low flow velocities the heating intensity usually causes a relatively small effect upon the fluid temperature.

Only for high shear rates typical for some ferrofluid applications (for example, in high-speed rotary seals and in loudspeakers) the viscous heating of the fluid can cause a significant rise of its temperature.

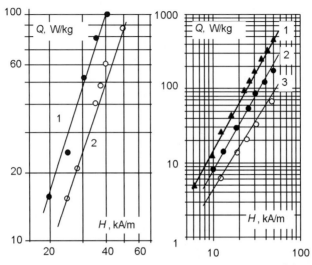

Fig. 1. Specific heat induced in magnetic fluids by a rotating field [3]. Left side: magnetite particles in kerosene, frequency $f = 1.75$ kHz, particle volume concentration $\varphi = 0.096$ (1) and $\varphi = 0.081$ (2). Right side: cobalt ferrite in kerosene, $\varphi = 0.086$, frequency $f = 1.75$ kHz (1), $f = 0.6$ kHz (2), $f = 0.2$ kHz (3)

Heating of the colloid due to internal rotations in the most pure form manifests in a rotating magnetic field. Despite the presence of anti-symmetric tangential stress, a liquid may remain quiescent, and only the second term in (3) is responsible for a rise of the fluid temperature. Figure 1 represents the heating intensity, which is measured employing magnetite and cobalt ferrite based ferrofluids [3]. From the presented results we see that in colloids of magnetically "hard" particles, the field-induced internal rotation causes a significant thermal effect. For example, in a cobalt ferrite based ferrofluid the rotating field 40 kA/m of frequency $\omega = 1.75$ kHz generates a heating density Q up to 600 W/kg [3]. In principle, the microconvection under the effect of field rotation induces also some changes in the fluid thermal conductivity. Microvortices alter the magnitude of heat flux along the temperature gradient and cause a transversal heat flux in the plane of rotation axis. Still, the numerical estimates show [4] that this magnetic analogue of Rigi-Leduc effect in colloids usually is negligible.

A detailed thermodynamic analysis of the energy conservation equation requires considering all thermodynamic parameters as functions of a magnetic field. However, from a practical viewpoint the corresponding corrections, except the situation when temperature is close to the Curie one, are negligibly small. A more important problem is the dependence of physical properties and transfer coefficients on the particle concentration and on the structure of colloids. The additivity law applies for the fluid density and the specific heat capacity; there is no observable effect of a magnetic field upon these. The transfer coefficients are affected by a magnetic field due to several reasons. It is well known that the magnetic braking of internal rotations causes a fluid magnetoviscosity. Internal rotations exert the Hall -type influence upon the heat and mass transfer coefficients as well. Still, the microconvective analog of the Rigi-Leduc effect of thermal conductivity is small [1]. More significant seem to be the magnetic field induced effects which are related to the aggregate forming and orientation.

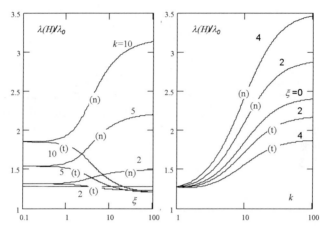

Fig. 2. Thermal conductivity of a diluted suspension of ellipsoidal particles (magnetite in kerosene, $\varphi=0.1$) under the effect of a magnetic field [5]. Lines marked with (n) represent the transient heat transfer with $\mathbf{B}\perp\nabla T$, lines with (t) correspond to $\mathbf{B}//\nabla T$. Parameter k is the relative elongation of particles, ξ is the Langevin parameter

Figure 2 shows the transient thermal conductivity of a diluted suspension of ellipsoidal particles calculated assuming the one-particle approximation [5]. In a qualitative sense this model represents the heat transfer through a colloid under the effect of magnetically oriented aggregates. The thermal conductivity of ferrite particles significantly exceeds that of a carrier liquid. Therefore, the conductive heat transfer through assemble of the non-spherical particles is higher than that of the spherical ones. The orientation of ellipsoids along the field direction during the colloid magnetization leads to appearance of an anisotropy of the transient heat transfer. From results presented in Fig. 2 we see that the magnetic field $\mathbf{B}$, which is oriented parallel to the temperature gradient ∇T, causes an increase in the thermal conductivity λ, whereas at $\mathbf{B} \perp \nabla T$ a reduction of λ should be observed. The one-particle approximation is valid for diluted colloids until the particle volume concentration does not exceed approximately the value $\varphi = 0.05$. The thermal conductivity λ of dense colloids of spherical particles can be calculated using the classical dependence [6]:

$$\frac{\lambda}{\lambda_0} = \frac{2\lambda_0 + \lambda_p - 2\varphi(\lambda_0 - \lambda_p)}{2\lambda_0 + \lambda_p + \varphi(\lambda_0 - \lambda_p)} \qquad (4)$$

Here λ_0 and λ_p are the thermal conductivity's of the carrier liquid and of particles.

The dependence of conductive heat transfer on a magnetic field in experiments employing stable ferrocolloids usually is not observed, but it is informative to note that most of the known experimental data suggest a more pronounced rise in thermal conductivity predicted by the relation (4). Obviously, this is due to the peculiarities of surfactant layers on particles. The presence in a colloid of various additional ingredients plays a more sufficient rule when the electrical conductivity of the colloid is considered.

3 Thermomagnetic Convection

To analyze the convective processes, the energy conservation equation should be considered together with the equation of a fluid motion. Let us consider the problem in a Bussines approximation, i.e. assuming the physical properties and the transfer coefficients (with exception of the density in the gravitation force $\rho \mathbf{g}$ and of the magnetization) constant. The equation of motion contains a new term of magnetostatic force:

$$\rho \frac{d\mathbf{v}}{dt} = -\nabla p + \eta \Delta \mathbf{v} + \rho \mathbf{g} + \mu_0 (\mathbf{M}\nabla)\mathbf{H} \qquad (5)$$

If the magnetic susceptibility χ is dependent solely upon the field and $\rho = const$, then under an approximation of equilibrium magnetization $\mathbf{M} = \chi \mathbf{H}$ both the gravitation and the magnetic forces are potential and there cannot arise internal convective flows in the fluid. The non-potentiality of bulk forces appears only if a fluid possesses the spatial non-uniformity of ρ and $\mathbf{M}$ due their dependence on temperature or on particle concentration φ. From (5) it follows that the free convection develops spontaneously if

$$\nabla \rho \times \mathbf{g} + \mu_0 \nabla \chi \times \nabla H^2 \neq 0 \qquad (6)$$

We can see that there exists a certain analogy between the thermogravitational and the thermomagnetic convection. The difference is only the spatial non-uniformity of magnetic acceleration ∇H^2 while the gravitation acceleration $\mathbf{g}$ is constant. There are many works published in which the magnetic convection under various field distributions are calculated and measured (see the review in Ref. [1]).

If (6) equals to zero, a problem of thermoconvective instability appears. The stability criterion is the Rayleigh number Ra. The equations (1) and (5), if they are linearized with respect to small perturbations of velocity, temperature and pressure, have the same form as those describing the conventional convection problems. The only difference lies in appearance of a different form of Ra (we assume the coordinate $\mathbf{x}$ being directed toward the vector $\mathbf{g}$) [7]:

$$Ra = \left(\beta_T \rho g + \mu_0 \alpha_T M \frac{dH}{dz}\right)\left[\frac{dT}{dz} - \frac{T}{\rho c_p}\left(\beta_T g \rho + \mu_0 \alpha_T M \frac{dH}{dz}\right)\right]\frac{c_p l^4}{\nu \lambda} \qquad (7)$$

Here β_T and α_T are the expansion and the pyromagnetic coefficients of the fluid (according to the definition they are positive), l is the characteristic length.

From (7) it follows that, without applying external temperature gradients, the fluid with respect to the adiabatic compression and magnetization is always stable (note, if $dT/dz = 0$, the Rayleigh number Ra is negative). The temperature gradients necessary to attain the critical value of Ra are so large that the adiabatic terms (the second part in the square brackets in (7)) usually can be neglected. In such approximation the Rayleigh number contains two additive terms, they reflect the thermogravitational (Ra_T) and the thermomagnetic (Rm_T) buoyancy forces:

$$Ra = Ra_T + Rm_T = \left(\beta_T \rho g + \mu_0 \alpha_T M \frac{dH}{dz}\right) \frac{dT}{dz} \frac{c_p l^4}{\nu \lambda} \qquad (8)$$

Convective instability can develop ($Ra > 0$) if the temperature gradient is directed along the gravitation or the magnetic force.

An interesting situation appears when the nonisothermic ferrofluid layer is subjected to a homogeneous magnetic field $\mathbf{B} = const$. From the field equation $\mathrm{div}\mathbf{H} = -\mathrm{div}\mathbf{M}$ it follows that thanks to the pyromagnetic properties of the fluid, a gradient of internal magnetic field in the layer appears. Instead of (8) the Ra now is the following:

$$Ra = Ra_T + Rm_T = \left(\beta_T \rho g + \mu_0 \frac{\alpha_T^2 M^2}{(1+\chi)} \frac{dT}{dz}\right) \frac{dT}{dz} \frac{c_p l^4}{\nu \lambda} \qquad (9)$$

(Here $\chi = \partial M/\partial H$ is the differential magnetic susceptibility.) As a result, the magnetic Rayleigh number Rm_T becomes square dependent on the temperature gradient and it is positive. It means that thermal perturbations of the magnetizing field always lead to the thermoconvective destabilization of the fluid.

The critical conditions for onset of the convection in a flat horizontal layer can be represented by a relation [8]

$$\frac{Ra^*}{R_0} = \frac{Rm^*}{Rm_0} \qquad (10)$$

Here $R_0 = 1708$ (flat layer of rigid boundaries), whereas Rm_0 is the critical Rm value when the gravitation is absent. The later depends on the magnetic properties of channel walls and is slightly influenced also by the nonlinearity of the fluid magnetization curve.

The general conclusions of thermoconvective stability theories (the pioneering one, obviously, is that published in Ref. [9]) are confirmed experimentally. Figure 3 represents the results [10] of measurements of the heat transfer through a horizontal layer of ferrofluid in the presence of a transversal magnetic field $B_z = const$. If $B = 0$, the convection appears (the Nusselt number Nu starts to exceed the value 1) when the layer is heated from below and the Rayleigh number reaches its critical value R_0. With a magnetic field applied the critical temperature gradient drops. Besides this, in accordance with (9) the convective

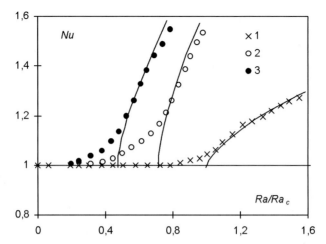

Fig. 3. Effect of a transversal magnetic field upon the heat transfer in infinite flat layer of magnetic fluid: 1 – heated from below, $H = 0.2$ – heated from above, $H = 40.8$ kA/m, 3 – heated from below, $H = 40.8$ kA/m [10]

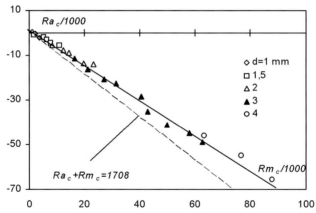

Fig. 4. Relation between Ra_c and Rm in a horizontal magnetic fluid layer of thickness d [10]. Dots represent the experimental results, solid curve – the results of theoretic calculations for the cell examined in experiments

instability sets on also when the layer is heated from above. Figure 4 represents the experimental values of Ra_c measured in the presence of a magnetic field. These results are in a good agreement with the theoretical dependence (10) if accounting for the critical values of both the R_0 and Rm_0. The Rm_0 is higher than R_0, (compare the points and the corresponding theoretical dependence in Fig. 4 with the curve $Ra_c + Rm_c = 1708$), it depends on magnetic properties of the channel walls [8,9]. The reason of that is a difference in mechanisms of the gravitational and the magnetic convective instabilities: the thermomagnetic instability is affected not only by perturbations of the fluid temperature but also by those of the magnetic field inside the fluid.

If the external magnetic field has a gradient non-parallel to the ∇T, inside the fluid sets on a non-threshold convection. Its intensity significantly exceeds the intensity of the thermogravitational convection and a significant enhancement of the heat transfer can be achieved. It means that magnetic control of the heat transfer in ferrofluids is an interesting problem of applications.

4 Mass Transfer

Considering stable colloids without chemical reactions, the mass conservation equation for two component systems (particles of a mass concentration ρ_i and a carried liquid) is the following:

$$\frac{d\rho_i}{dt} = -\text{div}\mathbf{j}_i \qquad (11)$$

Ultra-fine particles in magnetic fluids obey an intensive Brownian motion. Therefore, the mass transfer in colloids can be considered similar to that in molecular liquids involving the concept of a gradient diffusion. The mass diffusion coefficient of nanoparticles is determined by the relation $D = kT/f_v$, where f_v is the coefficient of the hydrodynamic drag force (for spherical particles of radius r the $f_v = 6\pi\rho\nu r$). The mass flux [1]

$$\mathbf{j}_i = -D\nabla\rho_i + \frac{m_g}{f_v}\left[-(\bar{V}_i - \bar{V}_0)\rho\mathbf{g} + \mu_0(\bar{M}_i - \bar{M}_0)\nabla H\right]\rho_i(1-n_i) - \rho_i(1-n_i)DS_T\nabla T \qquad (12)$$

contains a new barodifusion term. In addition to the conventional gravitational sedimentation of particles (m_g is the mass of the particle, $\bar{V}_i$ and $\bar{V}_0$ are the specific volume of the solid phase and that of the carrier liquid) it is necessary to take into account also for their magnetic sedimentation ($\bar{M}_i = M_i/\rho_i$ and $\bar{M}_0 = M_0/\rho_0$ are correspondingly the specific magnetization of the particles and the carrier). The mass diffusion coefficient of colloidal particles is several orders of magnitude less than that of molecules in liquids. Therefore, the gravitation and magnetic sedimentation processes in ferrofluids play a more significant rule than those in molecular liquids. The thermodiffusion term in (12) is retained for purpose of generality, S_T there is the Soret coefficient of particles, $n_i = \rho_i/\rho$ is the mass fraction of solid phase.

The conditions of the fluid convective instability in isothermal liquids also are determined by the relation (6). The corresponding solute Rayleigh number now is the following:

$$Ra = Ra_c + Rm_c = -\left(\beta_c\rho g + \mu_0\alpha_c M\frac{dH}{dz}\right)\frac{d\rho_i}{dz}\frac{l^4}{\eta D} \qquad (13)$$

(According to the definition, the solute expansion (β_c) and magnetic (α_c) coefficients are positive). It is important to note that there is a principal difference between the thermal and the solutal convective stability. For stable colloids the concentration boundary conditions are the unpermeability of surfaces. It means that the concentration gradients cannot be applied independently; they are formed internally by the sedimentation forces near walls in a form of diffusive boundary layers. From (12) and (13) it follows that the stratification of particles in colloids under the action of gravitation and/or external magnetic field gradients always causes an increase in the fluid convective stability. The influence of internal magnetic field gradient is quite different. If the fluid layer is placed in a homogeneous magnetic field $B = const$, the only reason of the fluid stratification is the gravitation force. The corresponding field gradient $\nabla H = -\nabla M$ is oriented opposite to the $\nabla \rho_i$. The magnetic term in (13) now becomes a square dependent on the concentration gradient and Rm_c is positive:

$$Rm_c = \left(\alpha_c M \frac{d\rho_i}{dz}\right)^2 \cdot \frac{\mu_0 l^4}{(1+\chi)\eta D} \tag{14}$$

That means that internal field gradients always cause a lowering of the convective stability of fluid layers.

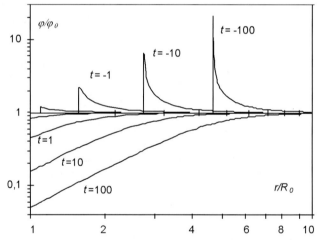

Fig. 5. Paramagnetic (positive t) and diamagnetic (negative t) separation of particles in the vicinity of a cylindrical current-carrying electric conductor of radius R_0[1]. Non-dimensional time t contains the characteristic magnetophoretic time R_0/u_0 with u_0 being the velocity of particle phoresis at the cylinder surface

From Maxwell equations it follows that the external magnetic field gradient cannot be constant, the magnetic force always is spatially non-homogeneous. As a result, the magnetic sedimentation forms a non-uniform distribution of particles not only near walls but also outside the boundary layers. As an example, in Fig. 5 are shown profiles of concentration in the vicinity of a cylinder when the particles are transferred by magnetophoresis under the effect of a radial gradient of the azimuthal magnetic field generated by axial electric current I

($H = 2\pi I/r$) [1]. In the case of a paramagnetic separation (magnetic particles of the colloid, $t > 0$) owing to a sharp acceleration of particles toward the cylinder, the wall region gradually depletes of a magnetic phase. In case of a diamagnetic separation (nonmagnetic particles in ferrofluid, $t < 0$) the withdrawal of a concentration front from the cylinder surface is observed. In both cases the gradients of concentration are oriented opposite to the magnetic driving force. It can be shown that this is a general law: the bulk gradients of concentration always are anti-parallel to the magnetic sedimentation force. Thus, the magnetic stratification of ferrocolloids is unstable. In Ref. [11] the convective stability of fluid in a coaxial gap is analyzed by taking into account not only for the magnetophoresis but also for the mass diffusion of particles. Despite the forming of concentration boundary layers near gap walls, the unstable bulk gradient of concentration still remains and, reaching a critical value of the solute Rayleigh number, the onset of a specific diffusion-magnetic convection is observed. This effect is confirmed experimentally [12].

If $\nabla \rho \times \nabla H \neq 0$, the solute-driven convection starts to develop monotonously and immediately after the moment when the magnetic field is switched on. The characteristics of isothermal magnetic convection in colloids are analyzed in Ref [13] considering the particle transfer near transversally magnetized cylinder. Numerical simulation of the problem shows that in the vicinity of the cylinder should develop a system of intensive convection rolls, the convection cause a deformation of the concentration profiles. This effect is investigated experimentally by performing holographic measurements of the unsteady distribution of particles around the cylinder [14]. A typical interferogram of the concentration field is displayed in Fig. 6. The experimental results coincide very well with the concentration profiles calculated under the assumption of a solute driven magnetic convection. Taking into account the numerically established fact that the concentration profiles formed under the effect of a convection, strongly differ from those valid for the motionless fluid, the results presented in Fig. 6 can be considered as an experimental confirmation of the theoretically predicted magnetic convection in isothermal fluids.

5 Heat and Mass Transfer Problems

The last term in (12) causes an interrelation of the heat and the mass transfer processes. Thermodiffusion in dispersions is a relatively new research problem. It is known that small particles in aerosols in comparison with the molecules in liquids have relatively high thermophoretic mobility. The Soret coefficient depends on Knudsen number, in rarefied gases it can be calculated taking into account for a slip velocity and a temperature drop on particle walls. The thermodiffusive properties of colloids are not investigated in details at the moment. The first experiments are evidence of an unusually high thermophoretic mobility of nanoparticles in ferrofluids. Using two different methods, the separation measurements in thermodiffusion column [15], and the analysis of a diffraction signal from optically induced particle grating in thin films [16,17], it is shown

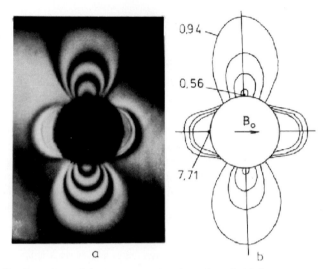

Fig. 6. Distribution of particle concentration in a ferrofluid (magnetite in kerosene, mean diameter of particles 8 nm) in the vicinity of a transversally magnetized ($B_0 = 65$ mT) cylinder (nickel wire, diameter 1 mm) [14]. (a) – experiment, duration of separation 20 min; (b) – the corresponding picture resulting from theory accounting for the magnetic solutal convection, the concentration difference between the neighboring interference fringes and lines is equal to $\Delta\varphi/\varphi_0 = 0,12$.

that surfacted ferrite particles in hydrocarbons are transferred toward decreasing temperatures. The measured Soret coefficients are several orders of magnitude higher than those usually found in molecular liquids, they reach values approximately equal to $S_T = +0,15$ K^{-1}. Electrically stabilized particles in ionic colloids usually have negative Soret coefficients [16,17].

To give a notion to the practical importance of thermodiffusive processes in ferrofluids, in Fig. 7 is presented a typical curve of particle separation in thermodiffusion column. The separation effect is very strong, even at relatively small temperature gradients a more than 50% difference in particle concentration in the channel ends during the first 4 - 5 hours can be achieved. In long time experiments, reaching the steady regime, almost complete educing of ferroparticles from the carrier liquid can be achieved [18].

The nature of the Soret effect in colloids is not well understood yet. Some general ideas of a theory of the slip velocity at a solid-liquid interface resulting from tangential temperature gradient are reviewed in Ref. [19]. Analyzing the enthalpy flux carried by forced convection of fluid across a porous barrier and applying Onsager's reciprocal theory for the slip velocity, in Ref. [20] it is predicted that free particles in surfacted colloids should move toward decreasing temperatures. According to the recent theoretic considerations, the Soret coefficient of surfacted particles is proportional to the Gibbs absorption and the length of surfactant molecules as well as to the size of particles [21]. The measured values S_T qualitatively well agree with this theory.

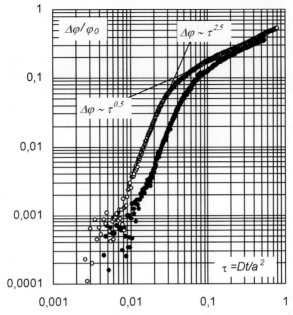

Fig. 7. Unsteady separation of particles (Mn-Zn ferrite in tetradecane,) in thermodiffusion column [30]. The column consists of a flat vertical channel of width $\delta=0.5$ mm and of height $L=780$ mm, difference of the wall temperatures is $\Delta T=10$ K. The nondimensional time is $\tau=Dt/\delta^2$. Dots correspond to the initial sample with $\varphi=0.092$, circles – to the diluted one with $\varphi=0.023$.

In recent experiments we have examined the thermophoretic properties of ferrofluid emulsions. Ferrofluid droplets in water are moving toward increasing temperature, the Soret coefficient reaches a value $S_T = -480$ K^{-1} [22]. We suspect that such a high thermodiffusive mobility of droplets is a result of Marangoni-type transfer. According to Ref. [23], the droplet velocity in a non-isothermal fluid follows the expression

$$u_\sigma = \frac{d}{2(3\eta + 2\eta_0)} \left(-\frac{\partial \sigma}{\partial T}\right) \nabla T \qquad (15)$$

Here η and η_0 are the viscosity's of the ferrofluid droplet and the surrounding liquid, $\partial \sigma / \partial T$ is the dependence of the surface tension on temperature and d is the diameter of the droplet. The experimental results agree qualitatively well with the value of S_T calculated from the dependence (15).

Theory [24] trays to explain the reason of negative Soret coefficients in ionic ferrofluids [16]. It is shown that the direction of thermodiffusive transfer of charged particles depends on electrochemical parameters of the colloid. For an infinitely thin Debye layer, independently of the surface potential value, the particle migration velocity is always directed opposite to the temperature gradient, $S_T > 0$. At finite thickness of the double layer this is a case only for small surface potentials. For high values of ζ-potential the theory predicts a reverse direction of particle transfer, starting some critical values of Debay length the Soret coefficient can be negative. The parameters of ionic ferrofluid used in the experiment [16] correspond to such requirements.

From (12) it follows that in colloids of a non-uniform magnetization specific mass transfer effects can be observed even if the external magnetic field is constant. Let us consider a homogeneous magnetic field $B = const$ directed along the temperature gradient. From equation $div\mathbf{H} = -div\mathbf{M}$ we obtain (we assume $M_0 = 0$)

$$\nabla H = -\nabla M_i = -\frac{\partial M_i}{\partial \rho_i}\nabla \rho_i - \frac{\partial M_i}{\partial T}\nabla T - \frac{\partial M_i}{\partial H}\nabla H \qquad (16)$$

The mass flux (12) with account for the equation (16) can be rewritten in the form of (for a simplicity the term of gravitation sedimentation is omitted)

$$j_i = -\left(D + \frac{\mu_0 \alpha_c M^2 m_g}{f_\nu(1+\chi)}(1+n_i)\right)\nabla \rho_i - \left(S_T - \frac{\mu_0 \alpha_T M^2 m_g}{f_\nu D \rho_i(1+\chi)}\right)\rho_i D(1-n_i)\nabla T \qquad (17)$$

Respectively, the magnetic stratification of ferroparticles under the influence of internal field gradients can be considered as an increase of the mass diffusion coefficient and as a reduction (in surfaced colloids) of the Soret coefficient. More detailed analysis leads to a conclusion, that the change in coefficient D should be observed also in the case when field is oriented normally to the temperature gradient [25,26]. The only difference is that now (at $\mathbf{B} \perp \nabla \rho_i$) the field causes a reduction of the mass diffusion coefficient. Both effects, the increase in D under the effect of a longitudinal field $\mathbf{B} \| \nabla \rho_i$ and its reduction $\mathbf{B} \perp \nabla \rho_i$, recently are confirmed experimentally [26,27], see Fig. 8.

The Soret effect of magnetic particles in the presence of a uniform magnetic field is considered in Ref. [28]. From the analysis of hydrodynamic Stokes problem for a spherical magnetic particle it follows, that if $\mathbf{B} \| \nabla T$, the nonuniformity of fluid magnetization causes a motion of particles in the direction of temperature gradient in accordance with the dependence (17). Contrary, in the presence of a transversal field $\mathbf{B} \perp \nabla T$, the particles are transferred toward decreasing temperatures. This so-called "magnetic Soret effect" and its anisotropy have been investigated experimentally. The separation measurements in thermodiffusion column [29,30] confirm the general predictions of hydrodynamic theory: in colloids with surfaced particles of positive Soret coefficients the transversal magnetic field $\mathbf{B} \perp \nabla T$ causes an increase in S_T, whereas at $\mathbf{B} \| \nabla T$ a strong reduction of S_T takes place (see Fig. 9).

Similar results are obtained also in the experiments on optical grating in thin ferrofluid layers [27]. But, it is interesting to note that the measured magnetic field effects always are significantly stronger than those predicted by the hydrodynamic theory. There might be several reasons of that. Firstly, the concentration profiles in nonisotermic fluids dvelop under the combined action of the thermodiffusion coefficient D_T and the mass diffusion coefficient D. Therefore, the measurd values of Soret coefficient $S_T = D_T/D$ reflect not only the direct influence of $\mathbf{B}$ on D_T but also the dependence $D = D(H)$. The second reason can be the polidispersity of colloids. In Ref. [31] are reported some calculations

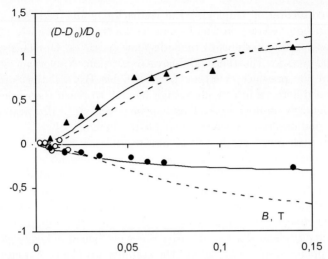

Fig. 8. Anisotropy of mass diffusion (meghemite particles in cyclohexane) in the presence of a magnetic field [26]. The diffusion coefficient is calculated analyzing the relaxation of a diffraction signal from optically induced grating structures. Triangles correspond to parallel field $\mathbf{B}//\nabla\rho_I$, dots and circles – to $\mathbf{B}\perp\nabla\rho_I$. Solid lines represent the calculation results based on thermodynamic theory [26], dashed lines – the corresponding results of hydrodynamic theory [28].

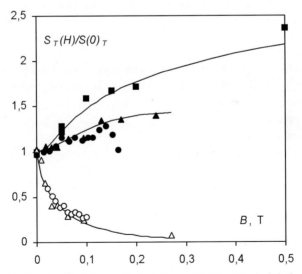

Fig. 9. Magnetic Soret effect in surfacted ferrocolloids. Black labels correspond to experiments with $\mathbf{B}\perp\nabla T$, white labels – to experiments with $\mathbf{B}//\nabla T$. Squares – magnetite in tetradecane, $\varphi=0.022$ [29], triangles – Mn-Zn ferrite in tetradecane, $\varphi=0.023$ [30], circles and dots – meghemite in cyclohexane, $\varphi=0.019$ [27]

of particle sedimentation employing the Batchelor model [32]. It is shown that the sedimentation velocity in polidisperse colloids is significantly higher than that in corresponding systems of monodisperse particles. Obvioulsly, a similar effect is valid also for the thermodiffusive separation. Numerical estimantions show that in the presence of strong parallel fields $\mathbf{B}//\nabla T$ the Soret induced concentration difference in colloids is high enough to reach the critical value of magnetic Rayleigh number (14). This can be a reason of disapearence of the difraction signal observed in some grating experiments.

The very high Soret coefficient of ferroparticles is not only an important practical problem for ferrofluid applications but brings up also new research tasks. The most interesting problems are: the combined thermal and Soret-driven convection in thermodiffusion columns [33], the stability of convective shear flow in channels under the effect of a transversal magnetic force in nonisothermic and stratified colloids, the stability of convection in vertical fluid layers with an adverse density gradient due to a negative Soret coefficient of heavy particles [34] etc. There appears also a new class of the stability problems of double diffusive magnetic convection in stratified colloids [35].

6 Conclusions

Magnetic convection in non-isothermic and stratified ferrocolloids can be used to develop new effective cooling systems (for loudspeakers, electric transformers, magnetically controlled thermosyphons) etc. Magnetic and thermodifusional stratification of colloids should be taken into account considering the long-time working ability of magnetofluid devices.

The research on separation in thermodiffusion columns reported in Section 5 is supported by BMBF and by Alexander von Humboldt Foundation.

References

1. E. Blums, A. Cebers, M. M. Maiorov: *Magnetic Fluids* (Walter de Gruyter, Berlin, New York, 1997)
2. R. E. Rosensweig: *Ferrohydrodynamics* (Cambridge University Press, Cambridge, 1985)
3. G. Kronkalns, M. M. Maiorov: Magnetohydrodynamics 3, 28 (1984) (in Russian)
4. V. G. Bashtovoy, A. N. Vislovich, B. E. Kashevsky: Zh. Prikl. Mat. Techn. Fiz. 3, 88 (1978) (in Russian)
5. M. A. Martsenyuk: "Thermal conductivity of a suspension of ellipsoidal particles in a magnetic field". In: *Proceedings of the 8^{ght} Riga MHD Conference* (Zinatne, Riga, 1975) **1,** pp. 108–109 (in Russian)
6. B. M. Tareev: Coll. Journ. 6, 545 (1940) (in Russian)
7. M. I. Shliomis: Fluid Dynamics 6, 957 (1973) (in Russian)
8. K. Gotoh, M. Yamada: J. Phys. Soc. Jap. **51**, 3042 (1982)
9. B. A. Finlayson: J. Fluid Mech. **40**, 753 (1970)
10. L. Schwab: Ķonvektion in Ferrofluiden. Ph.D. Thesis, Ludwig-Maximillians Universität München, München (1989)

11. A. Yu. Chukhrov: Magnetohydrodynamics **22**, 254 (1986)
12. S. Odenbach: Konvektion durch Diffusion in Ferrofluiden. Ph.D. Thesis, Ludwig-Maximillians Universität München, München (1993)
13. E. Blums, A. Rimsa, A. Yu. Chukhrov: Magnetohydrodynamics **23**, 139 (1987)
14. E. Blums, A. Yu. Chukhrov, A. Rimsa: Int. J. Heat and Mass Transfer **30**, 1607 (1987)
15. E. Blums, A. Mezulis, M. Maiorov, G. Kronkalns: J. Magn. Magn. Mater. **169**, 220 (1997)
16. J.Lenglet: Generation de second harmonique et diffusion Rayleigh forcee dans les colloides magnetiques. Ph.D. Thesis, University Paris 7, Paris (1996)
17. A. Mezulis: Mass transfer phenomena in non-isothermic magnetic colloids. Ph.D. Thesis, University Paris 7, Paris (1999)
18. T. Völker, E. Blums, S. Odenbach: Magnetohydrodynamics **37**, 274 (2001)
19. J. L. Anderson: Ann. Rev. Fluid Mech. **21**, 61 (1989)
20. B. V. Derjaguin, V. Churaev, N. V. Muller: *Surface Forces* (Consultants Bureau, New- York, 1987)
21. K. I. Morozov: "Soret effect in disperse systems". In: *Proceedings of the 3rd ASME/JSME Joint Fluids Engineering Conference, July 18-23, 1999, San Francisco*, CA (ASME, 1999), FEDSM99-7784
22. E. Blums, A. Mezulis, N. Buske, M. Maiorov: J. Magn. Magn. Mat. (to be published)
23. N. O. Young, J. S. Goldstein, M. J. Block: J. Fluid. Mech. **6**, 350 (1959)
24. K. I. Morozov: J. Magn. Magn. Mater. **201**, 248 (1999)
25. K. I. Morozov: J. Magn. Magn. Mater. **122**, 98 (1993)
26. J. C. Bacri, A. Cebers, A. Bourdon, A. Demouchy, G. Heegard, B. M. Kashevsky, R. Perzinsky: Phys. Rev. E **52**, 3936 (1995)
27. G. Demouchy, A. Bourdon, J.-C. Bacri, F. Da. Cruz, A. Mezulis, E. Blums: "Forced Rayleigh scattering determination of the Soret coefficient and of the Thermodiffusion mobility of ferrofluids under applied magnetic field" . In: *Proceedings of Fourth International PAMIR Conference "Magnetohydrodynamic at Dawn of Third Millenium"*, Presqu'ile de Giens, France, September 18-22, 2000 (LEGI, Grenoble, 2000), pp. 433–438
28. E. Blums: J. Magn. Magn. Mater. **149**, 111(1995)
29. E. Blums, A. Mezulis: "Thermal diffusion and particle separation in ferrocolloids". In: Transfer *Phenomena in Magnetohydrodynamic and Electroconducting Flows*, ed. by A. Alemany, Ph. Marty, J. P. Thibault (Kluwer, Dordrecht, 1998) pp. 1–18
30. E. Blums, S. Odenbach, A. Mezulis, M. Maiorov: Phys. Fluids **10**, 2155 (1998)
31. A. Yu. Chukhrov: "Gravitation Sedimentation of Polydisperse Magnetic Colloids".In: *Proceedings of 6th USSR Magnetic Fluid Conference* (Institute of Mechanics, Moscow State University, Moscow, 1991) pp. 157–158 (in Russian)
32. G. K. Batchelor: J. Fluid Mech. **119**, 379 (1982)
33. E. Blums, S. Odenbach: Magnetohydrodynamics **37**, 187 (2001)
34. M. M. Bou-Ali, O. Ecenarro, J. Madariaga, C. M. Santamaria: Phys. Rev. **59**, 1250 (1999)
35. M. I. Shliomis, M. Souhar: Europhys. Letters **49**, 55 (2000)

Part III

Rheological Properties

Statistical Physics of Non-dilute Ferrofluids

Andrey Zubarev

Department of Mathematical Physics, Ural State University, Lenin Av., 51, 620083 Ekaterinburg, Russia

Abstract. The influence of inner microscopical structures, consisting of various number of particles, on the macroscopic physical properties of ferrofluids is discussed. The main attention is focused on linear chain-like microclusters. The simplest model of the system, taking these chains as rod-like aggregates, where thermal fluctuations of the shape of these chains are ignored is considered. The boundaries of validity of this model are estimated and it is shown that - as a first approximation - it can be used for many real ferrofluids. The influence of the chains on magnetization and rheological properties of ferrofluids is evaluated. The analysis shows, that these aggregates induce physical phenomena which are impossible in magnetic liquids with separate particles and which are thus not described by the classical models for ferrofluids.

1 Introduction

One of the fundamental problems of physics of ferrofluids (magnetic liquids) is the determination of their macroscopical characteristics as a function of their inner composition, i.e. of shape, size distribution, physical properties and concentration of the magnetic particles and of the properties of the carrier liquid.

The classical theories for ferrofluids (see references and discussion in [1], [2]) deal with ideal models of very dilute systems in which any interactions between particles are negligible. However in real magnetic liquids these interactions often lead to long-range correlations between positions and orientations of the particles and to the appearance of bulk drop-like, linear chain-like and other heterogeneous aggregates, consisting of various number of particles. These inner structures play a fundamental role for the physical properties of magnetic fluids and, therefore, the study of the structures and internal states of ferrofluids gives a key for the understanding of the properties and the macroscopical behavior of these systems.

Since the particles in typical ferrofluids are small (mean diameter is about 10 nm), they are subject to intensive translational and rotational Brownian motion. Any inner structure in a ferrofluid is a result of a competition between thermal motion of the particles, their magnetic interaction and hydrodynamical forces, which are especially significant when the magnetic liquid is exposed to macroscopical flow. That is why the internal states of magnetic liquids can be understood and described only on the basis of an appropriate statistical theory. From the point of view of statistical physics the following main specifics of ferrofluids have to be considered.

The first important peculiarity of magnetic liquids is the magnetodipole interaction between the particles being non central and of long-range nature, decreasing only like r^{-3}.

Secondly, the increase of orientation of the magnetic moment of the ferroparticles under influence of a magnetic field leads to an increase of the effective attraction between them. This effect stimulates, and, moreover, can induce bulk condensation phase transitions in ensembles of particles and can lead to the appearance of dense drop-like aggregates observed in many experiments (see below). The condensation phase transition, induced by an external magnetic field, is a unique peculiarity of magnetic liquids and similarly of magnetorheological suspensions.

Next, unlike molecular systems, where all molecules are identical, real magnetic fluids are always polydisperse. The peak of the particle size distribution lies for typical ferrofluids in a region close to 10 nm. The biggest particles - with a diameter of more than 15 nm - having, as a rule, a very small concentration, can form various microstructures or can play the role of centers of cluster formation for the small ones. Therefore, the biggest particles, in spite of their small concentration, can play an important, often decisive role in determining the physical properties of ferrofluids and must be taken into account in theories for real systems.

The first statistical models of dense equilibrium homogeneous (without any heterogeneous clusters) ferrofluids were suggested in [3,4]. In these models magnetic interaction between the particles was taken into account with the help of the classical Weiss theory of a self-consistent field. With the help of these models [3,4] it was shown for the first time that interparticle interaction increases the equilibrium magnetization of a ferrofluid, and that a magnetic field stimulates the "gas-liquid" phase transition in an ensemble of particles. However, since these models are based on the Weiss approach, they predict spontaneous ferromagnetic order in a ferrofluid and that the critical temperature of the paramagnetic - ferromagnetic phase transition is higher than the temperature of the condensation transition. Therefore, in these theories, the dense ("liquid") phase must be always in a ferromagnetic state. But the ferromagnetic order in ferrofluids has been never observed in experiments. Moreover, there are fundamental reasons to doubt that spontaneous magnetization can occur in suspensions of dipole-dipole interacting particles. Indeed, as well known from magnetism in solid state physics, exchange forces create the ferromagnetic order; dipole-dipole interactions destroy the order. It is unclear, why and how the forces, destroying magnetic order in solid systems, should create it in liquid ones. It should be mentioned that various variants of self consistent effective fields were taken into account in [5]. As in the theories in [3,4], these models lead to the conclusion that spontaneous magnetization can occur in ferrofluids. This result is a direct consequence of the Weiss-like approximation used in these models. For the reasons just mentioned a model with a self-consistent field for the description of ferrofluids should be used with caution.

More advanced theories of homogeneous ferrofluids, using ideas of the Wertheim mean spherical model and the regular theory of thermodynamical perturbations, were suggested in [6] and [7], respectively. The results of calculations of the magnetization of ferrofluids using these theories are in good agreement with experiments and with each other. The models [6,7] predict a "gas - liquid" phase transition in ferrofluids and the existence of these phases in a paramagnetic state only.

The dynamical properties of non dilute homogeneous magnetic fluids, containing no heterogeneous "drops" or "chains", were studied in [8] on the base of the mathematically regular theory of perturbations [7]. The results of [8] show that interparticle interactions increase the characteristic times of magnetic and hydrodynamic relaxation of dense ferrofluids as well as their effective viscosity.

We just noted that real non dilute ferrofluids are often inhomogeneous - bulk (drop-like) as well as linear (chain-like) micro- and mesostructures appear in these systems under suitable conditions. The drop-like aggregates were observed in many experiments (see, for example [9-13]). The linear chains are too thin to be detected in most direct observations, however recent computer experiments [14-17] demonstrate that they can appear in ensembles of dipolar particles under suitable conditions. The appearance of such heterogeneous structures affects strongly the macroscopic characteristics of ferrofluids, especially their dynamical properties. For instance, experiments [18,19] show that typical commercial ferrofluids, like APG513A from Ferrofluidics, with a total hydrodynamic volume concentration of about 27 %, show an increase of the effective viscosity of up to 16 times compared with the zero field viscosity, even under influence of weak magnetic fields, being far from saturation of the magnetoviscous effect! The magnetoviscous effect is especially strong when shear rate is small (about $0.1\ \text{s}^{-1}$). When the shear rate increases, the viscosity decreases very fast. At the same time calculations of this effect, carried out on the base of classical theories of non interacting particles, predict a rise of viscosity about several per cent only under the conditions of [18,19]. Furthermore the dependence of the viscosity on shear rate is much weaker than in the experiments. This significant (several orders of magnitude) discrepancy between the experimental results and the predictions of the classical theories show clearly that a new physical effect, not present in dilute ferrofluids, appears in concentrated systems.

This effect can either be the formation of bulk (drop-like) or linear (chain-like) heterogeneous structures, appearing in dense ferrofluids due to dipole-dipole interactions between the particles. We just discussed briefly statistical models of the condensation phase transition in ferrofluids, which lead to the appearance of bulk-like dense "drops". The first theoretical model of chains in ferrofluids was suggested in [20]. However, this model is based on the theory of homogeneous fluctuations of density. Therefore, the structures, described there, are rather clouds of density than heterogeneous aggregates. The first model of chains as heterogeneous clusters was suggested in [21]. The chain length distribution was estimated in these works using methods of the kinetics of chemical reactions, modelling the combination and recombination of particles from chains. Unfortunately, the final results of the theory are cumbersome and it is not an easy thing to use them for calculations of the physical properties of magnetic fluids. Similar methods of chemical kinetics for the investigation of chain-like structures in ferrofluids in the absence of external field were used in [22].

Models of chains as heterogeneous fluctuations of density in magnetic fluids on the basis of statistical thermodynamics were suggested in [23-25]. Methods, developed for the theory of very long flexible macromolecules, with a tremendous number of monomers in the order of 10^4-10^9), were used in the last mentioned model. However it is difficult to expect such long linear clusters in real ferrofluids. The models in [23,24] use a very simple approximation of the chains as straight rod-like aggregates, i.e. thermal fluctuations of the shape of the chains are ignored in these models. In spite of the fact that this strong simplification can be used only when most of the chains are short enough, this allows us to understand the physical origin of group of phenomena in magnetic fluids and to estimate them, at least, by the order of magnitude. The main points and results of the models [23,24] are discussed in parts 2-4 of this chapter. It should be noted that effects driven by the flexibility of the chains are discussed in [26].

The formation and evolution kinetics of the "drops" as a result of the "gas-liquid" phase transition in a system of interacting, but single particles, was studied theoretically in [27,28]. As in [3,4] and [6,7], any linear chains are ignored in these models and the ferroparticles are treated as separate. These theories describe correctly the principal points of the phase transitions, but numerical experiments [14-17] and analytical models [25,29] show that linear chains appear in systems of dipole-dipole interacting particles long before the bulk condensation phenomena takes place. Therefore these chains must be necessarily taken into account in theories of phase transitions and other structural transformations in ferrofluids.

It is very difficult to study various kinds of inner structures in ferrofluids and their influence on the macroscopic properties of these systems in the framework of one model. That is why it is reasonable to start with separate analysis of each kind of these structures. In this paper some results of the theoretical study of effects of chain-like aggregates on the macroscopic dynamical properties of ferrofluids are presented. To avoid too cumbersome mathematics and to reach physically reasonable estimates, we use the following approximations.

1. We start with an analysis of monodisperse systems. A bidisperse model of a polydisperse ferrofluid is considered in the last part of the paper.

2. We assume that the considered particles are big enough and that the magnetic moment of each of them is frozen into its body. Estimates of the minimal size of the particle necessary to provide this approximation are given, for example, in [1] or [2]. One can show that the magnetic interaction of very small superparamagnetic particles with magnetic moments rotating freely inside their bodies, is too weak to provide any heterogeneous structures. Therefore, the assumption of magnetically hard particles is in agreement with our consideration of chain-like structures consisting of reasonably sized particles.

3. We neglect any interactions between the chains. One can show that, because of the fact that the chains can not overlap, particles from the different chains can not be situated in places, where their interaction is strong, since these positions, for every particle, are occupied by other particles from the same chain. Therefore interaction between particles of different chains is weaker than interaction between those from one chain. That is why the interparticle interaction inside one chain plays the main role in the formation of the chain structure. Thus, the model of non interacting chains seems to be a reasonable first approximation, at least when concentration of the interacting particles is not high. This assumption is not suitable for very dense systems, where the chains lose their individuality and are destroyed due to interaction between them [16].

4. We restrict ourselves to situations where the interaction of the magnetic particles with the external magnetic field is much smaller than the magnetic interaction of two neighboring particles in the chain. It means that the chains react with the field as whole aggregates. This restriction is valid in many real experimental situations.

2 The Chain Size Distribution

The influence of a chain on the macroscopic properties of a ferrofluid is as stronger as longer the chain is. On the other hand, the length of the chain is a magnitude that is determined by a competition between magnetic interaction of the particles and their thermal motion. Therefore our first problem is to determine the size distribution of the chains. In this section we consider equilibrium situations; a possible generalization of these results for systems in steady state far from equilibrium is discussed in section 4.

Consider a ferrofluid consisting of identical spherical particles with hydrodynamical radius (including surface layer) a and magnetic moment m. We denote the hydrodynamical volume concentration of the particles by φ. Let g_n be the number of the n-particle chains in a unit volume of the ferrofluid. In the equilibrium state this distribution function provides a minimum of the free energy F of the unit volume. Under the approximation discussed above this energy can be presented as (see, details in [23,24]):

$$F = kT \sum_{n=1}^{n_c} \left(g_n ln \frac{g_n}{e} + g_n f_n(\boldsymbol{\kappa}, \varepsilon) \right) \quad (1)$$

$$e = 2.72..., \quad \boldsymbol{\kappa} = \mu_0 \frac{m\boldsymbol{H}}{kT}, \quad \varepsilon = \frac{\mu_0 m^2}{16\pi a^3 kT}$$

Here kT is the absolute temperature in energetic units, f_n is the dimensionless free energy of the chain due to its inner structure, n_c is the maximal number of particles in the chain, estimated below. For equilibrium states one may put $n_c = \infty$ (see eq. (15)). The first term in the parentheses in equation (1) represents the entropy of the ideal gas of n-particle chains determined by their translation motion, $\boldsymbol{H}$ is the magnetic field inside the sample, μ_0 is the vacuum permeability,

κ and ε denote the dimensionless energies of the interaction of the magnetic particle with the field H and of the magnetodipole interaction of two neighboring particles in the chain, respectively.

We should minimize the free energy (1) under the obvious condition of normalization:

$$\sum_{n=1}^{n_c} n g_n = \frac{\varphi}{v}, \quad v = \frac{4}{3}\pi a^3 \tag{2}$$

After standard calculations one can obtain:

$$g_n = \frac{1}{v} exp\left(-f_n + \lambda n\right) \tag{3}$$

where λ is the Lagrange multiplier. To calculate it we've to substitute equation (3) into equation (2). This leads to a non linear equation with respect to λ which can be solved analytically or numerically. However first one needs to determine the "internal" free energy f_n. This energy can be presented as follows:

$$f_n = -\ln Z_n \tag{4}$$

$$Z_n = \int exp\left(\sum_i^n (\boldsymbol{\nu}_i \boldsymbol{\kappa}_i) - \sum_{i>j}^n U_d(\boldsymbol{\nu}_i, \boldsymbol{\nu}_j, \boldsymbol{r}_{ij})\right) d\Omega$$

Here $\boldsymbol{\nu}_i$ is the unit vector aligned along the magnetic moment of the $i-th$ particle, $\boldsymbol{r}_{ij}$ is the radius-vector, connecting the centers of the $i-th$ and $j-th$ particle, Ω includes all $\boldsymbol{\nu}_i$ and $\boldsymbol{r}_{ij}$ and U_d is the dimensionless potential of the dipole-dipole interaction.

As usual, the statistical integral Z_n can not be calculated in a general form. To reach reasonable estimates we may take into account only the interaction between nearest particles in the chain. This approximation was used in the models [22-26]. The interaction between nearest particles and throughout the nearest neighbors was taken into account in [21] and led to cumbersome final results. In the "nearest neighbor" approximation the statistical integral (4) is

$$Z_n = \int exp\left(\sum_i^n (\boldsymbol{\nu}_i \boldsymbol{\kappa}_i) - \sum_i^n U_d(\boldsymbol{\nu}_i, \boldsymbol{\nu}_{i+1}, \boldsymbol{r}_{i,i+1})\right) d\Omega \tag{5}$$

Even in this approximation the many-particle integral can not be decoupled to low-dimensional integrals for finite κ. That is why a mathematically strict calculation of this integral for arbitrary κ is impossible. Some estimates for the limiting situations $\kappa = 0, \kappa = \infty$ and extrapolated approximations for finite κ are suggested in [21,26].

Here we take into account that the appearance of chains in ferrofluids is possible only when the dimensionless parameter ε of the dipole-dipole interaction is significantly larger than unity. Since the potential U_d has a minimum when the vectors $\boldsymbol{\nu}_i, \boldsymbol{\nu}_{i+1}$, $\boldsymbol{r}_{ij}$ are parallel and $r_{i,i+1} = 2a$, we can replace

$exp\left(U_d(\boldsymbol{\nu}_i,\boldsymbol{\nu}_{i+1},\boldsymbol{r}_{i,i+1})\right)$ by $\delta(\boldsymbol{\nu}_i-\boldsymbol{\nu}_{i+1})\delta(\boldsymbol{\nu}_i-\boldsymbol{r}_{i,i+1})\delta(r_{i,i+1}-2a)$ as a first approximation. In this approximation we ignore thermal fluctuations of the chains and treat them as straight rod-like aggregates. This is justified when the contour length of the chain is less or about the persistent length l_0 of the chain. Using standard considerations of the theory of linear polymers, one can show that $l_0 \sim 2a\varepsilon$. Therefore the "rod" model of the chains can be considered as justified when the mean contour length $2a<n>$ of the chain is not more than $2a\varepsilon$, i.e. $<n>\le \varepsilon$ where $<n>$ is the mean number of particles in the chain.

In the framework of this model the dimensionless free energy f_n may be written down as:

$$f_n = -\left[\varepsilon(n-1) + ln\left(\frac{\sinh \kappa n}{\kappa n}\right)\right] \quad (6)$$

The first term in the square brackets represents the absolute value of the energy of magnetic interaction of particles inside the rod-like chain, estimated in the approximation of nearest neighbors. The second term corresponds to the Langevin energy of the rod in the dimensionless field κ. It should be noted that relation (6) is the upper estimate for the absolute magnitude of f_n.

Substituting equation (6) into (3) and, then, into (2), taking into account that one can put $n_c = \infty$ for equilibrium systems (see equation (15)), we get after some calculations (details in [23])

$$g_n = \frac{x_0^n}{v}\frac{\sinh \kappa n}{\kappa n}\exp(-\varepsilon), \quad (7)$$

where

$$x_0 = \frac{2y\cosh\kappa - \sinh\kappa - \sqrt{(2y\cosh\kappa - \sinh\kappa)^2 - 4y^2}}{2y}, \quad y = \kappa\varphi\exp(\varepsilon)$$

Figure 1 illustrates calculations of the volume concentration vng_n of the particles collected in the n-particle chains. This volume concentration, as function of n, has a maximum increasing with κ and ε. Similar maxima were obtained also in [29].

The mean number of particles in the chain is

$$<n> = \frac{\sum_n ng_n}{\sum_n g_n} = \frac{\varphi}{v\sum_n g_n} \quad (8)$$

Let φ_c be the volume concentration, corresponding to the condition $<n> = \varepsilon$. The plots of φ_c as function of κ for two values of ε are shown in Fig 2. If at a given ε the condition for the concentration fulfills $\varphi < \varphi_c$, then the model of rods can be used.

Using distribution function (7), one can estimate magnetization M of ferrofluid as in [23]:

$$M = m\sum_n nL(\kappa n)g_n \quad (9)$$

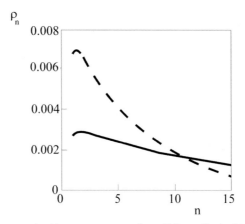

Fig. 1. The volume concentration $\rho_n = vng_n$ of particles collected in $n-$ particle chains vs. n for $\varphi = 0.05, \kappa = 1$; $\varepsilon = 4$ and 5 (dashed and solid lines, respectively)

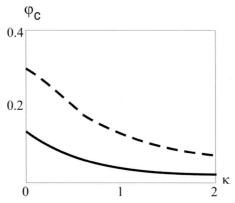

Fig. 2. The critical volume concentration φ_s of the particles for the model of straight aggregates vs the dimensionless magnetic field. Solid and dashed lines correspond to $\varepsilon = 5$ and $\varepsilon = 3$, respectively

where $L(x) = \coth(x) - x^{-1}$ is Langevin function. Comparison of calculations of the magnetization with the experiments [30] are shown in Fig. 3. One can see that the model of chains leads to reasonable results, at least when only short chains are expected.

3 Rheological Properties for Vanishing Shear Rate

Let us consider a ferrofluid subjected to a shear flow. We assume here that the shear rate is weak enough to neglect deviation effects from the equilibrium size and shape of the chains. As shown in [23], deformation of the chains under

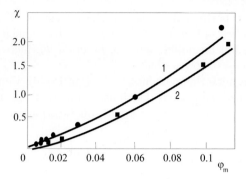

Fig. 3. Results of measurements [30](dots) and calculations according to equation (9)(lines) of the initial susceptibility χ of a ferrofluid vs. the magnetic volume concentration φ_m for two different systems. Their parameters are given in [30]. Circles and squares correspond to lines 1 and 2, respectively

hydrodynamical flow with shear rate $\dot{\gamma}$ is weak when

$$\dot{\gamma} \ll \frac{kT\varepsilon}{\eta_0 (2a)^3 <n>}$$

Here η_0 is the viscosity of the carrier liquid. This inequality is valid for most part of experimental situations.

It is well known since Einstein's classical work, that the influence of dispersed particles on the rheological properties of suspensions is connected with their perturbations of the macroscopic flow of the suspension. It is practically impossible to calculate the perturbations induced by the chains because of their complex shape. In order to get reasonable estimates, we model the n−particle chain, like in [23] and [24], as a prolate spheroid with semiminor axis a and semimajor one an. It is of principal importance that the volume of this spheroid is equal to the sum of the volumes of the particles in the chain. Therefore, the volume concentration of the model spheroids is equal to the total volume concentration of the particles.

Using the well known results of the theory of statistical hydromechanics of dilute suspensions of rigid spheroids (see, for example, [23,24,31]), we can write down expressions for the Cartesian components of the average viscous stress tensor $\hat{\sigma}$ as follows

$$\sigma_{ik} = 2\eta_0 \left[\gamma_{ik} + \sum_{n=1}^{n_c} \Phi_n(\gamma_{lm}, \omega_{lm}, <e_j e_s>_n, <e_l e_m e_j e_s>_n) g_n \right] \quad (10)$$

$$<...>_n = \int ...\phi_n(e)de, \quad i,k...s = x,y,z$$

The symbol $\hat{\ }$ here and below indicates tensor magnitudes, e is a unit vector aligned along axis of the chain, $\phi_n(e)$ is a distribution function over the orientations of the chain normalized to unity, γ_{ij} and ω_{ij} are the symmetrical and antisymmetrical parts of the tensor of the velocity gradient, Φ_n is a known and cumbersome linear function of γ_{ij} and ω_{ij}. Its explicit form is given in [23],[24],[31]. It should be noted that the function Φ_n increases fast with the number n.

In order to determine the moments $<...>_n$ in Eq.(6), we need to determine the non equilibrium orientational distribution function ϕ_n. This can be found using the Fokker-Planck equation for a non-spherical particle (see, for example, [23,24,31]) :

$$\frac{\partial \phi_n}{\partial t} = -\mathbf{I}(\phi_n \frac{d\mathbf{e}}{dt}) + D_n \mathbf{I}^2 \phi_n, \tag{11}$$

where

$$\frac{d\mathbf{e}}{dt} = -\frac{D_n}{kT}\mathbf{I}U_\kappa + \hat{\omega}\mathbf{e} + \lambda_n \hat{\gamma}\mathbf{e} - ((\mathbf{e}\hat{\gamma}\mathbf{e})\mathbf{e}), \tag{12}$$

$$\mathbf{I} = \left[\mathbf{e} \times \frac{\partial}{\partial \mathbf{e}}\right], U_\kappa = -kT(\mathbf{e}\boldsymbol{\kappa}n),$$

$$D_n = \frac{D_1}{n\delta_n}, \quad D_1 = \frac{kT}{6\eta_0 v},$$

Here D_n denotes the coefficient of rotational diffusion of a spheroid with an axis ratio equal to n, the parameter δ_n, increasing with n, is given in [23,24,31].

The exact solution of equation (11,12) is unknown. To find its approximate solution we use the ideas of the theory [32] and present the distribution function ϕ_n in the form of the following trial function:

$$\phi_n = \phi_n^o \left(1 + a_i \left(e_i - <e_i>_n^o\right) + b_{ik}\left(e_i e_k - <e_i e_k>_n^o\right)\right), \tag{13}$$

$$\phi_n^o(\mathbf{e}) = \frac{\kappa n}{4\pi \sinh \kappa n} \exp\left((\mathbf{e}\boldsymbol{\kappa})n\right), \quad <...>_n^o = \int ...\phi_n^o(\mathbf{e})d\mathbf{e}$$

Here ϕ_n^o is the equilibrium Boltzman orientational function for the n-particle rod in the magnetic field $\mathbf{H}$, $\mathbf{a}$ and $\hat{b}$ are a vector and a second rank tensor to be determined. It should be noted that the trial function, used in the theory [32], is similar to that in equation (13), but with $b_{ij} = 0$.

According to the ideas of Ref. [32], we determine the parameters a_i and b_{ij} using the equations for the first and second moments of the vector $\mathbf{e}$ of function (13). For this purpose equation (11) should be multiplied by components e_k and by the complex $e_i e_k - \frac{1}{3}\delta_{ik}$, and after integration over $\mathbf{e}$ we come to a set of equations for the first and second moments of the function ϕ_n. Substituting there ϕ_n in the form (13), we obtain a system of algebraic equations for a_i, b_{ij} (see details in [23,24,31]). If the state of the system is not far from equilibrium, we may take these equations in linear approximation in a_i, b_{ij} and the components components γ_{ij}, ω_{ij} of the velocity gradient. After solving this system and determination of a_i, b_{ij}, using (13), we can calculate the moments $<...>_n$ in equation (10), and therefore, estimate the stress tensor $\hat{\sigma}$ or, that is the same, all rheological characteristics of the ferrofluid in the framework of the discussed approximations.

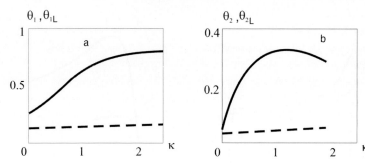

Fig. 4. Dimensionless stationary effective viscosities θ as functions of external magnetic field for $\varepsilon = 3, \varphi = 0.05$. Solid lines correspond to model with chains, dashed ones to model with separate particles. The field is aligned along (a) gradient of flow and (b) velocity

In general, the problem of the calculation of a_i and b_{ij} is cumbersome, but principally not difficult. The detailed calculations for two typical situations, when the field is parallel to (1) the gradient and (2) the velocity of a flow, are given in [24]. Some results for the dimensionless stationary effective viscosity $\theta_p = (\eta_p - \eta_0)/\eta_0)$ are shown in Fig.4 both for the model with chains and under assumption that all particles are free (the last case corresponds to the classical theories, discussed in [1,2]). The parameter η_p is the effective viscosity and the index $p = 1, 2$ corresponds to the mentioned orientations of external magnetic field. These plots show that the chains increase the effective viscosity significantly. The magnetoviscous effect is especially strong when the field is parallel to the gradient of velocity ($p = 1$). In this situation the viscosity increases with the field monotonously. When the field is parallel to the velocity ($p = 2$), the magnetoviscous effect is smaller, and, moreover, the viscosity depends on the field non monotonously with a maximum. The physical origin of the last result lies in the fact that the hydrodynamical resistance of the elongated n-particle aggregate, which is parallel to the velocity, is smaller than the total resistance of the n separate particles. When the field is small, the orientation of the chains is chaotic and the main role in the formation of the field dependence of viscosity is played by a blocking of the free rotation of the particles and chains in the shear flow (it is the same mechanism that creates the rotational viscosity in dilute ferrofluids with separate particles). However, when the field is large enough, the system of chains is highly ordered and an increase of the field leads to an increase of this order as well as to an increase of the number of long, highly oriented chains. This induces a decrease of viscosity. A similar effect of a decrease of viscosity under influence of a magnetic field parallel to the flow velocity, is well known for nematic liquid crystals. The maximum in the plot of viscosity vs. magnetic field in Fig. 4 is a result of a competition between these two mentioned mechanisms. One needs to stress that in the models of separate particles the effective viscosity of ferrofluids monotonously increases with magnetic field for all orientations.

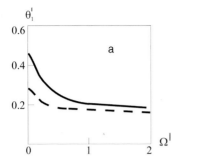
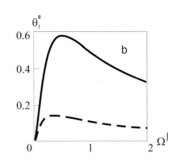

Fig. 5. (a) Real and (b) imaginary part of complex dimensionless effective viscosity vs frequency Ω of an oscillating flow ($\Omega' = \Omega/D_1$). The external field is aligned along the gradient of the flow velocity, $\varepsilon = 3, \varphi = 0.05$. Solid and dashed lines correspond to $\kappa = 1$ and $\kappa = 0.5$, respectively

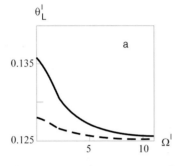
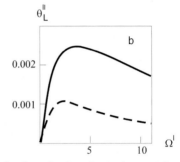

Fig. 6. The same curves as in Fig.5 for ferrofluid with single particles

Figure 5 shows calculations of the real θ'_1 and imaginary θ'' part of the dimensionless complex viscosity as a function of the frequency of the flow for a magnetic field parallel to the gradient of the flow velocity. For comparison, Fig.6 shows the results of calculations for the same magnitudes under the assumption that all particles are free. Comparing the results in Figs. 5 and 6, one can note a strong increase of both components of complex viscosity due to the chains and the fact that the frequency, corresponding to the maximum of θ'', is much smaller in the model with chains than in the model for free particles. It means that the time of hydrodynamical relaxation in ferrofluids with chains is significantly longer than the same time predicted by the classical models of separate particles.

Using this approach, one can estimate the normal components of the stress tensor under shear flow [24]. It is known that the normal stresses can lead to the so-called Weissenberg effect [33], observed usually in isotropic polymer liquids. The results of an experimental study of this effect in ferrofluids are discussed in [34]. It should be noted that this phenomenon can not take place in systems with separate spherical particles only. The fact that it is observed in ferrofluids is a result of the presence of anisotropic (chain-like) aggregates and qualitatively supports the suggested model.

In the discussed model it is possible to estimate the components of magnetization of a ferrofluid under conditions of a deformation flow. To do this, we should first to take into account that - by definition - the macroscopical magnetization is:

$$\boldsymbol{M} = m \sum_{n}^{\infty} n <\boldsymbol{e}>_n g_n, \qquad (14)$$

Let the magnetic field and the velocity gradient of the flow be aligned along the axis Oz, the velocity being aligned along axis Ox. For this situation, in linear approximation in the symmetrical γ and antisymmetrical ω parts of the velocity gradient tensor, the calculations give:

$$a_x = \omega A_1 + \gamma \Lambda_n A_2, \quad 2b_{xz} = \omega B_1 + \gamma \Lambda_n B_2$$

and

$$<e_x>_n = A_1 \left(<e_x^2> - n^o + B_1 <e_z e_x^2>_n^o\right) \omega + \Lambda_n \left(A_2 <e_x^2>_n^o + B_2 <e_z e_x^2>_n^o\right) \gamma$$

$$\Lambda_n = \frac{n^2 - 1}{n^2 + 1},$$

The parameters $A_{1,2}, B_{1,2}$ can be determined from a linear algebraic system of equations for a_x, b_{xz}, which is given in [24] (equations (16),(17) of this work). Substituting the last relation into (14), we get:

$$M_x = \lambda_1 \omega + \lambda_2 \gamma$$

$$\lambda_1 = m \sum_{n=1}^{n_c} n \left(A_1 <e_x^2>^o + B_1 <e_z e_x^2>^o\right) g_n,$$

$$\lambda_2 = m \sum_{n=1}^{n_c} n \Lambda_n \left(A_2 <e_x^2>^o + B_2 <e_z e_x^2>^o\right) g_n,$$

The calculations show that the parameters $\lambda_{1,2}$ are increasing functions of φ, ε and non monotonous functions of κ.

The same form for M_x with λ_1 and λ_2 is obtained in [35] from phenomenological considerations. Both microscopical and phenomenological approaches show that $\lambda_2 = 0$ when only separate spherical particles are present in the ferrofluid. This term can be non zero only due to the presence of non spherical aggregates in ferrofluid. We would like to note that the classical models for magnetic fluids, which deal with separate spherical particles and ignore any clusters, give $\lambda_2 = 0$. Specially organized experiments [35] demonstrate that for ferrofluids with non vanishing concentration of particles $\lambda_2 \neq 0$. This shows that elongated inner structures are present in real ferrofluids.

The experiments on Weissenberg effect and on the "γ-dependence" of magnetization in ferrofluids are in a qualitative agreement with the suggested model. Unfortunately, because of tremendous mathematical problems, appearing in the solution of the macroscopic equations of flow for ferrofluids under external magnetic field, it is difficult to give a precise quantitative comparison between the theory and experiments. Such a comparison is discussed in part 5 of the paper for the rheological properties of magnetic fluids.

4 Rheological Properties of Ferrofluids for Non Vanishing Shear Rate

In a non equilibrium situation the theorem of the minimum of the free energy, used in part 3 for the determination of the distribution function g_n is, strictly speaking, not valid. The search for a solution of equation (12) for situations far from equilibrium presents a very difficult mathematical problem. That is, why a generalization of the approach, developed for quasi equilibrium ferrofluids with chains, to non equilibrium situations (for example, when shear rate is not vanishing) is a very complicated problem. However these difficulties can be resolved for many actual situations when the inequality $\dot{\gamma}/D_n \ll 1$ ($\dot{\gamma}$, as usual, is shear rate) is held for all chains with non vanishing volume concentration vng_n. One can show that this inequality corresponds to the condition that the shear Pecklet number Pe, constructed for the chains, is small.

In this situation the distribution function ϕ_n in equation (11) is not far from its equilibrium magnitude and the trial function (13) can be used in linear approximation in $\dot{\gamma}$. At the same time, since Pe is small, the transport of the particles near the chains is almost the same as for the equilibrium state. Therefore, the condition of the minimum of the free energy can be used for an approximate estimate of g_n. At the same time it is necessary to keep in mind that hydrodynamical forces destroy too long chains in a ferrofluid under shear flow. The maximal number n_c of the particles in the non destroyed rod-like chain has been estimated in [19] as:

$$n_c \approx \sqrt{\varepsilon \frac{D_1}{\dot{\gamma}}} \qquad (15)$$

When $\dot{\gamma}$ is vanishing, n_c tends to infinity. This result was taken into account in the previous sections. When $\dot{\gamma}$ is significant, but the inequality $\dot{\gamma}/D_1 \ll 1$ is fulfilled, we should use the estimate (14) in equations (1),(2) and (10). Thus, in first approximation, when the inequality $\dot{\gamma}/D_n \ll 1$ is true for all chains, the influence of shear rate on the chain structure can be taken into account by a use of the estimate (15) in relations (1,2) and (10), remaining all other considerations of parts 2 and 3 without changes.

5 Comparison with Experiments

A strong increase of viscosity in dense ferrofluids under weak magnetic field, parallel to the gradient of flow, was detected in the experiments [18], [19]. Some results of these measurements are shown in Fig.7 (dots). These experiments were carried out with polydisperse magnetite ferrofluids containing about 27 vol% of the hydrodynamical volume of particles; the mean diameter of the magnetic core of the particles is about 10 nm. A diagram of size distribution for the particles is given in Fig. 8.

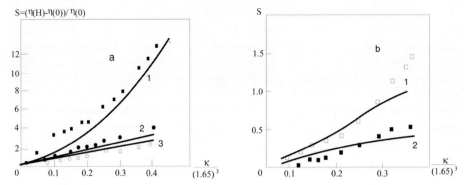

Fig. 7. Experiments [18] (dots) and calculations (lines) of the magnetoviscous parameter S. a) Squares corresponds to shear rate $\dot\gamma = 0.1s^{-1}$, line 1 to $\dot\gamma = 0$; black circles and line 2 to $\dot\gamma = 0.5s^{-1}$, white circles and line 3 to $\dot\gamma = 0.9s^{-1}$. b) light squares and line 1 correspond to $\dot\gamma = 1.05s^{-1}$, black squares and line 2 to $\dot\gamma = 5.23s^{-1}$

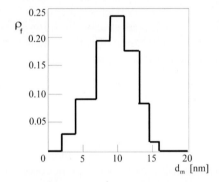

Fig. 8. Diagram of the size distribution for ferrofluid, used in experiments [18], [19]

One needs to stress that the 10 nm particles are superparamagnetic and the energy of magnetic interaction between them is much smaller than the thermal energy for room temperature. Therefore, these particles neither can influence significantly the magnetorheological properties of the ferrofluid, nor they can form any heterogeneous aggregates. It means that only the biggest particles, in spite of their small concentration, give rise to the observed strong magnetorheological effects. The analysis, given in Section 3, shows, that the physical origin of this phenomenon can be chain-like aggregates, formed by the biggest particles.

The problem of a theoretical description of chain-like structure in a polydisperse ferrofluid, even under the used strong simplification, is very complicated. To avoid cumbersome mathematics, we use a model of a bidisperse system consisting of the "small" particles with sizes being near the mean size of particles in the ferrofluid and "big" particles with sizes and volume concentration to be determined. For maximal simplification of the calculations we assume that there are only the "big" particles, which can form the chains. All the "small" particles are supposed to be free.

Using the approach, described in Sections 2 and 3, we calculated the parameter $S(H) = (\eta_1(H) - \eta_1(0))/\eta(0)$ of the magnetoviscous effect for vanishing shear rate. Then the magnetic diameter d as well as the hydrodynamic volume concentration φ of the "big" particles have been determined by comparing the results of the calculations with the experiments for the smallest shear rate (namely $\dot{\gamma} = 0.1 s^{-1}$, see Fig 7a). As a result we obtained $d = 16.5 nm, \varphi = 0.17$. If we look at the diagram in Fig. 8, these results seems to be reasonable. Then, using the consideration of section 4, we calculated S for several magnitudes of $\dot{\gamma}$, used in the experiments. The results of the calculations are given in Fig 7. The agreement between the theory and experiment can be recognized as good, especially taking into account that the model is extremely simple.

6 Conclusion and Discussions

The obtained results allow us to make the following qualitative conclusion. In real polydisperse ferrofluids large particles are present, which provide strong magnetoviscous and other dynamical effects, in spite on their very small concentration. The physical origin of these effects are heterogeneous (for example, chain-like) structures, formed by these particles. The magnetorheological effect is especially strong for small shear rate, when the structures are not destroyed. Increase of shear rate leads to disruption of the structures and to a fast decrease of effective viscosity.

The discussed simple model of the chains as rod-like aggregates allows us to understand the physical origin of such physical phenomena in real ferrofluids as non monotonic dependence of viscosity under magnetic field aligned along the velocity of the flow; the dependence of the components of the macroscopical magnetization on the symmetrical part of velocity gradient tensor; the Weissenberg effect; the strong magnetorheological effect. The estimates of the magnetoviscous effect in real ferrofluid are in a reasonable quantitative agreement with experiments.It should be stressed that these phenomena are impossible in systems with spherical particles only.

At the same time one needs to admit that the simple model of chain structures in ferrofluids opens many questions that can be a subject of further investigations.

First. What is the role of the flexibility of the chains in the formation of the macroscopic properties of ferrofluids? The matter is that the longest chains influence the dynamical properties of ferrofluids especially strong. However the long chains can not be rod like, they should be similar to worm-like micelles. Unfortunately, it is a very complicated problem to calculate the influence of these "micelles" on the physical properties of the magnetic liquids.

Second. Can the real long flexible chains be in a linear state or do they inevitably contain particles joined to the chains by side? In the last situation the chains are rather bulk than linear aggregates and their analysis requires a development of a special theoretical approach.

Third. What is the role of the "small" particles in the formation of the inner structure of ferrofluids? The point is that simple estimates show that the energy of magnetic interaction between typical magnetite particles with a magnetic diameter about 10 nm and those with 16-17 nm for room temperature is slightly more than kT. Therefore, the "small" particles can take part in the formation of the chains. The previous analysis [37] shows that their influence on the chain structure is not negligible, however this problem is worth a further detailed examination.

Forth. What is the role of the chain-chain interaction in ferrofluids? This question is especially intrigue, since many experiments (for example, [9-13]) demonstrate that the "gas-liquid" phase transition can take place in an ensemble of ferroparticles under suitable conditions. Theories [3,4,6,7,27,28] explain and describe the main points of these phase transitions, however they ignore any possibility of the formation of chain-like structures. Computer simulations [14-17] and theories [25,36] that take into account this possibility, conclude that the condensation phase transition in ferrofluids is impossible - when temperature decreases the chains become longer in the spatially homogeneous system. Thus, there is a qualitative contradiction between experimental and theoretical (including computer experiment) results on phase transitions in ferrofluids. A physical nature of this contradiction may lie in the following facts. First, the surface layers on the ferroparticles do not screen van-der-Waals interaction entirely. The combination of dipole-dipole forces and "tails" of the central van-der-Waals interaction can lead to a bulk phase transitions even in ferrofluids with chains [25]. The second factor, which can induce this transition, can be polydispersity of real ferrofluids. Indeed, the "small" particles, with magnetic diameter about 10 nm and high concentration, can create a strong osmotic pressure that pushes the "big" particles to regions where concentration of the "small" ones is low. As a result domains with high concentration of big particles, where the osmotic pressure compensate the pressure in an environment with small particles, can appear as thermodynamical stable phases. Both of these scenarios of the bulk separation (first, provided by the van-der-Waals tails and the second one, induced by the "small" particles) are worth a thorough examination.

Finishing, one can note that in spite of the mentioned unsolved problems, the simple model of non-interacting rod-like chains can be considered as a robust basis for a qualitative understanding and study of the dynamical properties of real non dilute ferrofluids. This model can serve also as a base for a development of more advanced theories of internal structures and macroscopic physical properties of magnetic liquids.

This work has been supported by grants of the Russian Basic Research Foundation, projects NN 02-00-17731, 02-01-6072, 01-01-00058, grant CRDF, REC-005 and grant of BMBF N RUS 00/196

References

1. M.I.Shliomis, Sov.Phys.Usp. **17**, 153 (1974)
2. E.Blums, A.Cebers, M.Majorov, *Magnetic Fluids* (de Gruyter, Berlin, New York, 1997)
3. A.O.Tsebers, Magnetohydrodynamics, **18**, 345 (1982)
4. K.Sano and M.Doi, J.Phys. Soc.Japan., **52**, 2810 (1983)
5. D.Wei, G.N.Patey and A. Perera, Phys.Rev.E, **47**, 506 (1993); H.Zhang and M.Widom, J.Magn.Magn.Mater.**122**, 119(1993); Phys.Rev.E **49**, 3591 (1994); Phys.Rev.B **51**,8951 (1995); B.Groh and S.Dietrich, Phys.Rev.E, **50**, 3814 (1994); Phys.Rev.E, **53**, 2509 (1996); Phys.Rev.E, **57**, 453518, (1998)
6. K.I.Morozov, Bull. Acad. Sci. Ussr, Phys. Ser. **51**, 32 (1987)
7. Yu.A.Buyevich and A.O.Ivanov. Phys. A, **190**, 276 (1992)
8. A.Yu.Zubarev and A.V.Yushkov, Zh.Eksp.Theor.Fiz **114**, 892 (1998) [JETP **87** 484 (1998)]
9. C.F.Hayes, J.Colloid. Interface Sci., **52**, 239 (1975)
10. E.A.Peterson and A.A.Krueger, J.Colloi. Interfase Sci., **62**, 27 (1977).
11. A.F.Pshenichnikov and I.Yu.Shurubor, Bull. Acad. Sci. USSR, Phys. Ser., **51**, 40 (1987)
12. J.C.Bacri, R.Perzynski, D.Salin, V.Cabuil and R.Massart,J. Colloid. Interface Sci., **132** 43 (1989)
13. P.K.Khizhenkov, V.L.Dorman and F.G.Bar'akhtar, Magnetohydrodynamics, **25**, 30 (1989)
14. J.M.Caillol, J.Chem. Phys., **98**, 9835 (1993)
15. M.E.Van Leeuwen and B.Smit, Phys. Rev.Let. **71**, 3991 (1993)
16. D.Levesque and J.J.Weis, Phys. Rev.E **49** 5131 (1994)
17. A.Satoh, R.W.Chantrel, S.Iand G.N.Coverdall, Coll., Intersace Sci., **181**, 422 (1996)
18. A.Zubarev, S.Odenbach, J. Fleicher J.Magn. Magn.Mat. (submitted as material of IX Int. Conf. on Magnetic Fluids)
19. S.Odenbach and H.Stork, J.Magn.Magn.Mat. **183**, 188 (1998)
20. P.G. de Gennes and P.A.Pincus, Phys. Condens. Mat. **11**, 189 (1970)
21. P.G.Jordan, Molecular Physics, **25**, 961(1973); ibid. **38**, 769 (1979)
22. A.Cebers, Magnitnaja Gidrodinamika, **2** 36 (1972)
23. A.Yu.Zubarev, L.Yu. Iskakova, J.Exp. Theoret. Physics, **80**, 857 (1995)
24. A.Yu.Zubarev and L.Yu. Iskakova, Phys. Rev. E, **61**, 5415 (2000)
25. M.A.Osipov, P.I.C.Teixeira and M.M. Telo da Gama, Phys.Rev.E., **54**, 2597 (1998). M. Tavares, M.M. Telo da Gama and M.A.Osipov, Phys. Rev. E., **56**, 6252 (1997), see, also, Errata, Phys. Rev. E, **57** 7367 (1998)
26. K.Morozov and M.Shliomis see p.204 of this volume.
27. A.O.Tsebers, Magnetohydrodynamics, **1** 3, (1994)
28. A.Yu.Zubarev and A.O.Ivanov, Phys. Rev. E, **55**,7192 (1997); A.O.Ivanov and A.Yu.Zubarev, Phys. Rev.E, **58**,7517 (1998)
29. J.M.Tavares, J.J.Weis and M.M.Telo da Gama, Phys. Review E, **59**, 4388 (1999)
30. M.Rasa, D.Bica, A.Philipse and M.Vekas, "Dilution Series Approach for Investigation of Microstructural properties and Particle Interaction in High Quality magnetic Fluids" (will be published)
31. V.N. Pokrovskij, *Statistical mechanics of Dilute Suspension*, (Nauka, Moscow, 1978)

32. M.A.Marstenjuk, Y.A.Raikher and M.I.Shliomis, Zh.Eksp. Teor Fiz **65** 834 (1974) [JETP, **38**, 413 (1974)]
33. D.D.Josef, G.S.Beavers and R.L.Fosdick, Arch. Rational Mech. Annals, **49**, 381 (1973)
34. K.Melzner and S.Odenbach, J. Magn. Magn. Mat, (submitted as material of IX Int. Conf. on Magnetic Fluids)
35. H.Muller and M. Liu see page 115 of this volume
36. S.Klapp and F.Forstmann, Phys. Rev. E., **60**, 3183 (1999).
37. A.O.Ivanov and S.Kantorovich, J.Magn.Magn.Mat. (submitted as material of IX Int. Conf. on Magnetic Fluids)

Magnetic Fluid as an Assembly of Flexible Chains

Konstantin I. Morozov[1] and Mark I. Shliomis[2]

[1] Institute of Continuous Media Mechanics, UB of Russian Academy of Sciences, 614013 Perm, Russia
[2] Department of Mechanical Engineering, Ben-Gurion University of the Negev, P.O.B. 653, Beer-Sheva 84105, Israel

Abstract. Dipolar chains formed in magnetic fluids out of colloidal magnetic grains have much in common with polymer molecules. An investigation of spatial and orientational intrachain correlations and elucidation of an important role of the chains flexibility lead us to natural and fruitful extension of basic concepts of polymer physics to the case of dipolar chains. Conformations of the chains (statistical coil, globule) in zero and infinitely strong external magnetic field are studied and the possible coil–globule phase transition is predicted and discussed.

1 Introduction

The phenomenon of formation of internal structure of magnetic fluids (MF) and its influence on the macroscopic properties of MF is one of the most exciting problems of physics of ferrocolloids. It is generally recognized that the origin of internal structure is due to dipolar interactions between magnetic grains composed MF. The basic features of the effects of dipolar interactions in MF were outlined long ago in the pioneering work of de Gennes and Pincus [1]. The authors had brilliantly foreseen the two possible consequences of magnetic interparticle interactions. First, in the zero external field they predicted *the phase separation* of a magnetic fluid into the dilute vapor and condensed liquid phases below some critical temperature in analogy with the case of simple fluids. A heuristic explanation of the phenomenon is following. The dipole $1/r^3$ pair potential (r the particle separation) being averaged over the dipole orientations becomes an attractive $1/r^6$-potential responsible for the van der Waals condensation of simple fluids. Second, at high external fields $\boldsymbol{H}$ the *association phenomenon* should appear: the dipoles align along the field direction, the head-to-tail configuration is energetically preferable and the magnetic grains tend to form *chains* along the direction of $\boldsymbol{H}$. Thus, the two types of effects of interparticle interactions in MF differ by their scales: the condensation manifests itself as *macroscopic* phenomenon whereas the grain association as *microscopic* one. De Gennes and Pincus discussed also the appearance of chain-like structure in zero external field. However, the relation between the condensation and the chain formation had remained unclear in this case. As we will see, just this question became the subject of the latest intensive investigations.

In the 70-s, the task of condensation of dipolar fluids in the *absence* of external field seemed to be well understood. All the theories of the system of interacting dipoles predicted the appearance of the gas-liquid phase transition under some critical temperature (see, e.g. [2] and references therein). Among the theories we note the analytical mean spherical model (MSM) of Wertheim [3] for the system of hard sphere dipoles. The thermodynamics for MSM was developed in [4] and the critical value of the coupling parameter $\lambda = m^2/d^3 k_B T$ (m is dipole moment, d is a grain diameter, k_B is Boltzmann's constant and T is the temperature) was determined. According to [4], this value is $\lambda_c = 4.445$. The condensation observed in Monte Carlo simulations [5] qualitatively corroborated the theoretical results.

The condensation of magnetic fluids in zero field as well as in external magnetic fields is reliably established experimentally. Starting from the probably first evidence, given by Hayes [6], it was observed in a numerous works for both types of MF - ionic [7–9] and surfaced ones [10–14]. It has been shown that an external magnetic field always promotes the phase transition. The droplets of concentrated "liquid" phase appear among the more dilute "gas" phase with decreasing of temperature or/and with growth of magnetic field. The real MF are polydisperse ones. So, there is no simple to determine experimentally the critical values $\lambda_c(H)$ of the beginning of the phase separation. The estimation of the value in strong magnetic filed is $\lambda_c(\infty) = 2.96$ [14]. In zero field the droplets of concentrated phase are spherical what means the *absence* of their spontaneous magnetization. The droplets have typical sizes of a few microns, i.e., they are *macroscopic* formations. Just the macroscopic scale of the droplets allows to identify them as new concentrated phase originating in the system. So, no wonder that the elongation of a droplet in external field is well described by the competition of the surface energy at the gas-liquid boundary and the magnetic energy of the droplet [7].

To describe the observable condensation of MF under influence of *external magnetic field* the statistical mechanics models [15–19] were proposed. Initially the dipolar interactions were taken into account within the framework of the classical Weiss model of ferromagnets [15,16] and antiferromagnetic model [17]. The both models quantitatively correctly describe the phase separation of MF however they possess the defect common for all the mean field theories. It is a prediction of spontaneous magnetized state of concentrated liquid phase in the absence of external field. This property was never observed in experiments with real MF. In contrast with the mean field theories, the more late models [18,19] took into account the orientational and spatial correlations of MF grains. In [18] the generalization of the MSM of Wertheim [3] onto the case of arbitrary external field was done. The mean spherical model has no the problem with spontaneous magnetized state of liquid phase [4]. The external magnetic field lowers the threshold of condensation from zero field value $\lambda_c = 4.445$ to infinite field value $\lambda_c(\infty) = 3.055$ [18] what is in agreement with mentioned estimation [14]. The analogous results were obtained within the framework of perturbation theory [19] which is more simple than MSM. The both theories describe perfectly the many properties of real MF such as the magnetization curve, temperature and concentration dependence of initial susceptibility [20–22].

Here it is necessary to note that the *usual* magnetic fluids used in practice are required to be stable against temperature and magnetic field action. In other words, in many cases the condensation of MF proves to be the undesirable phenomenon. Thus, the ordinary magnetic fluids and commercial MF especially are characterized by very low values of dipolar interactions, $\lambda \leq 1$. Often due to the special techniques only, the authors could increase value of λ up to $3 \div 4$ and observe the phase transition in MF [7–9,13,14]. As we will see, under these small values of λ the *association phenomenon* – the second consequence of dipolar interactions according to de Gennes and Pincus – is negligible. Another aspect of dealing for a long time with mentioned MF is the problem of synthesizing of MF with strong magnetic interactions. Only recently such ferrocolloids were successfully synthesized [23]. Now Nakatani and co-authors achieved the values of the coupling parameter $\lambda = 3 \div 10$ [24]. Just in such MF the association of grains should be essential. In contrast to condensation, to detect the particle association taking place on microscopic level is a serious experimental problem. We can judge about MF microstructure using only indirect data on *negative viscosity effect* and *magneto-vortical resonance* in MF under an oscillating magnetic field [25]. The data gave sure signals of the presence of a *wide spectrum* of magnetization relaxation times τ instead of only one Brownian time for single magnetic grains. Assuming that this spectrum originates from chains formed out of the grains, authors of [25] have reached a perfect agreement between the theoretical and experimental frequency dependencies of MF viscosity and dynamic susceptibility.

The intense interest to the systems with strong dipolar interactions arose however a bit earlier, at the beginning of 90-s, as a corollary of new unexpected data of numerical experiments. First, a surprising result of Monte Carlo simulations was the *absence* of vapor-liquid transition in the system of interacting dipoles. Instead, simulations revealed a phenomenon of formation of *chains* of head-to-tail aligned dipoles in the dipolar hard sphere fluid [26,27] as well as in the Stockmayer [28] and dipolar soft sphere fluid [29,30]. The evidence of the above transition did was given after all [31]. This transition proved to be rather unusual, namely, it occurs between highly associated vapor phase and a more normal dense liquid phase and accompanies by extremely small entropy and enthalpy changes.

Another exciting property of dipolar fluid revealed in simulations is that dipolar forces *alone* can create an orientationally ordered liquid state. Starting from the first demonstration of the phenomenon in molecular-dynamics simulations [32,33], this conclusion was confirmed later by Monte Carlo calculations [26,34–36]. In fact, there is no simple connection of spontaneous order of dipolar fluid and chain association phenomenon of aligned dipoles [35,37]. However the contraposition of liquid condensation and association feature seemed to be well established in the numerical experiments [26–31]. At the same time, the data of Monte Carlo simulations made in [38–40] with different conditions at the boundary of simulation cell similarly to old study [5] fully contradict to results [26–37] and predict at $\lambda > 3$ the *traditional* gas-liquid phase transition in the course of which *non-chain-shaped* clusters of magnetic grains are formed.

It is clear that these breathtaking simulation data initiated the new theoretical investigations of association phenomenon in dipolar fluids. We recall for readers that there are two basic statistical mechanics approaches to the problem of interparticle interactions. They are the technique of correlation functions and the association Frenkel-Band theory. The former is based upon the solution of the Ornstein-Zernike equation under some reasonable closure relation for the direct correlation function [2]. Such an approach was realized for the case of MF in [1]. Due to necessity to solve the integral equation, it is very complicated for analysis. Moreover, it contains only implicit information about structure of MF. What is why this approach is rarely used in theoretical works. The Frenkel-Band association theory treats the statistical system of interacted particles as chemically equilibrium mixture of non-interacting clusters of different size and the interparticle interactions are taken into account only inside clusters [41]. For the case of MF in strong magnetic field, this approach was realized in the pioneering works of Jordan [42,43]. In contrast to the correlation function approach, the microstructure of MF is a straightforward result of the association theory. Of course, the both theories do not give the identical results. However, they should have the same limit of low-concentrated MF what follows immediately from nature of the approximations [41]. Nevertheless, Jordan confirmed the main prediction of de Gennes and Pincus theory [1] about formation of magnetic grain chains parallel to external field. [It is interesting to note that the impulse for Jordan to develop his previous study of microstructure in MF [43] was the experimental data [6,10] where the mentioned condensation (not association of grains!) of MF into liquid droplets took place. As we have seen, the condensation and particle association are different phenomena connected with interparticle interactions. But even now both phenomena are confused with each other [44]].

In the papers Jordan took into account all the interparticle interactions in a chain of grains. Then, calculating the partition function of a chain, he had to make some additional assumptions about spatial and orientational configuration of dipoles. The final results are very cumbersome even in limiting cases of infinite external field and strong dipolar interactions, $\lambda \gg 1$. The most important on our mind conclusion of the Jordan's calculations is the account of dipolar interactions of *non-nearest* neighbors along the chain is reduced in fact to *re-normalization* of the *nearest-neighbors* interactions: $\lambda \to \lambda \zeta(3)$, where $\zeta(3) = 1.202$ is the Riemann ζ-function. Thus, it is possible to exclude the interactions of the non-nearest neighboring particles without any significant loss of information about the system. Just this reasonable approximation within the framework of association theory became the basis for latest theoretical works [45–53]. The critical review of number of these papers is given in the next Section where the association theory is considered in detail. Here we will mention only the main general results of cited papers.

The phenomenological theory [47], combined the van der Waals theory for liquid condensation with the association Jordan's theory, revealed a competition between liquid condensation and chain formation driven by anisotropic dipolar interactions. The analogous result was obtained in [46,48–50] where some phenomenological parameters of work [47] were estimated. An interesting conclusion of theoretical studies [48] is that a validity of association theory of noninteracting chains is possibly broader than it was assumed before [42,43] and is not limited only to the low-density phase of fluid. This result is indicated by the appropriate evaluation of the steric and dipolar interchain interactions [48]. Finally, authors [45,53] studied the magnetic and rheological properties of MF assuming the rod-like form of aggregates of magnetic grains.

Summarizing the review of recent theoretical works we see that all of them deal with only one characteristic of the particle chains - their length. Meanwhile a genuine understanding of the phenomenon of chain formation remains incomplete without studying the *spatial* and *orientational correlations* inside dipolar chain. As shown below, the association theory allows to investigate this aspect of the problem in detail and often analytically. Developing our approach, we establish an important role of chain *flexibility*; this attribute of dipolar chains has been remained without proper attention in previous works [45,46,48,49,51–53]. De Gennes and Pincus [1] were the first who noted an analogy between dipolar chains and polymer molecules. Actually, our investigation of the pair correlations and the chain flexibility leads us to natural extension of the basic concepts of polymer physics like the persistent length and Kuhn segment, coil, globule [54] etc. to the case of dipolar chains. The aim of this paper is study of the spatial and orientational *correlations* inside dipolar chains, determination of possible *forms* of the chains (statistical coil, globule), and investigation of the coil-globule *phase transition*. The paper is organized as follows. In Sect. 2 we consider the basic statements of an association theory of MF, define the concept of chain of magnetic grains, establish the main property of the proposed model of neighboring grain interactions and determine the averaged length of dipolar chains. The statistical properties of a single dipolar chain are studied in Sect. 3 where the important role of chain flexibility is established and main flexibility characteristics are determined. The influence of chain flexibility on structure of a chain and coil-globule transition are considered in Sect. 4. We make some concluding remarks in Sect. 5.

2 Association Theory of Magnetic Fluid

2.1 Definition of a Chain of Grains

Let us consider MF as an assembly of chains of different length formed out of magnetic grains in the absence of external field or in infinitely strong magnetic field. We assume that all the particles are identical spheres of diameter d, only neighboring grains in the chain interact with each other, and there is no interaction between grains of different chains. The interaction neighboring particles with numbers i and $i+1$ is the sum of hard-sphere $U_{HS}(r_{i,i+1})$ and dipole-dipole U_{dd} interactions:

$$U(i,i+1) = U_{HS} - \frac{m^2}{r_{i,i+1}^3}\left[3(\boldsymbol{e}_i \boldsymbol{n}_{i,i+1})(\boldsymbol{e}_{i+1}\boldsymbol{n}_{i,i+1}) - \boldsymbol{e}_i \boldsymbol{e}_{i+1}\right], \qquad (1)$$

where m is the magnetic moment, $\boldsymbol{e}_i$ and $\boldsymbol{n}_{i,i+1}$ are unit vectors along the dipole of particle i and the interparticle vector $\boldsymbol{r}_{i,i+1}$, respectively.

Let g_N be the number of N-particle chains per unit volume. The volume density F/V of the free energy of system includes ideal gas term and the intrachain contribution [45,46,48]

$$\frac{F}{Vk_BT} = \sum_{N=1}^{\infty} g_N \ln\frac{g_N v}{e} + g_N f_N, \qquad (2)$$

where $v = \pi d^3/6$ is the particle volume, $f_N = -\ln Z_N$ is the dimensionless "internal" free energy of N-particle chain, Z_N is its partition function.

Now it is necessary to give the definition of a chain. First, pair of grains are defined as bonded if their dipole potential for the head-to-tail configuration of dipoles exceeds thermal energy, i.e., the condition $2m^2/r^3 k_B T \geq 1$ is fulfilled. Second, calculating the partition function Z_N of N-particle chain, the integration over displacement vector $\boldsymbol{r}_{i,i+1}$ of two neighboring particles should be done over *half* of spherical volume admitted by mentioned above condition.

Let us comment both statements. Our choice of the threshold energy for cluster formation is the same as in [29,45]. Of course it is not unique and another definitions are also possible [26,30,34,51]. We note that this ambiguity takes place for finite values of $\lambda \geq 1$ and disappears when $\lambda \gg 1$. The latter case of large λ was the main approximation in [1,42,43,48,49]. In our consideration it is supposed that λ can vary from 0 to 10. What is why we had to make more precise the definition of a chain of grains and used for it mentioned physical on our mind limitation. As for the procedure of integration, this method proposed in [45] allows effectively to take into account the steric interactions with another grains in a chain. Of course, the given integration narrows down the real region of space admitted for the particle in a chain. However, it contains the main region around head-to-tail configuration of dipoles. The results of exact integration and integration over half of spherical volume prove to be practically coinciding starting from $\lambda \geq 3$, that is from the same beginning of the grain association as we will see later. In order to show the chosen method of integration we use below the prime near the symbol of integral.

Let us write the expression for partition function of N-particle chain in the absence of external field

$$Z_N = \int' \exp\left\{-\sum_{i=1}^{N-1} \frac{U_{dd}(i,i+1)}{k_B T}\right\} \prod_{i=1}^{N} \frac{d\boldsymbol{e}_i}{4\pi} \prod_{i=1}^{N-1} \frac{d\boldsymbol{r}_{i,i+1}}{v}, \qquad (3)$$

In spite of the clearness of this relation it is necessary to discuss one question of principle connected with (3). The point is that authors [51,52] unexpectedly introduced in the denominator of (3) the symmetry factor $N!$ "missed" in their previous considerations [48,49] as well as in [42,43,45]. They believe that such factor takes into account the indistinguishability of grains due to fact that magnetic particles diffuse through all the chains in the course of Monte Carlo calculations [35]. This mistaken statement has very long history. The exhaustive consideration of the question was given by Ehrenfest and Trkal in 1920 [55]. The detail information on this subject can be found in [56]. Here we only note that the presence of symmetry factor is not connected absolutely with simulation data because it is determined by writing of the Hamiltonian used in the theory. The authors [48,49,51,52] used the same Hamiltonian as we did (see Eq. (5) in [48] and Eq. (19) in [51]), i.e., as a sum along chain of dipolar interactions (1) between consecutive particles. It means that all the particles in the chain with $N \geq 3$ are numbered and particle with number i always follows beyond grain with number $i-1$. By the way, this grain numbering was taken into account by authors [48] explicitly (see Sect. II in [48]). Due to this numbering of all grains in the chain with $N \geq 3$ there is *no need* in additional factor $N!$ [55,56]. Thus the revising in [51,52] the previous approach [48,49] is incorrect. Nevertheless, there is the special case when the symmetry factor is necessary. It is the case of dimers, the clusters of two grains only. The phase volume of dimer is two times higher than it should be. Thus, due to mentioned indistinguishability of the grains the partition function Z_2 indeed should be divided by factor 2. However this factor is taken into account in our consideration *automatically* by the mentioned method of integration over half of spherical volume. Therefore, equation (3) do not have any additional factor at all.

Let us return to analysis of (3). The main advantage of proposed model is the *factorization* of the chain partition function – Z_N. Such a factorization takes place in the considered cases of zero and infinite external magnetic field but not for arbitrary field. This property of dipolar chain with interactions of neighboring grains is quite analogous to that for the chain of spins with the Heisenberg interactions [57]. In the case of zero field, it is immediately followed from the absence of a preferable direction in the system. Indeed, calculating Z_N one should integrate over variables e_i, $r_{i-1,i}$ starting from the last particle with number N ("tail" of the chain) to the first one ("head" of the chain) and taking at each step the direction e_{i-1} of previous grain as polar axis. The other essential fact is independence of the integral

$$I = \int' \exp\left\{\frac{m^2}{r^3 k_B T}(3(e_1 n)(e_2 n) - e_1 e_2)\right\} \frac{de_2}{4\pi} \frac{dr}{v}, \qquad (4)$$

on the orientation e_1 of previous particle. It is clear that partition function Z_2 of a dimer coincides with I. We also note that at large values of λ the integral I is also the absolute value of the *second virial coefficient*

$$b_2 = -\frac{1}{2}\int \{\exp[-U(12)/k_B T] - 1\} \frac{de_1}{4\pi} \frac{de_2}{4\pi} \frac{dr}{v}. \qquad (5)$$

Here is the dimensionless value of b_2 in units of particle volume. The integral is without of prime, i.e., the integration is fulfilled over all space. Finally we

achieve
$$Z_N = Z_2^{N-1}. \tag{6}$$

The factorization of the problem in the case of infinitely strong magnetic field is obvious.

The mentioned above setting of the problem is considered in number of works [46,48,49,51]. As we know, the factorization was noted only in [46]. Unfortunately, the results of this work are most likely mistaken (see below). Now we can calculate the distribution g_N of chains upon their length. Substituting (6) into (2) and using the condition of chemical equilibrium between chains of different sizes [42,43] or equivalently the minimization of the free energy of the system [45,48] we find
$$g_N = x^N/vZ_2, \tag{7}$$
where parameter x is expressed via normalization condition $\sum_{N=1}^{\infty} N g_N = n$ (n is the concentration of magnetic grains in MF) as
$$\frac{x}{(1-x)^2} = \Phi Z_2. \tag{8}$$

Here $\Phi = nv$ is the volume fraction of magnetic particles. From (7) and (8) we finally determine the average number of particles in the chain
$$\langle N \rangle = \frac{1}{2} + \sqrt{\frac{1}{4} + \Phi Z_2}. \tag{9}$$

We see that owing to factorization the problem of determination of the average length of the chains is reduced to calculation of the partition function of dimer. The asymptotic representation at $\lambda \gg 1$ of the dimer partition function in zero and infinite external field is
$$Z_2(0) = \frac{e^{2\lambda}}{3\lambda^3}\left(1 + \frac{8}{3\lambda} + \frac{23}{3\lambda^2} + \frac{229}{9\lambda^3} + \frac{5263}{54\lambda^4} + \frac{11536}{27\lambda^5} + \frac{57427}{27\lambda^6}\right), \tag{10}$$

$$Z_2(\infty) = \frac{e^{2\lambda}}{3\lambda^2}\left(1 + \frac{5}{3\lambda} + \frac{41}{12\lambda^2} + \frac{155}{18\lambda^3} + \frac{11195}{432\lambda^4} + \frac{39235}{432\lambda^5} + \frac{628145}{1728\lambda^6}\right). \tag{11}$$

These expansions are given up to terms of order of $O(\lambda^{-6})$ because of the slow convergence of the series. The main (first) term in the right side of (10) and (11) is a result of de Gennes and Pincus [1]. The other terms are new. We note that in [48] it was informed about mistaken value of the second virial coefficient b_2 (or Z_2) in [1]. This is however misunderstanding: the authors [48] used for b_2 the different notation than de Gennes and Pincus did. The calculated dependencies of both partition functions are shown in Fig. 1; note that in the figure and below the minimal value of λ is 0.5 according to our definition of a chain.

It is interesting that maximal deviations of calculated values from unity (i.e., from de Gennes and Pincus asymptotic values) takes place at small $\lambda \sim 2 \div 3$

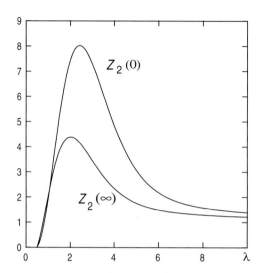

Fig. 1. Partition functions of dimer in zero and infinite external field in units of de Gennes and Pincus asymptotic values $e^{2\lambda}/3\lambda^3$ and $e^{2\lambda}/3\lambda^2$, respectively

however even at $\lambda = 10$ it achieves 50%. The expansions (10) and (11) describe the calculated values within the accuracy of 10% starting from $\lambda \sim 5 \div 6$.

Let us analyze relation (9) in some limiting cases. When $\lambda \gg 1$ we recover the result for average chain length in zero field [48,49]. Next we consider the case of low-concentrated magnetic fluid with strong dipolar interaction so as $\Phi b_2 \ll 1$. As it was noted above, both main theories should have the same limit of low concentration. From (9) and (5) it follows $\langle N \rangle = 1 - \Phi b_2$ (b_2 is negative). At the same time the theory of de Gennes and Pincus [1] predicts $\langle N \rangle_{GP} = 1 - 2\Phi b_2$. The origin of double difference of the coefficient is clear. In the approach [1] the steric interactions inside the chain were ignored. It results in double exceeding of contribution of neighboring particles interaction in partition function of a chain. As we mentioned above the factorization of the considered problem was noted in [46] where the grand partition function of the system was calculated. The author obtained for variable equivalent to our x the cubic expression (see equation (6) in [46]) instead of quadratic one (8). The character of Letter does not allow to check all the calculations made in [46]. Nevertheless, we argue that there is a mistake in his equation (5), containing the redundant coefficient 2. Without this factor the equation (6) in [46] reduces to the correct relation (8). We can support our point of view on [46] also *a posteriori*. Indeed, the relation (8) is a straightforward consequence of our formalism based on calculation of the partition function of a chain. It is clear that the final result should not depend on the choice of thermodynamic potential considered.

In this Section we developed the association theory of ideal chains of magnetic grains. The size distribution of dipolar chains obeys to equations (7) and (8). The average length of chains (9) increases $\sim \Phi \exp(2\lambda)$ under small values of this parameter and $\sim \sqrt{\Phi}\exp(\lambda)$ at high λ. The average value $\langle N \rangle$ in zero and infinite external field can easily calculated according to (9)-(11) and data shown in Fig. 1. Being the very important integral characteristics of the phenomenon of chain formation, the average length itself however does not contain an exhaustive information about system. Our understanding of the phenomenon would be incomplete without study of *statistical properties* of the chains, without their spatial and orientational correlations. Just these questions we start to investigate.

3 Statistical Properties of Ideal Chains

3.1 Correlation Functions of Dimers

Here we will see that the factorization is a general property of the considered model what allows to express any interparticle correlations in terms of spatial and orientational correlations inside dimers. First, let us define four orientational correlation functions

$$A = \langle e_1 e_2 \rangle, \quad B = \frac{\langle e_1 r_{12} \rangle}{d}, \quad C = \langle (e_1 e_2)^2 \rangle, \quad D = \frac{\langle (e_1 r_{12})^2 \rangle}{d^2}, \qquad (12)$$

and two spatial characteristics for a dimer:

$$E = \frac{\langle r_{12} \rangle}{d}, \qquad F = \frac{\langle r_{12}^2 \rangle}{d^2}. \qquad (13)$$

Averaging in these expressions is fulfilled over variables r_{12} and e_2 of second particle whereas the orientation e_1 of first grain is taken as polar axis. The polar angle between e_1 and the displacement vector r_{12} takes the values from zero to $\pi/2$ according to method of integration mentioned at the beginning of Sect. 2. Due to such integration over spatial variable the correlation function B is not zero. It is clear that in considered cases of zero and infinite external field all the correlations (12) do not depend on the orientation e_1. The correlations (12) together with (13) are dimensionless functions upon λ and easily calculated. For the case of zero external field they are depicted in Fig. 2. We note that the curve B lies higher the curve D. The maximal size of dimer (see curves E and F) suits to intermediate values of λ. Indeed, when $\lambda \leq 1$ the cluster of two grains falls into separate particles even for very small distance between them, $r \sim d$. At high values of λ, the mean interparticle distance $\langle r \rangle$ is of the order of d as well due to strong dipolar interactions. We note also that $\langle r \rangle$ never exceeds $2d$ (in fact, it is always less than $1.4d$) what means the numbering of the grains in a chain and verifies the chosen in Sect. 2 method of integration over space coordinate.

All the introduced dimer's variables tend to unity with increasing of coupling parameter. Their asymptotic representation at $\lambda \gg 1$ is

$$A = 1 - \frac{2}{\lambda} - \frac{2}{\lambda^2} - \frac{20}{3\lambda^3} - \frac{266}{9\lambda^4} - \frac{4241}{27\lambda^5} - \frac{77669}{81\lambda^6}, \qquad (14)$$

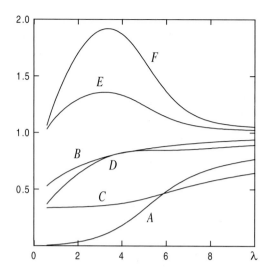

Fig. 2. Correlation functions (12) and spatial mean values (13) of dimers in zero field versus the coupling parameter λ

$$B = 1 - \frac{1}{2\lambda} - \frac{5}{9\lambda^2} - \frac{11}{6\lambda^3} - \frac{689}{81\lambda^4} - \frac{15593}{324\lambda^5} - \frac{229265}{729\lambda^6}, \qquad (15)$$

$$C = 1 - \frac{4}{\lambda} + \frac{4}{\lambda^2} + \frac{8}{3\lambda^3} + \frac{74}{9\lambda^4} + \frac{998}{27\lambda^5} + \frac{16880}{81\lambda^6}, \qquad (16)$$

$$D = 1 - \frac{1}{\lambda} - \frac{7}{18\lambda^2} - \frac{43}{27\lambda^3} - \frac{397}{54\lambda^4} - \frac{20185}{486\lambda^5} - \frac{398579}{1458\lambda^6}, \qquad (17)$$

$$E = 1 + \frac{1}{6\lambda} + \frac{4}{9\lambda^2} + \frac{44}{27\lambda^3} + \frac{1175}{162\lambda^4} + \frac{18007}{486\lambda^5} + \frac{153715}{729\lambda^6}, \qquad (18)$$

$$F = 1 + \frac{1}{3\lambda} + \frac{17}{18\lambda^2} + \frac{97}{27\lambda^3} + \frac{2657}{162\lambda^4} + \frac{41507}{486\lambda^5} + \frac{719573}{1458\lambda^6}. \qquad (19)$$

Within accuracy of a few percents these formulas describe the calculated values A, B, C when $\lambda \geq 5$ and D, E, F for $\lambda \geq 7$. We note that the terms $\sim \lambda^{-1}$ for A, B and E were found in fact in [42]. It is interesting, that even for the strong dipole interactions the correlations are still far from unity. For example, for $\lambda = 10$ the mean values of cosine and its square of dipole orientations are $A = 0.77$ and $C = 0.64$, respectively. It means that orientations of neighboring dipoles preserve a fairly high rotational mobility inside dimers.

Up to now we assumed that the external field is absent. Now, let us consider the opposite case of infinitely strong external field. The dimer correlations A_∞, B_∞ etc. in infinite field differ from their zero-field values. Evidently, $A_\infty = C_\infty = 1$. The calculated values of four remaining variables are depicted in Fig. 3. We indicate also their asymptotic expansions. They are

$$B_\infty = 1 + \frac{7}{36\lambda^2} + \frac{25}{36\lambda^3} + \frac{1787}{648\lambda^4} + \frac{1000}{81\lambda^5} + \frac{715597}{11664\lambda^6}, \qquad (20)$$

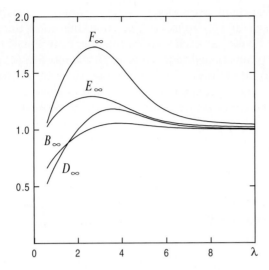

Fig. 3. Correlation functions B_∞, D_∞ and mean values E_∞, F_∞ of dimers in infinite field versus the coupling parameter λ

$$D_\infty = 1 + \frac{4}{9\lambda^2} + \frac{85}{54\lambda^3} + \frac{343}{54\lambda^4} + \frac{56255}{1944\lambda^5} + \frac{106571}{729\lambda^6}, \quad (21)$$

$$E_\infty = 1 + \frac{1}{6\lambda} + \frac{13}{36\lambda^2} + \frac{119}{108\lambda^3} + \frac{2675}{648\lambda^4} + \frac{34801}{1944\lambda^5} + \frac{508025}{5832\lambda^6}, \quad (22)$$

$$F_\infty = 1 + \frac{1}{3\lambda} + \frac{7}{9\lambda^2} + \frac{67}{27\lambda^3} + \frac{3109}{324\lambda^4} + \frac{10366}{243\lambda^5} + \frac{2472103}{11664\lambda^6}. \quad (23)$$

The results of calculation on formulas (20)–(22) with accuracy of $3 \div 5\,\%$ coincide with exact values of B_∞, D_∞, E_∞ at $\lambda \geq 5$. The accuracy of (23) is worse: the mentioned difference is achieved at higher $\lambda \geq 7$. As before, we note that terms $\sim \lambda^{-1}$ in (20) and (22) were determined in [42].

Six dimer variables (12) and (13) contain *all* physically significant statistical information about the system. In order to verify this property it is necessary to consider the characteristic intrachain correlations.

3.2 Calculation of Intrachain Correlations

For now we assume MF to be in zero external field till the opposite case of infinite field will be considered. First, we calculate two correlations typical of the system: $\langle e_{i;\alpha} e_{j;\beta} \rangle$ and $\langle r_{i,i+1;\alpha} r_{j,j+1;\beta} \rangle$. Here and below we denote the numbers of grain in a chain by Latin letters whereas the Greek ones indicate the components of orientation e_i and displacement $r_{i,i+1}$ vectors. The symbol $\langle ... \rangle$ means the averaging over spatial and orientational coordinates of all particles in a chain excepting the first particle – the "head" of a chain. The orientation e_1 of first grain is assumed to be fixed. In the case of zero external field, it determines a single preferable direction in the system and plays a role of external field for other particles in the chain. We average over the variables of particles starting from the "tail" of a chain, i.e., from Nth grain and then going to the grains with number $N-1$, $N-2$ etc. At each step we take the direction of previous grain as polar axis. Due to mentioned independence of the integral (4) as well as the dimer variables (12) and (13) on orientation of previous (first) particle, at such step of averaging we obtain one of the dimer functions Z_2, A, B, C, D or F. Just this signifies the factorization of any two-particle correlation.

a. Calculation of $\langle e_{i;\alpha} e_{j;\beta} \rangle$

Let us assume for definiteness that $1 \leq i < j \leq N$. The integration over variables of grains N, ..., $j+1$ is obvious: it reduces to multiplying by unity, $\langle ... \rangle_{N,...,j+1} = 1$. Here and further we indicate explicitly the numbers and the order of particles been integrated. Integration over particle j in the coordinate system connecting with particle $j-1$ gives

$$\langle e_{i;\alpha} e_{j;\beta} \rangle_j = A e_{i;\alpha} e_{j-1;\beta} . \tag{24}$$

Repeating this procedure $j-i$ times we find

$$\langle e_{i;\alpha} e_{j;\beta} \rangle_{j,j-1,...,i+1} = A^{j-i} e_{i;\alpha} e_{i;\beta} . \tag{25}$$

Now we consider mean value $\langle e_{i;\alpha} e_{i;\beta} \rangle$ which looks like initial quantity, but with $j=i$. Again we connect the coordinate system with particle $i-1$ and integrate over variables of grain i. The result is

$$\langle e_{i;\alpha} e_{i;\beta} \rangle_i = \frac{1-C}{2} \delta_{\alpha,\beta} + \frac{3C-1}{2} e_{i-1;\alpha} e_{i-1;\beta} , \tag{26}$$

where $\delta_{\alpha,\beta}$ is a unit tensor. The last expression is a recurrent one. After averaging over grains with numbers $i-1$, $i-2$, ..., 2 we have

$$\langle e_{i;\alpha} e_{i;\beta} \rangle_{i,i-1,...,2} = \frac{1}{3} \delta_{\alpha,\beta} + T^{i-1} \Delta_{\alpha,\beta} , \tag{27}$$

where the notation $T = (3C-1)/2$ is used and the irreducible second rank tensor $\Delta_{\alpha,\beta}$ is introduced according to relation

$$\Delta_{\alpha,\beta} = e_{1;\alpha} e_{1;\beta} - \frac{1}{3} \delta_{\alpha,\beta} . \tag{28}$$

Finally, combining expressions (25) and (28) the considered mean value is found in the form

$$\langle e_{i;\alpha} e_{j;\beta} \rangle = A^{j-i} \left(\frac{1}{3} \delta_{\alpha,\beta} + T^{i-1} \Delta_{\alpha,\beta} \right) . \tag{29}$$

b. Calculation of $\langle r_{i,i+1;\alpha} r_{j,j+1;\beta}\rangle$

There are two different cases, $j = i$ and $j > i$, which should be considered separately. First let us put $j = i$. We take $\boldsymbol{e}_i$ as polar axis and integrate over coordinates of grain $i+1$ (cf. with (26))

$$\langle r_{i,i+1;\alpha} r_{i,i+1;\beta}\rangle_{i+1} = \left(\frac{F-D}{2}\delta_{\alpha,\beta} + \frac{3D-F}{2}e_{i;\alpha}e_{i;\beta}\right)d^2 . \tag{30}$$

Using here equation (27) we deduce

$$\langle r_{i,i+1;\alpha} r_{i,i+1;\beta}\rangle = \left(\frac{F}{3}\delta_{\alpha,\beta} + \frac{3D-F}{2}T^{i-1}\Delta_{\alpha,\beta}\right)d^2 . \tag{31}$$

Second, let be $j > i$. In this case the integration over particle $j+1$ gives

$$\langle r_{i,i+1;\alpha} r_{j,j+1;\beta}\rangle_{j+1} = B d r_{i,i+1;\alpha} e_{j;\beta} . \tag{32}$$

Averaging the right side consecutively over particle with numbers $j, j-1,...,i+2$ we can write

$$\langle r_{i,i+1;\alpha} r_{j,j+1;\beta}\rangle_{j+1,j,...,i+2} = A^{j-i-1} B d r_{i,i+1;\alpha} e_{i+1;\beta} . \tag{33}$$

Integrating now over variables of particles $i+1,...,2$ we finally obtain

$$\langle r_{i,i+1;\alpha} r_{j,j+1;\beta}\rangle = A^{j-i} B^2 d^2 \left(\frac{1}{3}\delta_{\alpha,\beta} + T^{i-1}\Delta_{\alpha,\beta}\right), \ j > i . \tag{34}$$

The expression (29), (31), (34) are the basic mean values needed for calculation of any two-particle intrachain correlation.

3.3 Persistent Length of Dipolar Chains

We use the developed formalism for calculation of some characteristic correlations typical of dipolar chain. First, let us consider the correlation $\langle \boldsymbol{e}_i \boldsymbol{e}_{i+k}\rangle$ of orientations of dipoles i and $i+k$ in the chain. Due to relation (29), this value is multiplicative, i.e.,

$$\langle \boldsymbol{e}_i \boldsymbol{e}_{i+k}\rangle = A^k , \tag{35}$$

what means its exponential decay. From here, using an analogy with polymer theory [54], we determine the *persistent length* (in units of particle diameter d) as $L_p = -E/\ln A$. This quantity characterizes the flexibility of the chain. It is shown below in Fig. 4. The formulas (14) and (18) allow to determine the asymptotic behavior of the persistent length up to terms of order λ^{-4}

$$L_p = -\frac{E}{\ln A} = \frac{\lambda}{2} - \frac{11}{12} - \frac{23}{18\lambda} - \frac{128}{27\lambda^2} - \frac{9698}{405\lambda^3} - \frac{1423259}{9720\lambda^4} . \tag{36}$$

The asymptotic representation well describes the calculated values of L_p starting from $\lambda \geq 6$. The evaluation of persistent length for $\lambda = 12.25$ gives $L_p = 5.1$ what is in reasonable agreement with the simulation data $L_p \sim 7$ mentioned in [35]. As seen, the persistent length is rather short. Thus, dipolar chains may be considered as rigid ones only on small scales not exceeding $\lambda/2$.

3.4 The Kuhn Segment of Dipolar Chains

The most informative quantity of chain configuration is "end-to-end" vector $\boldsymbol{R} = \sum_{i=1}^{N-1} \boldsymbol{r}_{i,i+1}$ connecting centers of first and last particles in a chain of N grains. We calculate the mean values $\langle R_\parallel^2 \rangle$ and $\langle R_\perp^2 \rangle$ along the orientation $\boldsymbol{e}_1$ of one end of chain and in perpendicular direction. Using (29), (31) and (34) after cumbersome but simple calculations we obtain analytically

$$\frac{\langle R^2 \rangle}{d^2} = F(N-1) + \frac{2AB^2}{1-A}\left[N - 2 - AG_{N-2}(A)\right], \tag{37}$$

$$\frac{\langle R_\parallel^2 \rangle - \langle R_\perp^2 \rangle}{d^2} = \frac{3D-F}{2} G_{N-1}(T) + \frac{2AB^2}{A-T}\left[AG_{N-2}(A) - TG_{N-2}(T)\right], \tag{38}$$

where $R^2 = R_\parallel^2 + 2R_\perp^2$ and the function $G_N(x)$ is determined as $G_N(x) = (1 - x^N)/(1 - x)$.

Let us analyze the result (37), (38). In the case of dimers, $N = 2$, the obvious result $\langle R_\parallel^2 \rangle/d^2 = D$, $\langle R_\perp^2 \rangle/d^2 = (F-D)/2$ is recovered. When the chains are very short, $N \ll \lambda$, we have $\langle R_\parallel^2 \rangle = N^2 d^2$, $\langle R_\perp^2 \rangle = 0$. This is just the case of rod-like aggregates considered in [45,53]. This case, however, is *never put into effect* in reality. The fact is that both the energy gain accompanying every pairing of magnetic grains ($\sim \lambda k_B T$) and the chain correlation length ($\sim \lambda d$) are determined by the same coupling constant λ. Therefore, the mean number of grains in a chain $\langle N \rangle$ as well as its correlation length L_p increases with enlarging of λ, so that the condition $\langle N \rangle \ll \lambda$ is *never satisfied*. [From the formal point of view, $\langle N \rangle$ increases with growth of λ exponentially (see (9)) whereas $L_p \sim \lambda/2$ according to (36)]. Statistically, some number of very short chains ($N \ll \lambda$) are certainly existed, but their fraction is negligible. We note that recently authors [45,53] changed their own previous representations and admitted the deformation of dipolar chain [58].

For a long chain equations (37) and (38) give $(\langle R_\parallel^2 \rangle - \langle R_\perp^2 \rangle)/\langle R^2 \rangle = 0$ and

$$\langle R^2 \rangle = Nd^2 \left(F + \frac{2AB^2}{1-A}\right), \quad N \to \infty. \tag{39}$$

Such a N-dependence of the mean-squared end-to-end distance of a long chain is typical of ideal polymer chain [54]. Its local stiffness is often characterized – along with the persistent length L_p – by the so-called *Kuhn segment* L_K determined by the relation $\langle R^2 \rangle = N \langle r \rangle L_K d$, where $\langle r \rangle$ is the mean distance between neighboring grains in the chain. Both the quantities, L_p and L_K, are depicted in Fig. 4 as functions of the dipole parameter λ. When it is sufficiently large, we have

$$L_K = \left(F + \frac{2AB^2}{1-A}\right) E^{-1} = \lambda - \frac{19}{6} + \frac{2}{9\lambda} - \frac{715}{108\lambda^2} - \frac{8377}{216\lambda^3} - \frac{332995}{1296\lambda^4}. \tag{40}$$

As before, this expansion perfectly describes the calculated values of L_K at $\lambda \geq 6$. It is interesting that starting from $\lambda \simeq 5$ (i.e., just when the aggregation

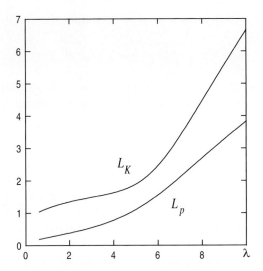

Fig. 4. The persistent length L_p and the Kuhn length L_K as a function of coupling parameter λ

phenomenon becomes appreciable) the ratio L_K/L_p is close to 2 as it takes place for many models of polymer molecules [54].

So, the linear along chain memory disappears at the distances of a few particles and the long dipolar chain with number of grains $N \gg L_K$ have practically the same characteristic sizes $\sqrt{\langle R_\parallel^2 \rangle}$ and $\sqrt{\langle R_\perp^2 \rangle}$. The long chain resembles *cloud* or *coil* of connected monomers and has quasi-spherical form. In fact however this conclusion proves to be valid even for sufficiently short chains with $N \geq L_K$ (see Sect. 3.5).

3.5 Case of Infinite External Field

Starting from Sect. 3.2 we assumed that the external field is absent. Now, let us consider the opposite case of infinitely strong external field. As noted above, the factorization of the problem in infinite field takes place as well as in zero external field. Thus, the general results (38) and (39) remain valid owing to the factorization only. Setting $A_\infty \to 1$ and $C_\infty \to 1$ we find the characteristic sizes of chain along the field and in perpendicular direction

$$\langle R_\parallel^2 \rangle = (N-1)\left[B_\infty^2(N-2) + D_\infty\right] d^2 , \qquad (41)$$

$$\langle R_\perp^2 \rangle = \frac{1}{2}(N-1)(F_\infty - D_\infty)d^2 . \qquad (42)$$

The size of a chain along field increases as $\sim N$ whereas the transverse size remains of the order $\sim \sqrt{N}$. The chain of magnetic grains has indeed *chain-like* form: it is elongated along external field.

3.6 Effective Form of Dipolar Chains

For chains of finite length it is convenient to introduce the variables

$$a = d + \sqrt{\langle R_\parallel^2 \rangle}, \quad b = d + \sqrt{\langle R_\perp^2 \rangle} \qquad (43)$$

that characterizes the chain scales in two directions. The effective form of the chain can be judged calculating the parameter of nonsphericity S defined as

$$S = (a^2 - b^2)/(a^2 + b^2). \qquad (44)$$

The parameter as a function of chain length is shown in Fig. 5 for MF with $\lambda = 10$ for cases of zero and infinity external field. As a chain length we use discrete variable N taking values from 2 to 29. Each diamond in Fig. 5 corresponds to specific value of N.

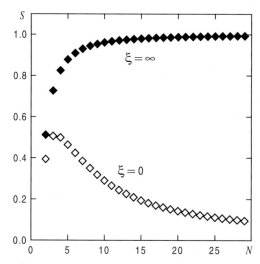

Fig. 5. Nonsphericity parameter S as a function of number of particles in chain in zero (white diamonds) and infinite (black ones) external field. The coupling parameter $\lambda = 10$

As it follows from analysis of the figure, even for a very strong dipole interactions the value of S for zero field (lower curve in Fig. 5) quickly decreases with growth of number of particles in the chain. In fact, *any* dipolar chain can be considered as a coil slightly elongated along one direction. The more number of grains in a chain, the smaller the coil elongation. At the same time in the case of infinite external field the nonsphericity parameter S increases fast with the growth of number of particles in a chain and approaches to unity (upper curve in Fig. 5). Does it mean that dipolar chain becomes stiffer in external field? Of course, it does not. The point is that each Kuhn segment is directed along the external field and all the Kuhn segments are connected with each other.

Summarizing the results of the Section 3 we can conclude that the ideal chain of magnetic grains is very flexible formation with short (of the order of a few particles) persistent length or the Kuhn segment. The chain looks like a quasi-spherical coil of connected links in the absence of external field. This coil is disentangled with increasing of the field and transforms into chain randomly and weakly bent relative to the field direction.

4 Nonideal Chains

Above we have considered the case of ideal chains. We made allowance only for interactions of neighboring particles belonging to the same chain ignoring interactions between 1) the non-nearest neighbors along the chain, 2) the distant segments of the same chain which are finding near each other owing to chain flexibility, and 3) the segments of different chains. The account of the first of them in the case of straight linear chains is reduced to re-normalization on 20% of the nearest-neighbors interactions (see Sect. 1). The chain flexibility should only reduce this value, so that one can regard these interactions as insignificant. The second type of interactions can form antiparallel side-by-side configurations of dipoles in zero field. The third type should be taken into account along with the second one because the friable coils of ideal chains (with typical volume of $\sim N^{3/2}$ in zero field) begin to overlap for strong dipolar interactions ($\lambda \geq 7$) already at small volume fractions ($\leq 1\%$) of ferroparticles. A due regard for these interactions can be carried out in the frame of the concept of *quasi-monomers* [54] treating every long flexible chain as a system of *disconnected* segments. We identify each a segment with the individual grain and take into account the interactions between segments in the approximation of the second virial coefficient b_m. [More precisely, one should examine the Kuhn segment or the segment of persistent length as an *effective monomer*. This strict consideration is similar to above-mentioned accounting for the next-nearest grains in the segment what re-normalizes slightly the grain-grain interactions. Such an approach, however, complicates considerably the calculations, what is why we use here more intuitive rather than the strict arguments]. b_m is not the usual second virial coefficient of interactions of two hard dipoles. The point is, even considering the chain as a system of disconnected grains, we should remember of the connection of all its segments. Thus, calculating b_m, we assume that the most energetically profitable positions corresponding to head-to-tail configurations are occupied already by the grains of main chain and then forbidden for any other interacting segments.

The way of determination of b_m can be divided on four steps. First, we right the usual second virial coefficient $b_2(\xi)$ of interacting dipoles in external magnetic field H

$$b_2(\xi) = -\frac{1}{2v}\left(\frac{\xi}{4\pi\sinh\xi}\right)^2 \int \{e^{-U(12)/k_\mathrm{B}T} - 1\}e^{\xi(e_1 h + e_2 h)} de_1 de_2 dr, \quad (45)$$

where $\xi = mH/k_BT$ is the Langevin parameter and $\boldsymbol{h}$ is unit vector along field. Obviously, the equation reduces to (5) in zero field. Second, we approximate (45) replacing the orientation of one particle by the field direction, $\boldsymbol{e}_1 \to \boldsymbol{h}$, so as

$$b_2(\xi) \approx -\frac{\xi}{8\pi v \sinh \xi} \int \{e^{-U(\langle 1 \rangle 2)/k_BT} - 1\} e^{\xi(\boldsymbol{e}_2 \boldsymbol{h})} d\boldsymbol{e}_2 d\boldsymbol{r} \,, \qquad (46)$$

where the symbol of averaging denotes the mentioned replacement. Equation (46) gives the exact values of the second virial coefficient (45) in the limiting cases of zero and infinite external field. The former follows from the independence of typical integral I (4) on orientation of first particle, the latter is evident. So, (46) is an interpolation formula for moderate values of external field. Our spot check for some values of λ and ξ shows that the difference between (45) and (46) does not exceed 10%. Third, in order to exclude the occupied states we restrict the possible values of polar angle θ between vectors $\boldsymbol{r}$ and $\boldsymbol{h}$ by the range $[\pi/3, 2\pi/3]$ what effectively takes into account the presence near particle 1 of two head-to-tail aligned neighboring dipoles. Fourth, we overcome the nonphysical logarithmic divergence owing to previous step by the cutting off the region of integration over interparticle distance r. We assume that r can vary up to its maximal value r_* defined by the physical limitation $m^2/r_*^3 k_B T = 1$ (within the spherical volume $r \leq r_*$ the dipole potential for the antiparallel orientation of dipoles exceeds thermal energy. Finally, we write the second virial coefficient of two interacting quasi-monomers in the form

$$b_m = -\frac{\xi}{4v \sinh \xi} \int_0^{r_*} dr \int_{\pi/3}^{2\pi/3} d\theta \int d\boldsymbol{e}_2 \{e^{-U(\langle 1 \rangle 2)/k_BT} - 1\} e^{\xi(\boldsymbol{e}_2 \boldsymbol{h})} r^2 \sin \theta \,. \qquad (47)$$

We have found b_m as a function of λ and the dimensionless external field ξ. We note that the asymptotic expansion at $\lambda \gg 1$ poorly describes the calculated values of b_m in the whole range of λ (up to 10) considered. What is why we do not perform it. The most interesting case is $b_m(\lambda, \xi) = 0$. It determines the *neutral curve* (see Fig. 6) corresponding to compensation of dipole and steric interactions of quasi-monomers.

This curve divides the plane of parameters $\lambda - \xi$ into two parts. Lower the curve the repulsive forces prevail and dipolar chains exist as chain-like formation - as coil in zero and weak fields, $\xi \ll 1$, and as almost linear formation in strong magnetic field $\xi \gg 1$. Indeed, under the field an antiparallel side-by-side configuration of dipoles belonging to two neighboring quasi-monomers becomes unprofitable: both them tend to orient along field. Vice versa, higher the curve the attraction forces predominate, the coils of grains become unstable against to interaction of distant along chain particles and any long chain collapses into dense *globule*. The indication that just long chains collapse is essential because the concept of quasi-monomers [54] is suitable only for such chains. In zero external field the *coil-globule phase transition* takes place at $\lambda = 3.2$ when b_m changes its sign. This value should be considered only as estimation due to

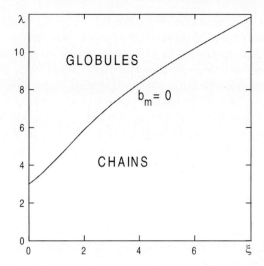

Fig. 6. Line of coil-globule transition at the plane coupling parameter λ - external field ξ

negligible number of long chains at so small values of coupling parameter (see Sect. 2).

The case of chains of finite length is much more complicated than studied one. Nevertheless, some general conclusions from the physical point of view can be done. Indeed, the interaction between distant along chain grains becomes important when chain length exceeds the value L_f equal to several of the Kuhn segments L_K (or persistent lengths). Roughly modelling the Kuhn segment by the hard rod and considering the dipolar chain as freely-jointed sequence of the segments, it is possible to estimate $L_f = 2L_K$. At $\lambda = 6$ this gives $L_f \approx 5$ (see Fig. 4). According to (9) the mean chain length $\langle N \rangle \leq L_f$ for volume fraction of magnetic grains $\Phi \leq 0.1$. Thus, the microstructure of such MF is represented mainly by the number of separate chains. At higher values of coupling parameter, $\lambda \geq 6$, the mean chain length exponentially grows. As a result, the majority of chains should transform into compact clusters - the prototypes of globules for the finite values of N. It is interesting that this qualitative physical picture well corresponds to the recent Monte Carlo data [39,40].

5 Conclusion

We studied here the spatial and orientational intrachain correlations in zero and infinitely strong external field and established the important role of flexibility in the description of conformational properties of dipolar chains. Among the problems which proved to be out of the scope of the present investigation we would like to mark the following ones. First, the problem of taking into account of the interchain interactions which can be significant due to the long-range nature of dipolar interactions. As it is shown in [58] the interchain correlation can noticeably change the mean length of a chain. Second, there is a problem of the unique description of both effects of dipolar interparticle interactions – the condensation and association phenomenon. On the language of present work it means the interconnection between condensation and globule formation. We assume that direct identification of both phenomena with each other is not appropriate. Most likely the globule formation precedes the liquid-gas phase transition and the globules themselves are the nuclei of future concentrated phase. Another interesting aspect of the problem is the study of both effects in the case of moderately concentrated magnetic fluids. In our approach it was convenient to divide mentally the microstructure formation on stages of (i) appearance of ideal chain and then (ii) its transformation to nonideal one due to chain flexibility. In reality both processes take place simultaneously. So, it is reasonable to wait for the formation in sufficiently concentrated magnetic fluids of extended network [37] instead of globule formation. Finally, there is the problem of description of thermodynamical and dynamical properties of magnetic fluids with developed microstructure. This work was begun in [45,48,49,53,58]. We will return to study of these intriguing questions later.

This work was supported by the Russian Fund for Fundamental Research (Project 02-03-33003) and the Israel Science Foundation (Grant 336/00).

References

1. P.G. de Gennes, P.A. Pincus: Phys. Kondens. Materie **11**, 189 (1970)
2. S.A. Adelman, J.M. Deutch: Anv. Chem. Phys. **31**, 103 (1975)
3. M.S. Wertheim: J. Chem. Phys. **55**, 4291 (1971)
4. W. Sutherland, G. Nienhuis, J.M. Deutch: Mol. Phys. **27**, 721 (1974)
5. K.-C. Ng, J. Valleau, G. Torrie, G. Patey: Mol. Phys. **38**, 781 (1979)
6. C.F. Hayes: J. Coll. Int. Sci. **52**, 239 (1975)
7. J.-C. Bacri, D. Salin: J. Phys. Lett. **43**, 649 (1982)
8. J.-C. Bacri, R. Perzynski, D. Salin, V. Cabuil, R. Massart: J. Magn. Magn. Mater. **85**, 27 (1990)
9. J.-C. Bacri, F. Boué, V. Cabuil, R. Perzynski: Colloids Surfaces A **80**, 11 (1993)
10. C.F. Hayes, S.R. Hwang: J. Coll. Int. Sci. **60**, 443 (1977)
11. E.A. Peterson, D.A. Krueger: J. Coll. Int. Sci., **62**, 24 (1977)
12. R.W. Chantrell, J. Sidhu, P.R. Bissel, P.A. Bates: J. Appl. Phys. **53**, 8341 (1982)
13. A.F. Pshenichnikov, I.Yu. Shurubor: Bull. Acad. Sci. USSR, Phys. Ser. **51**, 40 (1987)
14. A.F. Pshenichnikov, I.Yu. Shurubor: Magnetohydrodynamics **24**, 417 (1989)
15. A. Cebers: Magnetohydrodynamics **18**, 137 (1982)
16. K. Sano, M. Doi: J. Phys. Soc. Jpn. **52**, 2810 (1983)
17. K.I. Morozov: Magnetohydrodynamics **23**, 37 (1987)
18. K.I. Morozov: Bull. Acad. Sci. USSR, Phys. Ser. **51**, 32 (1987)
19. Yu.A. Buyevich, A.O. Ivanov: Physica A **190**, 276 (1992)
20. K.I. Morozov, A.V. Lebedev: J. Magn. Magn. Mater. **85**, 51 (1990)
21. A.F. Pshenichnikov, V.V. Mekhonoshin, A.V. Lebedev: J. Magn. Magn. Mater. **161**, 94 (1996)
22. A.O. Ivanov, O.B. Kuznetsova: Phys. Rev. E **64**, 041405 (2001)

23. I. Nakatani, M. Hijikata, K. Ozawa: J. Magn. Magn. Mater. **122**, 10 (1993)
24. H. Mamiya, I. Nakatani, T. Furubayashi: Phys. Rev. Lett. **84**, 6106 (2000)
25. F. Gazeau, C. Baravian, J.-C. Bacri, R. Perzynski, M.I. Shliomis: Phys. Rev. E **56**, 614 (1997)
26. J.J. Weis, D. Levesque: Phys. Rev. Lett. **71**, 2729 (1993)
27. J.M. Caillol: J. Chem. Phys. **98**, 9835 (1993)
28. M.E. van Leeuwen, B. Smit, Phys. Rev. Lett. **71**, 3991 (1993)
29. M.J. Stevens, G.S. Grest: Phys. Rev. Lett. **72**, 3686 (1994)
30. M.J. Stevens, G.S. Grest: Phys. Rev. E **51**, 5962 (1995)
31. P.J. Camp, J.C. Shelley, G.N. Patey: Phys. Rev. Lett. **84**, 115 (2000)
32. D. Wei, G.N. Patey: Phys. Rev. Lett. **68**, 2043 (1992)
33. D. Wei, G.N. Patey: Phys. Rev. A **46**, 7783 (1992)
34. J.J. Weis, D. Levesque: Phys. Rev. E **48**, 3728 (1993)
35. D. Levesque, J.J. Weis: Phys. Rev. E **49**, 5131 (1994)
36. M.J. Stevens, G.S. Grest: Phys. Rev. E **51**, 5976 (1995)
37. P.J. Camp, G.N. Patey: Phys. Rev. E **62**, 5403 (2000)
38. A.F. Pshenichnikov, V.V. Mekhonoshin: J. Magn. Magn. Mater. **213**, 357 (2000)
39. A.F. Pshenichnikov, V.V. Mekhonoshin: JETP Lett. **72**, 182 (2000)
40. A.F. Pshenichnikov, V.V. Mekhonoshin: Eur. Phys. J. E **6**, 399 (2001)
41. T.L. Hill: *Statistical Mechanics* (McGraw-Hill, New York 1956)
42. P.C. Jordan: Mol. Phys. **25**, 961 (1973)
43. P.C. Jordan: Mol. Phys. **38**, 769 (1979)
44. B. Jeyadevan, I. Nakatani: J. Magn. Magn. Mater. **201**, 62 (1999)
45. A.Yu. Zubarev, L.Yu. Iskakova: JETP **80**, 857 (1995)
46. R.P. Sear: Phys. Rev. Lett. **76**, 2310 (1996)
47. R. van Roij: Phys. Rev. Lett. **76**, 3348 (1996)
48. M.A. Osipov, P.I.C. Teixeira, M.M. Telo da Gama: Phys. Rev. E **54**, 2597 (1996)
49. J.M. Tavares, M.M. Telo da Gama, M.A. Osipov: Phys. Rev. E **56**, R6252 (1997)
50. Y. Levin: Phys. Rev. Lett. **83**, 1159 (1999)
51. J.M. Tavares, J.J. Weis, M.M. Telo da Gama: Phys. Rev. E **59**, 4388 (1999)
52. M.M. Telo da Gama, J.M. Tavares: Comp. Phys. Communic. **121-122**, 256 (1999)
53. A.Yu. Zubarev, L.Yu. Iskakova: Phys. Rev. E **61**, 5415 (2000)
54. A.Yu. Grosberg, A.R. Khokhlov: *Statistical Physics of Macromolecules* (AIP, Woodbury, New York 1994)
55. P. Ehrenfest, V. Trkal: Proc. Amst. Acad. **23**, 162 (1920)
56. J.E. Mayer, M. Goeppert Mayer: *Statistical Mechanics* 2nd edn. (John Wiley & Sons, New York 1977)
57. M.E. Fisher: Am. J. Phys. **32**, 343 (1964)
58. A.Yu. Zubarev: Lecture Notes in Physics (2002) (this issue)

Supplementary Glossary

$A = \langle \boldsymbol{e}_1 \boldsymbol{e}_2 \rangle$	dimer correlation function
$a = d + \sqrt{\langle R_\parallel^2 \rangle}$	longitudinal size of chain
$B = \langle \boldsymbol{e}_1 \boldsymbol{r}_{12} \rangle / d$	dimer correlation function
B_∞	dimer correlation function B in infinite field
$b = d + \sqrt{\langle R_\perp^2 \rangle}$	transverse size of chain
b_2	second virial coefficient of dipoles
b_m	second virial coefficient of quasi-monomers
$C = \langle (\boldsymbol{e}_1 \boldsymbol{e}_2)^2 \rangle$	dimer correlation function
$D = \langle (\boldsymbol{e}_1 \boldsymbol{r}_{12})^2 \rangle / d^2$	dimer correlation function
D_∞	dimer correlation function D in infinite field
d	particle diameter
$E = \langle r \rangle / d$	dimensionless interparticle distance
E_∞	dimer correlation function E in infinite field
$\boldsymbol{e}_i$	unit vector along dipole i
$F = \langle r^2 \rangle / d^2$	dimensionless square of interparticle distance
F_∞	dimer correlation function F in infinite field
g_N	number of N-particle chains per unit volume
H	external magnetic field
$\boldsymbol{h}$	unit vector along field
k_B	Boltzmann's constant
L_K	Kuhn segment length
L_p	persistent length
m	particle magnetic moment
N	number of grains in chain
n	number of grains per unit volume
$\boldsymbol{R} = \sum_{i=1}^{N-1} \boldsymbol{r}_{i,i+1}$	"end-to-end" vector of N-particle chain
$\langle R_\parallel^2 \rangle$	longitudinal mean square of "end-to-end" vector
$\langle R_\perp^2 \rangle$	transverse mean square of "end-to-end" vector
$\langle r \rangle$	mean interparticle distance
$S = (a^2 - b^2)/(a^2 + b^2)$	nonsphericity parameter
T	temperature
v	grain volume
Z_2	dimer partition function
$\lambda = m^2 / d^3 k_\mathrm{B} T$	coupling parameter
$\xi = mH / k_\mathrm{B} T$	Langevin parameter
$\Phi = \pi n d^3 / 6$	volume fraction of magnetic grains

Magnetoviscous Effects in Ferrofluids

Stefan Odenbach and Steffen Thurm

ZARM, University of Bremen, Am Fallturm, D-28359 Bremen, Germany

Abstract. The appearance of field and shear dependent changes of viscosity in ferrofluids opens possibilities for future applications e.g. in damping technologies. To enhance the effects, it is necessary to understand the observed magnitudes of magnetoviscosity in commercial ferrofluids from a microscopic point of view. Starting from experimental results, it is described how the magnetoviscous effects can be explained by chain formation of a small fraction of large particles in the fluid. With a dedicated experiment ferrofluids are separated into fractions with high and low amount of such particles. The rheological characterization of the fractions prove the aforementioned model. Furthermore it leads to additional information concerning viscoelasticity of the suspensions in a magnetic field.

1 Introduction

The phenomenon of field dependent changes of viscosity of a suspension of magnetic nanoparticles is known since more than 30 years. In 1968 *McTague* [1] discovered an increase of viscosity of a ferrofluid containing Co-particles in the presence of a magnetic field. His experiments showed a dependence of the viscosity increase on the magnetic field strength as well as on the field's direction relative to the flow. In his paper as well as in an accompanying paper by *Hall* and *Busenberg* [2] the effect was explained by a hindrance of rotation of the suspended particles due to the action of the magnetic field. This concept should be shortly explained here, since it has fundamental importance for almost all discussions about magnetoviscous effects in suspensions of nanosized particles.

Assume a ferrofluid under influence of a shear flow, as shown in figure 1. In such a situation the magnetic particles will rotate in the flow due to the action of mechanic torque produced by viscous friction in the fluid. If a magnetic field H is applied to the fluid, the magnetic moment m of the particles will align with the field direction. In a situation where field direction and vorticity of the flow are collinear (figure 1a) the magnetic alignment will only lead to the fact that the magnetic moment of the particles becomes collinear with the direction of vorticity. An influence on the motion of the particle and therefore on the flow of the fluid as a whole does not appear. The situation changes if vorticity and field direction are perpendicular. Under these conditions the mechanic torque will force a misalignment of the magnetic moment of the particle and the field direction, provided the magnetic moment's direction in the particle is fixed. An angle between the mutual directions of the magnetic moment and the field will immediately give rise to a magnetic torque, trying to realign m and H. This

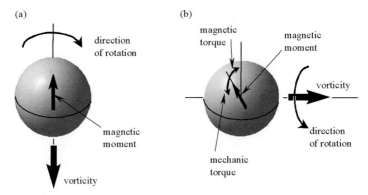

Fig. 1. On the origin of a field dependent change of viscosity in a suspension of magnetic nanoparticles. Detailed explanation is given in the text.

torque acts opposite to the mechanic torque and causes thus a hindrance of the free rotation of the particle in the flow. This increases the flow resistance and thus the fluid exhibits an increased viscosity.

A rigorous theoretical analysis of the phenomenon has been given by *M. Shliomis* [3] four years after the experimental discovery. Taking into account Brownian motion of the particles he derived an expression describing the change of viscosity - called rotational viscosity η_r - as a function of the strength and direction of the magnetic field H in the form

$$\eta_r = \frac{3}{2}\phi'\eta_0 \frac{\alpha - \tanh\alpha}{\alpha + \tanh\alpha} <\sin^2\beta>, \qquad (1)$$

where η_0 denotes the viscosity of the fluid in the absence of a magnetic field, ϕ' the volume concentration of the magnetic particles including the surfactant and β is the angle between vorticity and field direction; $<...>$ denotes the spatial average. The parameter α is the ratio of magnetic and thermal energy of the particles

$$\alpha = \frac{\mu_0 m H}{kT}, \qquad (2)$$

where μ_0 denotes the vacuum permeability, m the particle's magnetic moment, k Boltzmann's constant and T the absolute temperature. Equation (1) has been derived under the assumption of two important limitations. First of all it has to be assumed, as already discussed in the qualitative description above, that the magnetic moment of the particles is spatially fixed within the particle. This assumption leads to a question concerning the relaxation of magnetization of the fluid. As discussed in *P. Fannin's* contribution [4] the magnetization can either relax by the Néel process, i.e. by a change of the magnetic moment's direction in the particle, or by the Brownian process, that means by a rotation of the particle in the flow. The relaxation takes place by the process with the shorter characteristic relaxation time. As shown in [4] both times depend on the size of the particles but the Brownian time scales only linear with the particles' volume while the Néel time depends exponential on the volume. Thus, small particles will relax by the Néel process while particles above a certain critical diameter follow the Brownian process. For the discussion of rotational viscosity this means, that only particles with a diameter above the critical one, usually called magnetically hard, will contribute to the changes of viscosity.

The second fundamental assumption in [3] has been the restriction to highly diluted systems, neglecting any interaction between the magnetic particles. For *McTague's* experiments both assumptions were fulfilled. He used Co-particles with a diameter about 10 nm being sufficiently larger than the critical diameter of 6 nm. These particles are thus magnetically hard and a respective suspension is expected to show rotational viscosity. Moreover the volume concentration of the suspensions in [1] were as low as 0.05 vol% - a high dilution preventing significant interparticle interaction.

A few month before *McTague's* experiment was published, *Rosensweig* [5] reported a viscosity change in concentrated magnetite suspensions. The increase of the fluids' viscosity found in these experiments was relatively high and reached about 200 % of the zero field value. A result principally encouraging further investigations in this direction especially with a focus towards technical applications using the phenomenon. Nonetheless, these results were not discussed furthermore and especially they were never compared with the theoretical description in [3].

2 Magnetoviscous Effects in Concentrated Suspensions

As mentioned, the microscopic explanation of rotational viscosity has been developed and proved with experiments using highly diluted suspensions of magnetically hard cobalt particles. Nevertheless, magnetic fluids used for technical applications and for numerous experiments in basic research, contain magnetite particles with magnetic volume concentrations in the order of 7-10 vol%. Thus, an investigation of the viscosity of such ferrofluids in the presence of magnetic fields is an interesting topic in the field of ferrofluid research. In an experiment using a relative high shear rate of $\dot{\gamma} = 500$ s^{-1} *Ambacher et al.* [6] investigated the field dependent increase of viscosity in a magnetite ferrofluid containing 7.2 vol% of the magnetic component.

Figure 2 shows the measured results leading to a maximum increase of viscosity of about 40 %. Comparing the experiment with the theory in [3] requires information about the size distribution of particles, since only particles with a diameter larger than about 13 nm - being the critical diameter for magnetite - can contribute to the viscosity changes. As seen from the inset in figure 2, these particles represent only a small fraction of the overall magnetic concentration of the ferrofluid. Calculating the rotational viscosity with these values, one obtains the dashed line in figure 2, being in obvious quantitative disagreement with the experimental result. Nevertheless a fit of equation (1) to the measured data shows that the general behavior of the viscous changes fits qualitatively well. This leads to the assumption that the concept of hindrance of rotation holds

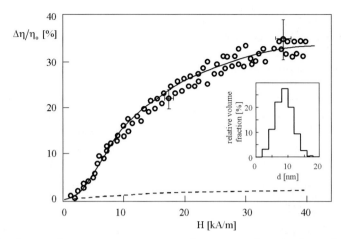

Fig. 2. The change of viscosity in a commercial magnetite ferrofluid together with the theoretical prediction following equation (1) calculated for the particle size distribution shown in the inset (dashed line) and with a fit of equation (1) (solid line).

even when the absolute values can not be explained in terms of individual particles rotating in the flow. Moreover, the fit parameters leading to an apparent mean particle size of about 16 nm for the full particle concentration indicate that interaction of particles, forcing formation of agglomerates, has significant importance in the concentrated suspension investigated.

To avoid confusion we will call the changes of viscosity in concentrated suspensions, where interparticle interaction plays a significant role "magnetoviscous effect", while the term "rotational viscosity" will be reserved for the case of highly diluted suspensions with negligible particle interaction.

A dominant contribution of interparticle interaction to the magnetoviscous effect leads immediately to the question what kind of microstructure is responsible for the observed effects. To get a deeper insight into the behavior of the fluid under shear and magnetic field influence we developed a specialized rheometer for the investigation of ferrofluids [7], shown in figure 3.

The core part of the rheometer is a modified cone-plate flow cell with a moving plate and a cone attached to the torque sensor. The connection between cone and torque sensor is provided by a long rod guided in an air bearing to prevent spurious friction due to the bearings. The long rod provides a sufficient spatial separation between the sensor and the magnetic field region, ensuring that no disturbing influence of the field on the measurement occurs. The rheometer allows investigations of magnetic fluids under stationary as well as time dependent load.

Using this device we have investigated [8,9] the dependence of the magnetoviscous effect on shear rate and fluid composition. Figure 4 shows a typical change of viscosity with field strength for various shear rates for the ferrofluid APG513A containing 7.2 vol% of magnetite particles with 10 nm mean diameter in an ester. Obviously a strong magnetoviscous effect is observed at low shear

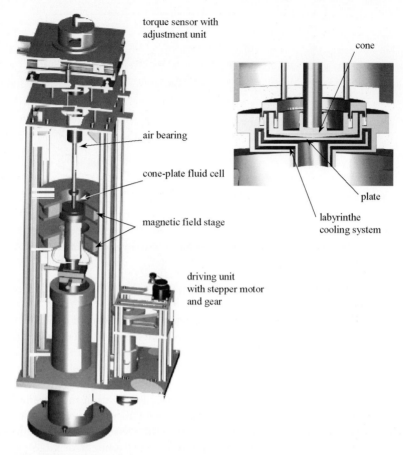

Fig. 3. The rheometer used for the investigation of magnetic fluids in the experiments presented here. Details are given in the text.

rate diminishing successively with increasing load. The absolute values of the magnetoviscous effect are again - as in [6] - significantly higher than expected from the theory in [3]. Furthermore, the field dependent shear thinning is an effect which is not predicted in [3] at all. These observations led to the assumption that the formation of chains or clusters of particles under the influence of a magnetic field has to be a leading component in the description of magnetoviscosity. The hindrance of rotation of these agglomerates is assumed to give rise to the strong increase of viscosity with magnetic field strength. Furthermore the rupture of the chains in a shear flow can be the basis for an explanation of the observed shear thinning [8,10].

Within the frame of this model one has to observe that only relatively large magnetite particles with a diameter of more than about 13 nm show an interparticle interaction being strong enough to contribute significantly to such chain formation. That means that only a small fraction of the overall magnetic con-

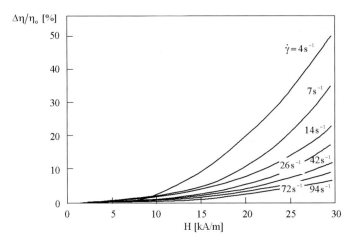

Fig. 4. The field and shear rate dependence of the viscosity of APG513A

centration in a magnetite ferrofluid is assumed to be of dominant importance for the magnetoviscous effects.

To obtain a first check of this hypothesis we have performed experiments using ferrofluids from a common production batch having identical overall magnetic concentration but being subjected to different purification in strong magnetic field gradients [9]. The variation of the purification process leads to differences in the concentration of large magnetic particles. Table 1 gives an overview of the properties of these fluids as obtained from magnetization measurement. Increasing ordinal number of the fluids indicates an increasing content of large particles. Comparing the magnetoviscous effect in these fluids we obtain a significant decrease of magnetoviscosity with decreasing amount of large particles as illustrated in Fig. 5.

Table 1. Properties of the fluids with variable content of large particles

Fluid	$M_s [kAm^{-1}]$	$d_{mean} [nm]$
$F1$	32.41	8.3
$F2$	32.34	8.8
$F3$	31.54	9.2
$F4$	32.17	9.2
$F5$	32.06	10.1

This result documents clearly that the magnetoviscous effect in commercial magnetite ferrofluids is dominated by the interaction of the small amount of large particles generated during the synthesis. On this basis a model [11] has

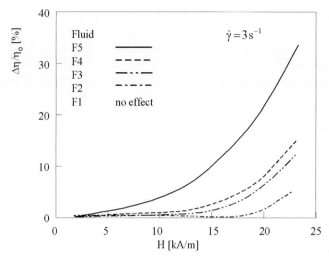

Fig. 5. The magnetoviscous effect for fluids with different content of large particles

been developed describing such ferrofluids as bidisperse systems containing a large fraction of small particles with weak interparticle interaction and a small fraction of large particles forming chains in the presence of a magnetic field. Under the assumption that the chains are rigid and straight and that interaction between chains and between chains and small particles can be neglected one can calculate the size distribution function of the chains from a minimization of the free energy of the system. Using this distribution function, an expression for the stress tensor can be obtained, allowing to fit the experimental data on magnetoviscous effects.

3 Controlled Changes of the Fluid Composition

We have shown in the previous chapter, that the size distribution of suspended particles is of fundamental importance for the viscous behavior of ferrofluids under influence of external magnetic fields and shear stress. Previous experiments [9] and a corresponding model [11] proved that the amount of large particles is of major importance for the effects observed.

To investigate the influence of the size distribution, and in particular of the amount of large particles, on the viscous behavior, it is necessary to change directly and continuously the fraction of these particles by a controlled separation process. Therefore we developed a separation device for ferrofluids, which bases on forced diffusion of magnetic particles due to a strong magnetic field gradient.

3.1 The Theory of Magnetic Driven Diffusion Processes

The development of a magnetic separation device, allowing controlled separation of a ferrofluid, requires precise information on numerous experimental parameters like diffusion times or required magnetic field gradient, as well as on fluid parameters like viscosity, interparticle forces and - as the most important factor - the particle size distribution. First step of the layout of the separation device was a numerical simulation of the diffusion process to provide information about expected diffusion times dependent on the magnetic field gradient, which can vary in strength and spatial structure. Later this approximation of field strength and geometry was used as input for a computer simulation to develop the shape of the pole shoes, which are the key parts of the experimental setup.

The numerical simulation bases on the assumption that the magnetic particles do not interact with each other. This is a safe approximation, because interparticle forces cause chain or cluster formation, which in turn accelerates the process of diffusion. In principle two different theoretical approaches exist to model the process of diffusion in magnetic fluids.

Close to equilibrium, i.e. for weak driving forces leading to diffusion in the system, an approach from the point of view of irreversible thermodynamics can be chosen. Detailed information for this approach can be found in *Prigogine* [12], *de Groot, Mazur* [13], *Blums et al.* [14] and *Odenbach* [15].

For strong driving forces - as they are required for the separation process in focus of our experiment - an approach based on statistical physics has to be used. The individual velocities and trajectories of all particles are obviously not important, only the integral change of the spatial concentration distribution with time has to be calculated. The fundamental equations describing this problem can be found e.g. in *Gerber et al.* [16], who simulated numerically the process of HGMS (*H*igh *G*radient *M*agnetic *S*eparation). Starting point is the equation of continuity, describing the temporal changes of the concentration distribution of the magnetic particles

$$\partial c/\partial t + \nabla \boldsymbol{J} = 0 \qquad (3)$$

here the total flux of particles $\boldsymbol{J}$ consists of a diffusive part, $\boldsymbol{J}_d = -D\nabla c$ and a part $\boldsymbol{J}_f = \boldsymbol{v}c$ driven by external forces, where $\boldsymbol{v}$ denotes the drift velocity of the particles. Substituting these expressions into (3) gives

$$\partial c/\partial t = \nabla(D\nabla c) - \nabla(\boldsymbol{v}c). \qquad (4)$$

Some general approximations are necessary to solve (4). First it is assumed that the diffusion coefficient D is independent from the concentration, and that the fluid can be considered as an ideal diluted system of spherical particles. Therefore the Nernst-Einstein relation $D = ukT$ can be used, with the mobility of spherical particles $u = 1/(6\pi\eta r)$, the Stokes' particle radius r, and the viscosity of the fluid η. The drift velocity $\boldsymbol{v}$ can be expressed by $\boldsymbol{v} = u\boldsymbol{F}$, so (4) becomes

$$\partial c/\partial t = D\nabla^2 c - \nabla(u\boldsymbol{F}c). \qquad (5)$$

The force $\boldsymbol{F}$ incorporates all forces driving diffusion in the system. The major contribution to this force is the magnetic force $\boldsymbol{F}_m = \mu_0 m \boldsymbol{\nabla} H$, which is exerted by the field gradient.

Beside the magnetic effect one could principally account e.g. for the gravitational force $\boldsymbol{F}_g$ or the viscous force $\boldsymbol{F}_v$ due to flow of the background fluid.

The first one is neglected since its influence is about two orders of magnitude smaller than the magnetically induced drift in reasonable field gradients. The latter is excluded by the assumption that the fluid as a whole is at rest during the separation process.

Further effects could appear due to van der Waals or magnetic interaction leading to an agglomeration of the particles. This would increase the magnetic moment entering the magnetic driving force and would thus enhance the drift velocity leading to an acceleration of the separation process. Since the simulation is thought to give an approximation of the diffusion process allowing the design of a separation device enabling separation of the large particles in the fluid on reasonable time scales, effects like the mentioned agglomeration will only enhance the performance of a system based on the above assumptions.

Finally (5) can be written as

$$\partial c/\partial t = 1/\left(6\pi\eta r\right)\left[kT\left(\nabla^2 c\right) - \nabla\left(\mu_0 V_{mag} M_0 \nabla H c\right)\right] \quad (6)$$

where m has been replaced by $m = M_o V_{mag}$ with V_{mag} denoting the volume of the magnetic particle and M_o being the spontaneous magnetization of the magnetic material – in our case magnetite with $M_o = 4.5 \cdot 10^5$ A/m.

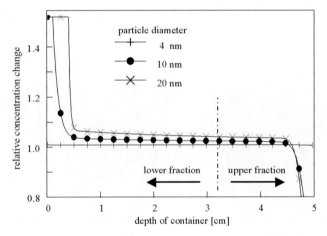

Fig. 6. Numerically calculated time dependent concentration profile for three different particle diameters after a separation time of 1 week. The height of fluid in the container is 5 cm

Using (6) one can calculate the concentration distribution of magnetic particles as a function of time for different particle sizes. From a comparison of the resulting distributions it can be judged whether a separation of a fluid in fractions with different content of large particles and mainly constant content of small ones is possible.

Fig. 6 shows the calculated time dependent concentration profile for three different particle diameters after a separation time of one week for a 5 cm deep container. The field gradient applied in the direction of diffusion is constant over the depth of the container at a value of 10^7 A/m^2. As a boundary condition it has been assumed, that the concentration at the lower wall of the container can not exceed 150 % of the original concentration.

One can see that the magnetic field gradient acts mainly on the largest particles. Particles which are smaller than 10 nm hardly react on the field gradient, thus we are able to concentrate just the fraction of bigger particles in the region of high magnetic field. The parameters separation time and field gradient control the accumulation and depletion of bigger particles in the lower and upper region of the container respectively.

3.2 The Experimental Setup

One of the key requirements for the separation device was the possibility to work with fluid volumes up to 300 ml. This large amount of fluid is necessary to enable successive separation steps with corresponding rheological investigations. This constraint and the available free space between the pole shoes of an high field electromagnet restricted significantly the geometrical parameters for the separation container. The final geometry of the setup was chosen by an iteration of the shape of the pole shoes, the height of the container and corresponding diffusion times. The result is a trapezoid shaped container with a height of 5 cm as shown schematically in Fig. 7.

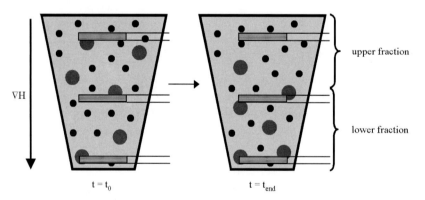

Fig. 7. Principle sketch of the experimental setup

The container is located in the gap between two tailor made pole shoes of an electromagnet. The pole shoes are designed in a way that they provide a constant magnetic field gradient of 10^7 A/m^2 over the whole fluid volume. This ensures a uniform diffusion of particles from the top to the bottom of the container as already shown in the simulation (see Fig. 6).

The container is equipped with an in-situ concentration measuring system, consisting of three coils, located in different heights. The inductance of the coils is directly proportional to the local concentration of magnetic particles. Therefore it is possible to measure a time dependent concentration profile in the container. In principle this information can be used to simulate numerically the diffusion process with consideration of chain and cluster formation.

The whole setup is temperature controlled to an accuracy of 0.1 K. This ensures that no convective flow due to temperature differences can disturb the diffusion process. Furthermore, the temperature stability is a technical requirement for the concentration monitoring using the sensor coils. The inductance of the coils depends not only on the concentration of magnetic particles, but also on temperature and the strength of the applied magnetic field. By keeping the external magnetic field and temperature constant during the measurements, the inductance of the coils corresponds directly to the local concentration of magnetic particles.

4 Results

The fluid used in the experiments discussed hereafter has been a sample of APG513A from Ferrofluidics, with a saturation magnetization of 32 kA/m, an initial susceptibility of 1.4 and a dynamic viscosity of 150 mPas. The magnetic separation ran over a period of 30 days. After this time period the separation was stopped, and the fluid was removed in two fractions from the container as indicated in Fig. 7.

4.1 Inductance Measurements during Separation Process

Fig. 8 shows the change of concentration for the upper and lower region of the container. It is calculated as the difference between the concentration signals of the sensor coil in the center and the upper and lower sensor coils respectively. The middle coil shows more or less no change in concentration. This result is expected, because the overall concentration in the container has to be constant, and according to numerical results the middle coil should not react before an advanced state of separation. Thus this coil can be used as a reference concentration as mentioned above. The upper coil shows a decrease in concentration, which is as strong as the increase of concentration of the lower coil. It can be seen, that saturation of separation is not reached after 30 days. In this context saturation means, that all bigger particles are in the lower region of the container, and the continuing process of diffusion is too slow to cause measurable changes of the signal of the coils. So one can expect, that a small amount of large particles is still present in the upper fraction and thus a certain magnetoviscous effect is expected for this part of the fluid too. The concentration difference between the lower and upper fraction of the fluid as obtained from the sensor signals is round about 6 percent.

4.2 Magnetization Measurements

Magnetization measurements provide information about saturation magnetization and initial susceptibility. The saturation magnetization M_s corresponds to

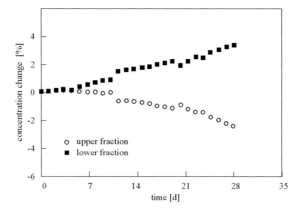

Fig. 8. Time dependent change of concentration during separation process

the concentration of magnetic material in the fluid, and the initial susceptibility to the mean particle diameter [17]. Thus after a separation process M_s should be increased in the sample taken from the lower part of the container and decreased in that extracted from the upper one, while the initial susceptibility χ_i should become smaller in the upper fraction, and larger in the lower fraction all compared with the values obtained for the original fluid before separation. Tab. 2 shows the results for the original sample of APG513A, and for both fractions after the separation process. The concentration difference between the upper and lower fraction of the fluid is about 10 percent. This value is somewhat higher than the one obtained from the in situ determination of the concentration changes using the sensor coils. The difference is simply caused by the fact that the magnetization measurement gives an information about the mean concentration of the respective fraction of the fluid, while the sensor coils just measure a concentration value at a certain depth inside the container.

Calculating the mean saturation magnetization for both fractions, and comparing it with the original fluid, we can find a loss of particles due to adhesion of particles at the container walls in particular in the lower part of the container, where the bigger particles and agglomerates are collected. This loss has no critical influence on the final characterization and discussion of results, but one should keep in mind, that the changes of the magnetoviscous effect discussed afterwards for the lower fraction would even be greater if these particles would also be removed from the container together with the lower fraction.

4.3 Rheological Measurements

Rheological measurements are especially suited for the investigation of the influence of the microstructural characteristics of ferrofluids on their viscous behavior. The discussion of the results obtained with these measurements is splitted into two parts. First we will focus on the magnetoviscous effect, measured with the rotational mode of the ferrofluid-rheometer described above. The elemental influence of bigger particles on the change of viscosity under influence of shear flow

Table 2. Magnetization data from the original APG513A and both fractions after the separation process

Fluid name: APG513A	M_s [kA/m]	χ_i	d_{mean} [nm]
original fluid	32.6	1.37	12.0
upper fraction	30.0	1.16	11.6
lower fraction	33.1	1.42	12.1

and magnetic fields is the central aspect of this discussion. In the second part we will present some results obtained in the oscillating mode of the rheometer, leading to information about magneto-viscoelastic effects.

The magnetoviscous effect. As seen from the experiments in [9], the investigation of field and shear rate dependent changes of viscosity in a ferrofluid provide an excellent tool to get an insight into the microscopic reasons for the rheological properties of the suspensions. The model of chain formation of bigger particles – being under test with the experiments described here – leads to a couple of phenomena being expected to appear in a ferrofluid separated as described above. First of all, in zero field, the dependence between stress and shear rate in both fractions should be linear. That means the fluid should behave Newtonian, since no field induced interparticle interactions leading to chains or clusters are present. The ferrofluid behaves like a normal suspension of non-magnetic particles.

Applying magnetic fields, the upper fraction containing only a negligible amount of large particles is expected to show weak magnetoviscous effects since the majority of particles in this fraction is too small to cause significant effects due to interaction.

In contrast the lower fraction should react noticeable on increasing magnetic field strength. These dependencies are illustrated in Fig. 9.

For the upper fraction - represented by open symbols - the relation between stress and shear is linear. This fraction shows mainly newtonian behavior and the existing weak dependence from magnetic field strength and shear rate will only be seen in a field dependent plot (see Fig. 10). This is a direct validation of the assumption, that just the bigger particles cause chain and cluster formation, which in turn gives rise to significant changes of the viscosity.

The lower fraction is represented by the filled symbols in Fig. 9 where different symbols indicate different stength' of the applied magnetic field. Starting with the curve for zero field, the similarity between the lower and upper fraction is obvious. This is expected for $H = 0$ since the only difference between the fluids is a slight change in number and size of the suspended particles. To understand this one can apply Rosensweig's equation for the dependence of viscosity on the

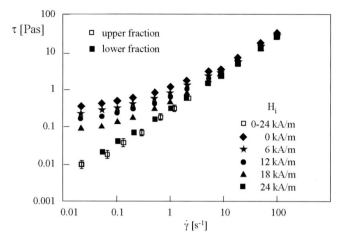

Fig. 9. Stress versus shear-rate for different magnetic fields. Both fractions and the original fluid are plotted

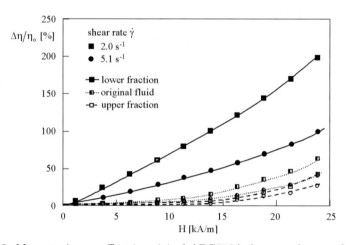

Fig. 10. Magnetoviscous effect in original APG513A, lower and upper fraction

volume fraction of magnetic particles in moderately concentrated suspensions [17],

$$\frac{\eta}{\eta_0} = \left(1 - \frac{5}{2}\tilde{\phi} + \left(\frac{5}{2}\tilde{\phi}_c - 1\right)\left(\frac{\tilde{\phi}}{\tilde{\phi}_c}\right)^2\right)^{-1}, \qquad (7)$$

where $\tilde{\phi}$ denotes the volume fraction of particles including their surfactant and $\tilde{\phi}_c$ is a critical volume fraction for which the suspension's viscosity diverges - this value is usually chosen as $\tilde{\phi}_c = 0.74$. It is easy to see that a difference of volume concentration of about 10% - as it was found for the two fractions of our separation experiment - leads to a viscosity difference of about 7% only.

With increasing magnetic field strength an increase of viscosity - the magnetoviscous effect - can been seen from the change of the slope of the stress-shear relation. Furthermore one observes that this difference diminishes with increasing shear rate. So presenting the magnetoviscous effect in the commonly used form by plotting the change of viscosity as a function of field strength for various shear rates leads to the well known behavior as shown in Fig. 10. As discussed earlier these changes of viscosity can be explained by the formation and rupture of chains formed by the large particles in the suspension in the presence of a magnetic field. We compare here the change of viscosity for the original fluid with both fractions for two shear rates as a function of magnetic field strength.

As expected, the magnetoviscous effect increases in all three fluids with decreasing shear rate. Furthermore the influence of the concentration of bigger particles is directly visible. The fluid with a large concentration of bigger particles has a much stronger magnetoviscous effect than the original fluid, and *vice versa* in the upper fraction with a small concentration of bigger particles.

Viscoelastic effects. The formation of chainlike clusters gives rise to the assumption, that viscoelastic effects should occur in ferrofluids under the influence of magnetic fields. A first evidence for such effects was provided from *Odenbach et al.* [18] with the investigation of the Weissenberg-Effect in ferrofluids exposed to magnetic fields. Using rheometrical investigations, the application of an oscillating load gives direct access to information about viscoelastic properties. The advantage of the oscillating mode is, that measurements with small amplitudes does not change the "zero-shear-rate" structure of the fluid.

As an example we'll discuss here the phase shift between shear and stress, being a characteristic measure for the appearance of viscoelasticity. For an elastic body this phase shift should be zero, while it is $\pi/2$ for a Newtonian liquid. Correspondingly viscoelastic liquids exhibit phase shifts between 0 and $\pi/2$.

It was shown in Fig. 9, that the upper fraction shows a linear relation between stress and shear rate, indicating newtonian behavior. This is confirmed by the phase shift $\delta = \pi/2$, being constant over the whole range of oscillation frequencies ω as shown in Fig. 11. In contrast the lower fraction with many bigger particles should have a phase shift between 0 degree and $\pi/2$, dependent on the length of the chains formed by the particles. This length itself depends on shear rate as well as on the strength of the applied magnetic field.

The lower fraction shows just under zero field no phase shift- i.e. it behaves Newtonian for vanishing field. If a field is applied, a phase shift between 0 and $\pi/2$ is observed. This phase shift decreases continuously with decreasing shear rate and thus with increasing length of the particle chains. The phase shift of the original fluid takes values in between the phase shifts of the upper and lower fraction. The higher number of bigger particles in the lower fraction obviously enhances the appearance of viscoelasticity, and a reduction of the number of bigger particles in the upper fraction finally leads to a pure Newtonian system.

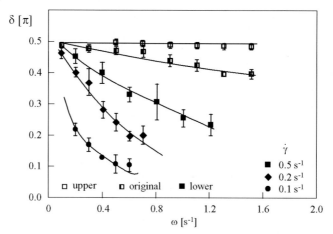

Fig. 11. The phase shift between shear and stress for the original fluid and both fractions obtained by separation as a function of the frequency of the oscillating load for different shear rates. To obtain the given data points, measured values in a field range between 3 kA/m and 17 kA/m have been averaged. The error bars give the standard deviation of this averaging process.

Acknowledgements

Support of this work by DFG (Grant.No. Od18/2 and Od18/3) is gratefully acknowledged.

References

1. J. P. McTague: *Magnetoviscosity of magnetic colloids.* J. Chem. Phys., vol.51, no.1 (1969)
2. W.F. Hall, S. N. Busenberg: *Viscosity of magnetic suspension.* J. Chem. Phys., vol.51, no.1 (1969)
3. M. I. Shliomis: *Effective viscosity of magnetic suspensioms.* Soviet Phys. JETP, vol.34, no.6 (1972)
4. P. Fannin: *Magnetic spectroscopy as an aide in understanding magnetic fluids.* Springer LNP this issue (Berlin, New York 2002)
5. R.E. Rosensweig, R. Kaiser, G. Miskolczy: *Viscosity of magnetic fluid in a magnetic field.* J. Colloid Interface Sci., vol.29, no.4 (1969)
6. O. Ambacher, S. Odenbach, K. Stierstadt: *Rotational viscosity in ferrofluids.* Z. Phys. B-Condensed matter, vol.86 (1992)
7. S. Odenbach, T. Rylewicz, M. Heyen: *A rheometer dedicated for the investigation of viscoelastic effects in commercial magnetic fluids.* J. Magn. and Magn. Mat., vol. 201 (1999)
8. S. Odenbach, H. Störk: *Shear dependence of field-induced contributions to the viscosity of magnetic fluids at low shear rates.* J. Magn. and Magn. Mat., vol.183 (1998)
9. S. Odenbach, K. Raj: *The influence of large particles and agglomerates on the magnetoviscous effect in ferrofluids.* Magnetohydrodynamics, vol.36, no.4 (2000)

10. S. Odenbach: *Magnetoviscous effects in ferrofluids.* Springer LNP m71 (Berlin, New York 2002)
11. A. Yu.Zubarev, S. Odenbach, J. Fleischer: *To the theory of dynamical properties of polydisperse magnetic fluids. II. Effect of chain-like aggregates.* submitted to Physical Review E (2001)
12. I. Prigogine: *Thermodynamics of irreversible processes.* (3. edition, John Wiley and Sons, New York 1967)
13. S.R. de Groot, P. Mazur: *Grundlagen der Thermodynamik irreversibler Prozesse.* BI 162/162 a (Berlin 1962)
14. E. Blums, Yu.A. Mikhailov, R. Ozols: *Heat and Mass transfer in MHD flows.* Series in theoretical and applied mechanics, 3. volume, ed. by R.K.T. Hsieh (World Scientific Publishing, Singapore 1987)
15. S. Odenbach: *Konvektion durch Diffusion in Ferrofluiden.* PhD Thesis, Ludwig-Maximilians University, Munich (1993)
16. R. Gerber, M. Takayasu, F.J. Friedlaender: *Generalization of HGMS theory: the capture of ultra-fine particles.* IEEE Trans. Magn., vol.mag-19, no.5 (1983)
17. R.E. Rosensweig, R. E. *Ferrohydrodynamics.* Cambridge University Press (Cambridge, New York 1985)
18. S. Odenbach, T. Rylewicz , H. Rath: *Investigation of the Weissenberg effect in suspensions of magnetic nanoparticles.* Physics of Fluids 11 (1999) 2901

Magnetorheology: Fluids, Structures and Rheology

G. Bossis[1], O. Volkova[1], S. Lacis[2], and A. Meunier[1]

[1] LPMC, UMR 6622, University of Nice, Parc Valrose, 06108 Nice Cedex 2, France
[2] Department of Physics, University of Latvia, Zellu str.8, LV-1586, Riga, Latvia

Abstract. Magnetorheological suspensions are complex fluids which show a transition from a liquid behavior to a solid one upon application of a magnetic field. This transition is due to the the attractive dipolar forces between the particles which have been magnetized by the applied field. The formation of a network of particles or aggregates throughout the suspension is the basic phenomena which is responsible for the strength of the solid phase. In this paper we shall give an overview on the fluids and their properties and we shall especially emphasize the interplay between magnetic forces which are responsible for the gelling of the suspension and on the other hand of hydrodynamic and thermal forces which contribute to break this gel and allow the suspension to flow. The combination of these three forces gives rise to a very rich rheology whose many aspects are still not understood.

1 Introduction

Since Rabinow and Winslow's [1,2] discoveries in the 1940's, magnetorheology (MR) and electrorheology (ER) have emerged as a multidisciplinary field whose importance has considerably increased these last ten years. The rheology of these fluids is very attractive since it can be monitored by the application of a field, either magnetic or electric. The most important advantages of these fluids over conventional mechanical interfaces is their ability to achieve a wide range of viscosity (several orders of magnitude) in a fraction of millisecond. This provides an efficient way to control force or torque transmission and many applications dealing with actuation, damping, robotics have been patented and are coming on the marketplace [3]. The basic phenomena in electro or magnetorheology is the ability to control the structure of a biphasic fluid. The two phases are usually made on one hand of solid particles in the micrometer range and on the other hand of a carrier fluid. The application of a field polarizes the particles and induces their transient aggregation, hence an increase of the viscosity. In some other cases the two phases can be two immiscible fluids and the application of the field will change the shape and size of the droplets of the dispersed phase which will modify the rheology [4,5]. Some intermediate cases can exist where the carrier phase is itself a suspension of nanoparticles, a typical case being a ferrofluid. Furthermore new materials where the two phases are solids, like solids particles embedded in a rubber matrix belong to the same category of smart composites whose rheology can be controlled by an external field [6–8].

All these materials have common features regarding the relation between the shape and size of the domains of the dispersed phase and their rheology, even if the microscopic interaction between the constituents of the dispersed phase can be very different, depending on their size (polymer, nanoparticle, micronic particle) and other physical characteristics like conductivity, permittivity, magnetic permeability, properties of the interface and so on. In this paper devoted to MR fluids we shall of course focus on magnetic interactions but many devolpments are also relevant to ER fluids who had received more attention in the past years with several review papers [9–14]. The interest in MR fluids has raised more recently and review papers are scarce in this domain [15–17]. Our goal is to give a brief overview of research on MR fluids and their rheology and also to give some guide lines to beginners in this domain.

In section 2 we shall describe the main phenomena, give some orders of magnitude and present the different kinds of fluids which are used with their advantages and disadvantages. Section 3 contains a presentation of the field induced structures. Section 4 is centered on the rheology of magnetic fluids, with a description of the apparatus and methods (section 4.1) a comparison of experiments and models in relation to the yield stress (section 4.2) and to the shear rate dependency (section 4.3).

2 Overview of MR Suspensions

2.1 General Description of the Interactions in an MR Suspension

If an isolated particle of relative magnetic permeability μ_p surrounded by a fluid of relative permeability μ_f is placed in an external magnetic field H_0, this particle will acquire a magnetic moment: $\boldsymbol{m} = 4\pi\mu_0\mu_f \beta a^3 \boldsymbol{H_0}$; a is the radius of the particle, μ_0 the permeability of vacuum and $\beta = (\mu_p - \mu_f)/(\mu_p + 2\mu_f)$. It is worth noting that this formula also holds when the permeability of the carrier fluid is larger than the one of the particle; in this case $\beta < 0$ and the magnetization vector is opposed to the field. The interaction energy between two dipoles of moment m is given by:

$$W = \frac{1}{4\pi\mu_0\mu_f} \left(\frac{\boldsymbol{m}_\alpha \boldsymbol{m}_\beta}{r^3} - \frac{3(\boldsymbol{m}_\alpha \boldsymbol{r})(\boldsymbol{m}_\beta \boldsymbol{r})}{r^5} \right) \quad (1)$$

where $\boldsymbol{r}$ is the separation vector between the centers of the two particles. This energy is minimum when the two dipoles are aligned with r and maximum when they are perpendicular leading to a preferential aggregation as chains of particles aligned on the direction of the field. The formation of aggregates of particles will depend on the ratio of this interaction energy to kT. Taking as reference the energy of two dipoles in repulsive configuration gives:

$$\lambda = \frac{1}{4\pi\mu_0\mu_f} \frac{m^2}{r^3} \frac{1}{kT} = \frac{\pi\mu_0\mu_f \beta^2 a^3 \boldsymbol{H_0}^2}{2kT} \quad (2)$$

For particles of diameter $1\mu m$ with large permeability ($\beta \approx 1$) and $T = 300K$, we obtain from (2) $\lambda = 1$ for $H = 127$ A/m (or $H = 1.6$ Oersted in cgs units). It means that for usual magnetic fields the magnetic forces dominate the Brownian forces. The situation is quite different if we consider ferrofluids which are magnetic fluids with particle diameters of about 100 Angstrom; then in the same conditions we have $\lambda = 1$ for $H = 1600$ Oersted. It means that for usual fields the magnetic forces will always be dominated by Brownian forces and we cannot expect to change significantly the viscosity of a ferrofluid by applying a magnetic field. Furthermore this criterion on λ has to be reinforced by the fact that the efficiency of thermal forces to break a chain of spheres increases with the length of the chain. Despite these restrictions which seem to prevent the finding of a noticeable ER or MR effect with suspensions of colloidal particles there are some experimental evidences for large changes of viscosity in some ferrofluids [18]; we shall discuss this point in more details in the next section. The quantity λ is the key quantity which, together with the volume fraction, $\Phi = Nv_p/V$, will determine the equilibrium structure of a suspension of monodisperse particles as a function of the applied field. We shall come back to this important point in section 3.

In order to obtain all the quantities which will rule the behavior of the suspension it is useful to start from the equation of motion of one particle and to put it in a dimensionless form. This is what we would do to calculate the trajectories of the particles by using Brownian or Stokesian dynamics [19–21]. For a given particle we can write:

$$m\frac{d\boldsymbol{v}}{dt} = \boldsymbol{F}^H + \boldsymbol{F}^{\text{ext}} + \boldsymbol{F}^I + \boldsymbol{F}^B \tag{3}$$

The first term $\boldsymbol{F}^H$ is the hydrodynamic force on the test particle coming from the hydrodynamic friction and is proportional to $-\xi(\boldsymbol{v} - \boldsymbol{v}^0(x))$, where $\xi = 6\pi\mu a$ with μ the viscosity of the suspending fluid and $\boldsymbol{v}^0(x)$, the imposed velocity field at the location x of the particle. The term $\boldsymbol{F}^{\text{ext}}$ is the hydrodynamic force due to the symmetric part of the velocity gradient tensor. In the case of a pure shear characterized by the shear rate $\dot\gamma$ this force scales as $6\pi\mu\dot\gamma a^2$. The third force $\boldsymbol{F}^I$ is the interparticle force coming from the dipole-dipole interaction and given by minus the gradient of (1). For two particles α and β the force on α will be:

$$\boldsymbol{F}^I_\alpha = 12\pi\mu_0\mu_f a^2 \beta^2 H_0^2 \left(\frac{a}{r}\right)^4 \left[(2\cos^2\theta_{\alpha\beta} - \sin^2\theta_{\alpha\beta})\boldsymbol{e}_r + \sin 2\theta_{\alpha\beta}\boldsymbol{e}_\theta\right] \tag{4}$$

The meaning of the different vectors is defined in Fig.1. For two spheres placed side by side ($r = 2a$, $\theta_{\alpha\beta} = 90°$) we have $F^I_\alpha = f_d = -(3/4)\pi\mu_0\mu_f a^2\beta^2 H_0^2$ which is repulsive as expected; we shall take $-f_d$ as the scaling factor of the interparticle force. The last term in (3) is the Brownian random force which scales as kT/a. In general we can neglect the inertial force (the time needed to reach a constant velocity: $\tau = m/\xi$ is $1.7\mu s$ for an iron particle of radius 1 micron in water; it is much smaller than the other characteristic times) so we can neglect the left hand side of (3). Dividing all the terms of (3) by $6\pi\mu\dot\gamma a^2$ and rearranging we obtain:

$$\frac{(\boldsymbol{v} - \boldsymbol{v}^0)}{\dot\gamma a} = \frac{[\boldsymbol{F}^I]}{Mn} + \frac{[\boldsymbol{F}^B]}{Pe} + [\boldsymbol{F}^{\text{ext}}] \tag{5}$$

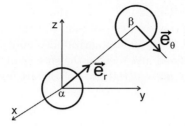

Fig. 1. Two particles in a magnetic field

The brackets mean that the forces have been divided by their scaling factor. Mn is the Mason number:

$$Mn = -\frac{6\pi\mu\dot{\gamma}a^2}{f_d} = \frac{8\mu\dot{\gamma}}{\mu_0\mu_f\beta^2 H_0^2} \tag{6}$$

It expresses the ratio of shear to magnetic forces. Note that in other papers the definition can differ by a multiplicative factor. $Pe = 6\pi\mu\dot{\gamma}a^3/kT$ is the Peclet number and expresses the ratio of shear forces to Brownian forces. For particles larger than one micron and reasonable shear rates, the Peclet number is large (for instance $Pe=4.5\cdot 10^6$ for a particle of radius one micron in water with $\dot{\gamma} = 1s^{-1}$) and the Brownian force can be neglected; nevertheless we have to keep in mind that the influence of the Brownian motion can still be important even at quite high Peclet number.

The dimensionless equation of motion depends on two quantities (actually the three quantities defined above are related by $Mn\lambda = 2\ Pe/3$) so, for a given suspension, all the trajectories and hence all the properties-and in particular the viscosity - will be the same for the same values of Mn and λ. Of course this equivalence only applies for systems of particles starting in the same initial conditions, that is to say with the same volume fraction Φ and the same initial configurations. This last point is usually not critical if we are only interested by equilibrium properties, so we can say that for monosized hard spheres with particles having the same magnetic permeability, the viscosity (normalized by the one of the carrier fluid) will depend only on three quantities which are Φ, Mn, λ.

The efficiency of an MR fluid is firstly judged through its yield stress, τ_y, which measures the strength of the structure formed by the application of the field. The restoring force per unit surface, which resists to the deformation of the structure, is given by the derivative of the magnetic energy per unit volume relatively to the strain $\gamma : \tau = -dW/d\gamma$. The yield stress represents the maximum of the stress versus strain: $\tau_y = \max(\tau)$ since above a critical shear strain, γ_c, the gel-like structure will break. This definition of the yield stress through the energy of the field allows to understand why it is easier to get a larger yield stress with MR fluids than with ER fluids : the vacuum magnetostatic energy density: $\mu_o H^2$ for $H = 3000$ Oe is an order of magnitude larger than the electrostatic energy density $\epsilon_o E^2$ for a field $E = 3$ kV/mm (close to the breakdown field).

It does not mean that ER fluids will never show yield stress as high as in MR fluids, because it also depends on the efficiency of the polarization mechanisms. For instance ER fluids having ionic polarizability may show high yield stress if the ions can remain confined inside or on the surface of the particles. This is an other issue, let us simply present the state of the art for MR fluids.

2.2 MR Fluids

Strong fluids. A simple and quite efficient MR fluid is obtained by dispersing iron powder in oil with the use of some surfactant - for instance stearic acid - in order to prevent irreversible aggregation. Magnetorheological suspensions made of iron synthesized from iron carbonyl precursors and known as carbonyl iron, have been extensively used, first by Rabinow [1], also by Shulman et al [22] and later on by many other teams. These particles are used for producing high permeability circuits, they are quite spherical and can be found rather easily in the range 2-10μm. The magnetization saturation of iron is large ($\mu_0 M_s = 2.1$ Tesla) and its remanent magnetization very low. The efficiency of this fluid is measured through its yield stress which is, for a volume fraction of 30%, typically 50 kPa for a field $H = 4000$ Oersted [23]. An other study report a yield stress of 100kPa for an induction of 1 Tesla and a volume fraction $\Phi = 0.36$ [24]. The magnetic induction B is given by:

$$\boldsymbol{B} = \mu_0(\boldsymbol{M} + \boldsymbol{H}) \quad \text{with} \quad \boldsymbol{M} = \chi \boldsymbol{H} \tag{7}$$

The magnetic induction can be measured with the help of a sensing coil and of a fluxmeter and is safer to use for comparisons between different fluids because it is a rather well conserved quantity inside a magnetic circuit. On the contrary the magnetic field H inside the fluid is not directly measurable; it can be deduced from the measurement of the field with a Hall gauge placed in an air gap close to the fluid; we shall come back to this point in section 4.1. This problem as well as the presence of additives which can modify the state of aggregation or impose a non zero gap between particles, also the lack of knowledge of the polydispersity, are factors which make difficult the comparison of the results between different authors. Density of iron is large (ϱ=7.87 g/cm^3) and a particle of radius one micron in water has a sedimentation velocity of 5.4cm per hour. In order to prevent the sedimentation and potential redispersion difficulties, it is usual to add a gel forming additive. The gel must have a low yield stress in order to flow easily under a small agitation. Nanometer silica particles can be used in this aim because they easily form a physical gel at low volume fraction (2-3%) and the yield stress in the absence of field is only a few Pascal. An other problem-even more important than sedimentation- is the corrosion of iron particles. It has been found [25] that under permanent use in a damper valve and in the presence of a magnetic field, the fluid thickens progressively and can show an increase of the off state (zero field) viscosity by a factor of 3 after 600000 cycles. This thickening has been identified as due to the removal of thin oxide fragments from the surface of the particles. The operating conditions in a damper (high shear rate: 10^4 to

10^5 s^{-1}, strong attractive forces between the particles in the presence of the field, possible corrosion due to dissolved oxygen) favor the fragmentation process and must be prevented by a proper surface protection of the particles. The new fluid made by Lord Corporation no longer show any appreciable thickening even after 2 million cycles. Progress in the strength of the fluids has been realized by using materials with a higher saturation magnetization like for instance cobalt-iron alloys ($\mu_0 M_s$ = 2.45 Tesla for about 50%Fe, 50%Co) which shows a higher yield stress than iron: 70kPa instead of 50kPa for H=4000 Oersted [23]. Commercial fluids available from Lord Corporation have typically yield stresses between 50 and 70kPa for a volume fraction between 0.35 and 0.4 and a field of 300kA/m (3770 Oersted). The off state viscosity is often higher than 1Pas at shear rate of $50s^{-1}$. The composition of these fluids should still progress in order to have a lower off state viscosity and still keeping a good protection against corrosion and a low sedimentation rate. At last it is worth mentioning that as proved by R.T.Foister [26] mixing two different sizes of carbon iron particles improves considerably - for the same volume fraction - the on/off ratio of stresses (they report an improvement of 2.7 times over the monosized suspension at $\Phi = 0.55$ and $B = 1$Tesla). This increase is quite understandable since it is known that the maximum packing fraction is increased by mixing two sizes such that the small spheres can be placed in the holes between the large spheres; consequently the viscosity diverges at a higher volume fraction, which means that, at the same volume fraction, the off state viscosity is lower than the one of a monosized suspension. More surprisingly they also observe - for the same volume fraction - an increase of the yield stress in the on state of about 25% for the bidisperse suspension which has no clear explanation.

Colloidal MR fluids. If iron based particles with radii larger than 1μm, give fluids with high strength, they have the inconvenience to be abrasive. Since the shear force, which will push the particles against the walls or against each other, is proportional to the square of the size of the particles, it would be advantageous to reduce this size, which would reduce both abrasion, sedimentation and fragmentation. On the other hand a too low value of the parameter λ would not authorize the formation of clusters, so there is a compromise to find. For a particle of diameter 0.1μm we need a field of about 50 Oersted in order to have a value of unity for λ. It seems difficult to go well below this limit but nevertheless Kormann et al [18] have shown that a suspension of soft ferrites with an average diameter of about 30nm and a volume fraction of 23% can give a dynamic yield stress of 2kPa for an induction of 0.2Tesla. If we suppose a permeability of about 2 and so $H = 1000$ Oersted we obtain $\lambda = 10$ which seems to be too low to explain this quite large effect on viscosity. It is quite likely that the nanometric particles are partially aggregated even in zero field; it happens quite naturally since the particles are monodomain and carry a permanent dipole. The average dipole of these clusters is zero in the absence of a field but the persistence length of the correlations between the orientations of the dipoles of adjacent particles can be much larger than the particle diameter and explain the large change of viscosity with the field (for magnetite the attractive dipolar energy between two particles at contact overcomes the thermal energy for a diameter larger than 10nm [27,28]). When the field is turned on, the dipoles align on the field, the cluster becomes magnetized and, thanks to its larger dimension, will attract other clusters and form the gel like structure. The zero field viscosity of this fluid is quite high (about 50Pas at $0.1s^{-1}$ [29]) which supports the hypothesis of a zero field aggregation. In any event such a fluid does not sediment and is not abrasive so it could be useful if not too high yield stress is needed. An other attempt to use nanoparticles has been presented recently with carbonyl iron [30]. The particles were obtained from decomposition of vapors of iron pentacarbonyl with an average diameter of 26nm. For a volume fraction of about 16% it gives an average increase of yield stress of 7kPA. The experiment was conducted with a MR damper and the value of the field is not given so it is not possible to compare with the ferrite particles but, here too, we see that this fluid can be interesting for applications. It is worth noting that this nanosized fluid presents a quite large yield stress at zero field (3.6kPA) indicating a permanent aggregation. This is not surprising if we suppose that these particles are still monodomain, then, due to their larger magnetization saturation compared to ferrite (by a factor of 4) the dipolar interactions will lead to this aggregation.

Concerning the use of colloidal particles it seems more interesting to use larger particles (for instance $0.1 \mu m$) in order to avoid the presence of a single ferromagnetic domain but still keeping a large enough value of λ.

Other types of MR composites. For most of the applications the fluids described above will be best adapted. Nevertheless numerous other fluids can be used with special goals. We shall describe some of them without pretending to be exhaustive.

An interesting fluid with regard to the influence of the Brownian forces on the rheology is the one made of a suspension of monodisperse non magnetic colloidal particles in a ferrofluid. As it can be seen from (1), the interparticle force is proportional to the square of the dipole. If the particle is not magnetic ($\mu_p=1$) but the suspending fluid is magnetic: $\mu_f >1$, then the induced moment is opposite to the field (β <0) but the interparticle force (4) is still attractive. The existence of a magnetorheological effect in this kind of system was demonstrated [31,32], and using different sizes of silica particles it was shown that the yield stress decreases when the size of the particles becomes smaller [33], the rheology of these inverse ferrofluids has also been recently investigated [34,35].

Magnetic particles are usually conductive particles and so not suitable for electrorheology, nevertheless by coating the magnetic particles with an isolating layer, it becomes possible to apply both an electric field and a magnetic field. Surprisingly the resulting yield stress $\tau_y^{H,E}$ was found larger than the sum of each separated yield stress: $\tau_y^{H,E} > \tau_y^H + \tau_y^E$ [36,37]. This synergistic effect is explained in the following way [38]: in ER fluids cancellation of the field on the electrodes introduces dipole images of the particles with respect to the plane of the electrodes; the attractive interaction between original dipoles and their images drift the chains of particles towards the electrodes and in a shear the weak point of the structure is in the middle of the cell. On the contrary for magnetic particles in the presence of a magnetic field and non magnetic electrodes the chain do not attach on the electrodes and the density of particles is larger at the center of the cell, then in the presence of the shear the fracture zone is located on the electrodes. In the presence of both electric and magnetic field the chain like structure will be more homogeneous throughout the cell without weak points, hence the synergistic effect. Nevertheless the use of a non magnetic layer decreases the magnetic force between particles, so it is not clear at all that the synergistic effect will compensate the decrease of magnetic stress. In [38] the results are presented for carbonyl iron coated with titanium oxide and a weight fraction of 40%. The thickness of the titanium layer is unknown but the volume fraction of iron is in any event larger than 7%. The magnetic stress is found to be 400 Pa for $H = 2000$ Oe and close to 1 kPa by adding an electric field of 1 kV/mm. By comparison for the same field, and without coating, the yield stress is 5kPa for a volume fraction of 15% [39]. As the proportionality between the yield stress and the volume fraction is quite well verified in this range of volume fraction [35], we should expect a yield stress of at least 2.5 kPa for the magnetic stress alone. It appears that, at least for this fluid, even the synergistic effect does not compensate the loss of magnetic stress due to the presence of the titanium oxide layer; progresses can likely be made by decreasing the thickness of the insulating layer.

The most promising type of MR fluid is simply obtained by using a ferrofluid as carrier liquid together with carbonyl iron particles. An increase in yield stress by a factor of four was reported in this situation [40]. The reason for this enhancement is easy to understand since the force between two particles, when mediated by a fluid of relative permeability μ_f, is multiplied by this permeability (cf. Eq. (4)). Also, due to local alignment and network formation of the nanodipoles between the ferromagnetic particles, there is no sedimentation. The remaining problem for industrial use could be chemical stability of the ferrofluid.

Besides MR fluids, new types of magnetic composites are appearing which are the solid counterparts of ferrofluid and magnetorheological fluids, namely magnetic gels and magnetic elastomers. Magnetic gels are chemically cross-linked polymers network swollen by a ferrofluid [41]. Placed in a gradient of magnetic field they will deform and change of shape; this ability to elongate and contract makes them looking like artificial muscle [42]. Nevertheless, if they can develop large change of shape, it is because their zero field elastic modulus is weak; in practice the energy which is developed is quite low. The second class: magnetic elastomers, are made by dispersing iron particles in a polymer and applying a magnetic field prior cross linking. T. Shiga et al [43] reported an increase in shear modulus of about 10kPa for a 28% volume fraction of iron particles (diameter about 100μm) at a strain $\gamma = 0.1$ and H=59kA/m. At lower strain and with carbonyl iron particles of average diameter 3-4μm, Jolly et al [6] obtained a much higher field induced shear modulus: 0.56 MPa for $\Phi = 0.3$, $\gamma = 0.01$, $\nu = 2Hz$ and flux density corresponding to saturation (0.8 Tesla). The investigation of the same kind of material under traction instead of shear gives comparatively a still higher increase in Young modulus (0.6MPa for $\gamma = 0.05$, $\Phi = 0.25$ and $H = 123$kA/m) [8]. The applications of these elastomers for semiactive damping of vibrations have been patented by several companies and their performance can still be improved by a better control of the column formation before curing. This remark introduces the next section which deals with the prediction of the structures formed by the magnetic particles in the presence of a magnetic field.

3 Structuration of a MR Fluid by a Magnetic Field

When a magnetorhological fluid is submitted to a magnetic field the particles will aggregate and form different kind of structures depending on the initial volume fraction, and on the parameter λ (Eq. (2)). Actually it is only true if we reach the equilibrium state and, as the particles are at the limit of the colloidal range, the way to increase the field in order to reach equilibrium is important. Increasing the field too quickly will give a kind of labyrinthine structure, whereas increasing the field slowly gives well separated columns [44,45]; also large volume fractions (typically above 5%) always tend to form labyrinthine structures [46]. Increasing the field slowly allows to observe a complex behavior of the transmitted light, with first a decrease of the transmission followed by an increase which ends up with a higher transmission than in the initial stage at $\lambda = 0$ [47]. Exactly the same behavior was found in ER suspensions [48,49]. The transmitted light first decreases because chains of particles, like cylinders, scatter the light more efficiently than individual particles [49]; then at higher fields the chains gather into thick fibers and the space between fibers form channels of clear liquid which transmit the light without attenuation, hence the large increase of transmitted light. The interpretation of the transmitted light versus λ in terms of critical fields defining different phases is rather ambiguous. What is not controversial is that the phase separation begins with the formation of individual chains and ends up with domains of particles which look like fibers and are well separated at low volume fraction. The problem of structure equilibrium being quite crucial, it seems that a good method is not only to rise the field slowly but also to interrupt it during a short time corresponding to the time needed for a particle to diffuse on one radius. Not only this procedure allows to better reach the equilibrium state but also it allows to precisely determine the critical values of λ between the isotropic and nematic phase and between the nematic phase and the columnar phase [49]. Instead of measuring only the transmitted light, it can be helpful to measure also the scattered light as a function of the angle of observation. In order to have access to the chain length the best geometry is the one where the light beam is perpendicular to the chains; then the light scattered in the plane perpendicular to the chains is proportional to the square of the length of the chains [50,51]. Nevertheless the information on chain length is indirect, and, for magnetic suspensions, it is restricted to low volume fraction due to the strong absorption of magnetic suspensions and the impossibility to realize refractive index matching. The prediction of the average chain length as a function of λ

has been derived by A. Cebers [52] from the calculation of the free energy of a perfect gas of chains of spheres. Analogous results have been obtained more recently with the same model but using a simpler derivation which can be easily extended to interacting chains. The result for the average number of particles per chain is [53]:

$$<n> = y \frac{\sqrt{1+4y}-1}{1+2y-\sqrt{1+4y}} \quad \text{with} \quad y = \Phi e^{2\lambda} \tag{8}$$

A good agreement was found between the predictions of such model and experiments on an ER fluid made of monodisperse silica particles [48]. On the other hand, if we do not look for the equilibrium distribution of chains but rather for their kinetic of formation it is possible to describe reasonably well the first stage of the chaining process with the help of the Smoluchowski equation [54]. In a monolayer the aggregation process can be followed with a microscope and the image analysis shows that the average length follows a power law:$< n(t) > = 1 + kt^\nu$; an exponent $\nu = 0.6$ was found for $\lambda = 31$ and $\lambda = 0.009$ [54] but a range of exponents ($0.37 < \nu < 0.6$) was also obtained depending on the experimental conditions [45]. Actually the Smoluchowski equation predicts an exponent $\nu = 0.5$ for a constant cross section of collision and a mobility of the aggregate decreasing as the invert of its length [55] (which is not exactly verified since the mobility of a slender body of length L, along its main axis, decreases as $(\ln(L/a) - 0.5)/2\pi\eta L$ [56]).

The second step of the aggregation process is the coarsening of chains into columns which is still not well understood. We just want to point out a few results to emphasize the nature of the problems raised by this coarsening process either in ER or MR fluids. First the field of a periodic chain of dipoles of infinite length dies off exponentially as $\cos(\pi z/a)\exp(-\pi\varrho/a)$ where ϱ is the distance in the perpendicular x-y plane and 2a the period along z. So, except at very short distances, two infinite chains will not feel each other. Now let us see the situation for two rigid dipolar chains of finite length. If we consider two parallel rigid rods (Fig.2a) separated by a distance d and carrying a dipolar density: m'=m/2a, the interaction energy between the two rods is:

$$W = \frac{1}{4\pi\mu_0\mu_f} \frac{m^2}{4a^2} \left(\frac{1}{\sqrt{d^2+z_{Aa}^2}} + \frac{1}{\sqrt{d^2+z_{Bb}^2}} - \frac{1}{\sqrt{d^2+z_{Ba}^2}} - \frac{1}{\sqrt{d^2+z_{Ab}^2}} \right) \tag{9}$$

The energy of interaction is the one of equivalent charges (plus and minus) located at the end of the rods. If we do not consider very large shifts between chains, this energy corresponds to a repulsive force, but, except for the very special case of two chains of the same length and at the same height, there is also a torque which will rotate the rods. As the chains are close to each other they will touch during the rotation well before shifting enough to connect end by end. After two particles of each chain have joined the coalescence occurs by a "zippering motion" [45] because of the short ranged attractive interaction between two chains shifted by a radius. Taking L as the order of magnitude of z_{Ba} in Eq. (9) and normalizing the torque K by kT we have $K/kT \approx \lambda a d/L^2$.

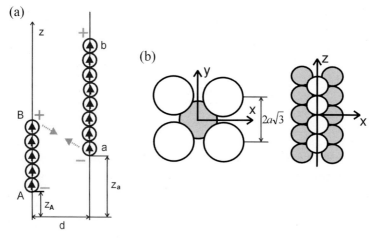

Fig. 2. Chains of spheres aligned by the field (a): Shifted chains at a distance d (b): Body Cubic tetragonal structure

Besides this simple mechanism of aggregation between chains, Halsey et al [57] have examined the interaction between chains due to the fluctuations of positions of the individual particles. The idea behind is that enhancement of fluctuations in one dimensional structures should give rise to a fluctuating field whose root mean square is slowly decreasing with the distance (as $1/d^2$). The resulting interaction energy coming from dipole induced dipole mechanisms similar to Van der Waals energy is attractive with an order of magnitude: $W_f/kT \approx (L/a)(a/d)^5$ and does not depend on λ. If we compare this energy with torque we see that, at large distances, where these theories are valid, (for instance d=10a) and moderate field ($\lambda = 10$) the energy of the fluctuating field overcomes the torque only for $L/a > 200$. At low field both mechanisms are present in the coarsening of the suspension but at high fields the torque induced aggregation is likely to be the dominant one. Actually an other clue that other mechanisms than thermal fluctuations can explain the coarsening process is that numerical simulations carried out without any Brownian motion also show this coarsening [58,59]. The time evolution of the thickening of the columns can be directly measured by light scattering in a configuration where the beam (Oz) is perpendicular to the field (Ox). In this case the formation of chains or fibers gives a strong scattering in the direction perpendicular to the field (Oy). The light scattered along Oy presents a maximum for a scattering wave vector q_{max} and $L = 2\pi q_{max}^{-1}$ gives the characteristic distance between aggregates and so indirectly their size. The growth of L in time was found to be $L(t) = L(0)(1+(t/\tau)^\alpha)$ with $\alpha = 0.42$ for a 11 wt. % suspension of $0.7\mu m$ silica spheres in electric field [51]. In magnetic colloidal suspensions most of the experiments are realized with a thin cell made of two microscope slides separated by a spacer and a field perpendicular to the slides [44,46]. The important difference relatively to electric field is that, with a magnetic field, the boundaries conditions do not require to use dipole images in order to cancel the field on the electrodes; then a chain filling the gap is a finite chain instead of an infinite one as in the case of electric field. We have seen in Eq. (9) that such finite chains repel each other, so looking from above we could expect to see the cross sections of individual chains. Actually there will be a compromise between the coarsening coming from phase separation and this long range repulsive energy which will prevent the fibers to join each other. It is worth noting that in ER suspensions the equilibrium state for high values of λ does not consist of individual fibers but of a unique domain with a crystal like structure where the particles are placed on a Body Cubic Tetragonal (BCT) lattice. In this lattice one chain of particles is surrounded by four chains shifted by a radius (Fig. 2b). Such a structure has been proved to have the lowest energy [60] and has been observed both by laser beam diffraction [61] end recently by confocal microscopy [62]. In practice, due to transport limitation, the size of the domains will remain small. Coming back to the case of magnetic suspensions, the individual fibers of particles can be easily seen with a microscope. As their shape can be quite irregular - especially at volume fraction larger than 5% - and also because their apparent size can depend on image thresholding - it is the average distance between fibers rather than their diameter which is measured by microscope observation [46,63]. The other way is to use the light scattered by the fibers with the beam (Oz) parallel to the field. It will give rings in the x-y plane and the position of the first ring will give the average distance between fibers [64]. As the field induced by equivalent end charges depends on the length of the aggregates - which is equal to the thickness of the cell at high λ -, then we expect that the distance, d, between domains will also depends on the thickness of the cell h. It is quite usual to fit the experimental results with a power law: $d = Ch^v$ but it appears that the power can vary from 0.37 [64] to 0.67 [63]. The spreading of the results is mainly due to the fact that there is no power law between d and h as demonstrated by a model taking into account the repulsive interaction between the fibers and the surface energy. This model assumes that there are no particles between the domains, also for the sake of simplicity only two kinds of domains are considered: ellipsoidal aggregates or stripes of particles; the first case can approximate the domains created in a unidirectional field, whereas the second one is appropriated for structures created in a rotating field or in a shear flow. If we call M_a the total dipole of an aggregate, we have to distinguish the dipoles which are on the surface of the aggregate from those which are in the bulk because the Lorentz field (coming from the surface of a virtual cavity around a given particle) will be different on the surface of the aggregate: indeed on the surface half of the surrounding particles are missing and so we can divide the Lorentz field, H_L, by a factor of 2. We shall have:

$$\mathbf{M_a} = N_B \mathbf{m_B} + N_S \mathbf{m_S} \qquad (10)$$
$$\text{with} \quad \mathbf{m_B} = \alpha(\mathbf{H_0} + \mathbf{H_D} + \mathbf{H_L})$$
$$\text{and} \quad \mathbf{m_s} = \alpha(\mathbf{H_0} + \mathbf{H_D} + \mathbf{H_L}/2)$$

The polarisability is $\alpha = 4\pi\mu_0\beta a^3$ (we suppose in the following that $\mu_f = 1$) N_B, m_B, N_S, m_S are respectively the numbers of particles and the magnetic moments of a given particle in the bulk of the aggregate and on its surface. H_0 is the external applied field, $H_D = -N_D I_a/\mu_0$ the field coming from the equivalent charges on the surfaces of the aggregates, $H_L = I_a/3\mu_0$ is the Lorentz field, and $I_a = M_a/V_a$ the magnetisation inside the domain of volume V_a. The demagnetisation factor, N_D, depends on the geometry of the domains [65]. The total magnetic energy per unit volume of the suspension is:

$$U_m/V = -1/2 N_a M_a H_0/V = -1/2 \varphi I_a H_0 \qquad (11)$$

The apparent volume fraction of aggregates $\varphi = N_a V_a/V$ should not be confused with Φ_a which is the volume fraction inside the aggregate, neither with the average volume fraction Φ. These three volume fractions are related by $\varphi = \Phi/\Phi_a$. The energy per unit volume is obtained as a function of φ and $d^* = d/h$ appearing in N_D

The magnetic energy can be minimized with respect to d^* in order to obtain the average distance between the aggregates but it will be for a given volume fraction inside the aggregate: Φ_a which is unknown. At high fields it is reasonable to suppose that Φ_a is close to the volume fraction of the BCT structure ($\Phi_a = 0.69$) [63] but it is possible to obtain its value directly by saying that at high field the total pressure inside the aggregates is close to zero: $P_m + P_{osm} = 0$ where $P_m = -dU_m/dV_a$ is the magnetic pressure and P_{osm} the osmotic pressure. Taking for the osmotic pressure $P_{osm} = 1.85(NkT/V)/(0.64-\Phi_a)$ which fits well the hard sphere pressure at high volume fraction [66] we shall finally have the two following equations [67]:

$$-\nu_p \left.\frac{\partial(U_m/V)}{\partial \varphi}\right|_d + 1.85 \frac{\Phi}{0.64\varphi - \Phi} = 0 \quad \text{and} \quad \left.\frac{\partial(U_m/V)}{\partial d}\right|_\varphi = 0 \qquad (12)$$

These two equations can be solved numerically for the two unknowns, d/h and φ (or Φ_a), and the solution depends on the initial volume fraction Φ and on λ. An example of the structure obtained in unidirectional field is shown in Fig.3a; the experimental data for d versus h are plotted in Fig.3b together with the solution of Eq. (12) for $\Phi = 0.045$ and $\lambda = 273$. The agreement is quite good and, in this range, is close to $d \propto h^{0.5}$. In this derivation the surface energy appears through the change of Lorentz field, δH_L, on the surface; it is proportional to the square of the magnetization and to a/h since the radius of the particles determines the thickness of the layer where the field differs from its bulk value [67].

Some detailed calculation, either with dipolar [68] or multipolar [69] interactions on different lattices have shown that the surface energy can be quite sensitive to the type of lattice, but the magnetic suspensions are not monodisperse and we do not expect a perfect ordered surface so this "homogeneous" approximation is likely appropriated. If the surface energy becomes negligible, then the energy depends only on d/h and φ and the solution of Eq. (12) no longer depends on a/h; in other words d increases linearly with h for a given Φ and λ. Although the equations (12) could be used to predict the critical field

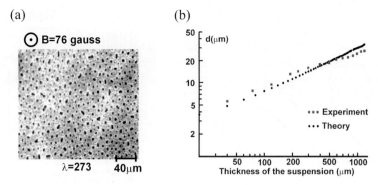

Fig. 3. Field induced phase separation. (a): top view of the structure. (b): average distance between aggregates versus the thickness of the cell: ○ Eq. (12); □ experiments

for phase separation (the value of λ above which the free energy becomes minimum for a value of φ different from unity) it is not believed to give the right answer because, as stated above, these equations skip the existence of the intermediate phase made of a gas of chains. To our knowledge such a theory is still missing since theories [70,71,49] rely on the hypothesis of a transition from a diluted isotropic phase to a condensed one, and use the osmotic pressure of a hard spheres liquid instead of the one corresponding to ellipsoids or rods.

Before ending with field induced structures it is worth saying a few words about structures in a rotating magnetic field. Let us consider a magnetic field rotating in the x-y plane: $\boldsymbol{H_0} = H_0(\boldsymbol{e_x}\cos\omega t + \boldsymbol{e_y}\sin\omega t)$ at a frequency high enough such that the dipolar force between two particles can be averaged over a period without significant change of the interparticle distance. Then the interaction energy corresponding to two dipoles $\boldsymbol{m_\alpha} = \boldsymbol{m_\beta} = m(\boldsymbol{e_x}\cos\omega t + \boldsymbol{e_y}\sin\omega t)$ rotating in phase with the field is still given by Eq. (1) whose average on a period gives: $W = m^2/2r^3(3\cos^2\theta - 1)$ where θ is the angle between Oz and r (cf. Fig. 1). This energy is negative for $\theta = \pi/2$ and positive for $\theta = 0$. It means that the particles will tend to gather into disks whose axis will be along Oz; these disks will also tend to separate from each other along the z axis due the repulsive force for $\theta = 0$. This disk shaped structure was observed with dielectric particles of diameter $10\mu m$ in a rotating electric field [72]. In the case of a magnetic colloidal suspension, the experiment made in a thin cell with the rotating field perpendicular to the plane of the cell shows a set of stripes which are the projections of the disk-shaped aggregates (Fig. 4) [73]. The thickness of these stripes being an order of magnitude larger than the diameter of the particles it seems that there are composed of many sheets of particles. If the field is rotating at a low enough frequency, chains of particles have enough time to form and rotate. This regime can give interesting informations on the kinetics of chain formation and chain rupture [74]. In the following section we shall see that in some cases the information on structure can help to explain the rheology.

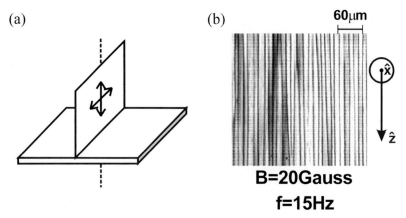

Fig. 4. Structures in a rotating field. (a): geometry; (b): top view of stripe structure

4 Rheology of Magnetic Fluids

4.1 Rheometry

The rheometers used to study the effect of a magnetic field on the rheology of MR fluids are usually conventional rheometers which have been adapted to install either coils or a magnetic circuit in soft iron. The installation of coils depend on the space available around the rotating tool. A pair of coils in Helmotz configuration allows to have a quite well defined field on a large diameter and minimizes field gradient but the maximum field attainable remains small (a few hundred Oersted). Using a single coil around the rotating tool can allow to reach 1000 Oersted but the coil must be cooled, furthermore it is difficult to control if the gap between the tools is correctly filled. For fields larger than about 1000 Oersted a magnetic circuit is needed. Either the field is parallel to the axis of rotation and the polar pieces have the same axis of symmetry (Fig. 5a); either the geometry is a cylindrical Couette cell and the field is everywhere perpendicular to the lines of flow(Fig. 5b) [75]. In this last case the field being radial is not constant throughout the gap and it can be 25% higher on the inner wall of the yoke than on the outer one [29]; this will attract the particles on the inner wall and decrease the apparent viscosity. Furthermore the height of the cylinder must be less than its radius in order to avoid the magnetic saturation of the central iron rod and this can introduce serious end effects.

In practice the parallel plate or the cone plate geometry is the most convenient device for high field measurements [76]. In each case the gap inside the magnetic circuit is composed of three parts: a fluid (length l_s) and an air path in series and an air path (l) in parallel (the different lengths are shown in Fig. 5). The length of the yoke is l_y. From Ampère's theorem $NI - H_y l_y = H_g l = H_s l_s + H(l - l_s)$ and flux conservation: $B_s = \mu_s H_s = \mu_0 H$ we readily get [75]:

$$\frac{B_s}{B_g} = \left(1 + \left(\frac{1}{\mu_s} - 1\right)\frac{l_s}{l}\right) \tag{13}$$

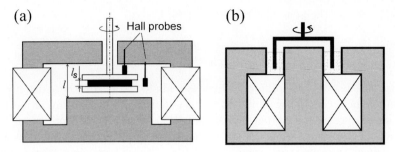

Fig. 5. Magnetic circuits for rheometry. (a): parallel plate geometry; (b): cylindrical Couette geometry

where B_g is the field measured with a Hall probe placed between the poles besides the magnetic fluid. From Eq. (13) we can see that the induction measured besides the fluid can be quite different from the one inside the suspension and that the correction can only be done if we know the permeability of the suspension. A way to overcome this problem is to place the Hall probe between the suspension and the yoke in order to measure directly B_s. Also with two Hall probes placed as shown in Fig. 5a it is possible to measure simultaneously the induction inside the suspension and its permeability by the mean of Eq. (13). Lastly, if the field is imposed only with an external coil, the induction is practically the one measured in the absence of the suspension (and the field inside the suspension is $H = H_0/\mu_s(H)$).

The parallel-plate or cone-plate geometries are widely used to study the rheology of magnetic suspensions. In the cone plate geometry the shear rate is constant inside the suspension but the variable thickness can induce a variable structure as discussed in the preceding section. In the parallel plate geometry the gap is constant and easy to change, allowing to get rid of slipping velocities on the plates but on the other hand the shear rate is not constant (since $\dot{\gamma} = \omega r/h$). For a fluid characterized by a Bingham law: $\tau = \tau_y + \eta\dot{\gamma}$ the torque recorded by the rheometer should be related to the shear rate $\dot{\gamma}_R = \omega R/h$ on the rim of the disk by:

$$T = \int_0^R r\tau(r)2\pi r dr = \frac{2\pi R^3}{3}\tau_y + \frac{\pi R^3}{2}\eta\dot{\gamma}_R \qquad (14)$$

The software of the rheometer will use the apparatus constant $2/\pi R^3$ to relate the stress to the torque independently of the rheological law, so multiplying (14) by this constant and extrapolating at zero shear rate we obtain $\tau = 4\tau_y/3$. We see that the apparent yield stress given by the rheometer in a parallel plates geometry is overestimated by a factor $4/3$. This problem does not exist with a cone-plate geometry neither with a sliding plate geometry [77] but this last technique only gives the static yield stress. In the parallel plates geometry it is necessary to use the general formulation [78]:

$$\tau(\dot{\gamma}_R) = \frac{T}{2\pi R^3}\left(3 + \frac{\dot{\gamma}_R}{T}\frac{dT}{d\dot{\gamma}_R}\right) \qquad (15)$$

Also the use of two different thickness allows to check the existence of a slipping velocity on the wall and to correct it if any. The determination of the yield stress, τ_y, as a function of the induction is the main property of a MR suspension. Experimentally the static yield stress is obtained by extrapolating at zero shear rate the stress versus logarithm of shear rate curve (there must be data until 10^{-3} s^{-1}) or by detecting the change of slope on a stress-strain curve. But the yield stress is not a well defined quantity and, in many cases, the yield of a suspension does not represent a bulk property of the fluid but the interactions between the wall and the particle [77,35]. A more reliable quantity is obtained by fitting the rheogram with a Bingham law or a Casson law: $\tau^{1/2} = \tau_d^{1/2} + (\eta\dot{\gamma})^{1/2}$ or Hershel-Bukley: $\tau = \tau_d + (K\dot{\gamma})^p$. The values of τ_d obtained by fitting these different laws are quite similar and represent what is usually called the dynamic yield stress. This stress can be viewed as the one which is needed to continuously separate the particles against the attractive magnetic forces in the low shear limit [20,79].

4.2 Models for the Yield Stress

The models aiming to predict the yield stress have to deal with two scales. Firstly, the particle scale which will give the force between two particles as a function of their physical properties and respective positions. Secondly the knowledge of the mesostructure and of its deformation at the scale of the device is needed. Two situations can be distinguished depending on the permeability of the particles. If the permeability is high ($\alpha = \mu_p/\mu_f \gg 1$) it is the particle scale which is important because the forces strongly depend on the interparticle gap. If α is low, then we can use the dipolar approximation and the separation between the particles does not change significantly the energy; this is mostly the shape of the mesostructure (cylinders, sheets...) and its inclination relatively to the field which will change the energy and generate the yield stress.

High permeability: $\alpha \gg 1$

The standard model for the structure is based on a cubic network of infinite chains of particles aligned in the direction of the field. When the material is strained, the chains are supposed to deform affinely with the strain (Fig. 6a); in other words the motion of the particles takes place only along the velocity lines. The result is that the distance between any pair of neighbors in the chains is the same and increase at the same rate with the strain $\gamma = \tan(\theta)$.

Some more realistic structural models have been considered [80,81] but it appears that in most cases the standard model gives a good prediction of the yield stress [82]. The field dependence of the yield stress is usually represented by a power law $\tau_y \sim H^n$ with $1 < n < 2$. The case $n = 2$ is the case of a linear magnetic material which is found at low fields or for particulates of low

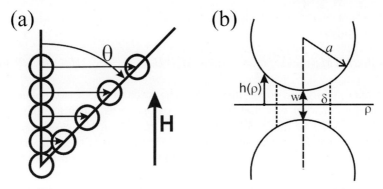

Fig. 6. Modelling the yield stress. (a) affine deformation of a chain; (b) gap between two particles; for $\varrho < \delta$, $H_g = M_s$

permeability [83]. The effect of saturation on the yield stress can be described by a simple model which can be applied both to ER [84] and MR [82] fluids. In this model the permeability of the particles is high enough to neglect the field H_i inside the particles. Then we can distinguish two domains inside the gap between two particles (Fig. 6b): the pole region with $\varrho < \delta$ (ϱ is the polar coordinate) where the field is given by the saturation magnetization: $H_g = M_s$ and the other with $\varrho < \delta$ where the field is given by $H_g = H(a + 0.5w)/h(\varrho)$ where H is the average field in the suspension, a, the radius of the particles and $h(\varrho)$ the distance between the plane of symmetry and the surface of the sphere: $h(\varrho) \cong 0.5 * (w + \varrho^2)$ with w the minimum gap between the two spheres and $\epsilon = w/a$. The distance δ is obtained by equating H_g and M_s for $\varrho = \delta$. If $H/Ms \ll 1$ the radial force between the two spheres which is the integral of the field on the plane separating the two spheres can be developed as:

$$F_r = \frac{\mu_0}{2} \int_0^a (H_g - H)^2 2\pi\varrho d\varrho \approx \pi a^2 \mu_0 M_s^2 \left(\frac{H}{M_s} - \frac{\epsilon}{2}\right) + \pi a^2 \mu_0 M_s^2 \frac{H}{M_s} \quad (16)$$

The first part of right hand side of Eq. (16) corresponds to $\varrho < \delta$ and the second part to $\varrho > \delta$. Knowing the force between the two spheres, the shear stress as a function of the strain $\gamma = tg(\theta) \approx \sin(\theta)$ is given by:

$$\tau(\gamma) = N\frac{F_r}{L^2} \sin(\theta) \cos^2\theta = \frac{3}{2}\Phi\frac{F_r}{\pi a^2}\frac{\gamma}{1+\gamma^2} \quad (17)$$

The quantity N/L^2 is the number of chains by unit surface and Φ is the volume fraction of solid particles. The $\cos^2(\theta)$ comes from the projection of the field squared on the unit vector joining the two spheres and the $\sin(\theta)$ of the projection of the radial force on the direction of shear. From Eqs. (16), (17) and using $\epsilon = 2\left(\sqrt{1+\gamma^2} - 1\right)$ we obtain an expression for the shear modulus as a function of the strain which is valid for $\gamma \ll 1$. The slope of this curve for $\gamma = 0$

gives the shear modulus and the maximum for $\gamma = \gamma_c$ gives the yield stress which are respectively:

$$G = 3\mu_0 \Phi H M_s \quad \text{and} \quad \tau_y = 2.31\Phi\mu_0 M_s^{1/2} H^{3/2}; \gamma_c = \sqrt{2H/(1.5M_s + 6H)} \tag{18}$$

It is also possible to solve directly for the force between two spheres in the presence of a constant external field or for two spheres which are part of a periodic chain by using finite element analysis [82,85]. In the case of an infinite chain, the unit cell contains one particle and the volume of the cylindrical cell is obtained from the volume fraction. The magnetic induction on the outer cylindrical boundary is imposed to be the average induction. Then the field can be calculated on the nodes of the mesh and the force is still given by the left-hand side of Eq. (16).

Both finite element and analytical expression predicts a yield stress proportional to the volume fraction and that is also what is observed experimentally at not too high volume fractions ($\Phi < 0.2 - 0.3$) [86,22] but an increase faster than linearly [35,87] is observed at higher volume fraction. This behavior can be expected if we believe that thick aggregates are more difficult to break than individual chains; it can also be related to non affine motion which leads to rupture of aggregates at higher angle-and so higher stresses [32,81].

Concerning the field dependence on the yield stress, several groups have found a power law with an exponent close to 3/2 as predicted by Eq. (18) [86,88,24], but some data show a linear dependence [39], or an exponent 1.27 - but at quite low fields and with steel spheres [90] -, and we can find results [77] or simulations [89] showing a continuous decrease of the slope with the field, so it is rather difficult to conclude at this stage.

A way to check the theories is to use steel spheres whose magnetic properties can easily be measured in the bulk material. The magnetization curve is well fitted by a Frolich-Kennelly curve: $M = M_s(\mu_i - 1)H/(M_s + (\mu_i - 1)H)$ with $M_s = 1360$ kA/m and $\mu_i = 250$. The calculation of the radial force F_r between two spheres is performed by using either finite elements or Eqs. (16), (17). The experiment consists in shearing chains of seven steel spheres placed on a ring inside a rheometer and glued at each extremity [90]. The results are shown in Fig. 7a for a field $H = 14.3 kA/m$ and an equivalent volume fraction of 15%. The finite element calculation is closer to the experimental result but the analytical solution gives the right behavior, even if the predicted yield stress is higher than the experimental one.

The maximum yield stress at high induction will be given by magnetic saturation since each particle will acquire the dipole: $m_s = v_p M_s$; then using affine displacement of the particles, the shear stress is obtained from the derivative of the energy relatively to the strain hence the yield stress and the shear modulus [6]:

$$\tau_y^s = 0.1143\zeta(3)\Phi\mu_0 M_s^2 \quad \text{and} \quad G^s = 0.294\zeta(3)\Phi\mu_0 M_s^2 \tag{19}$$

For $\Phi = 0.3$ and $\mu_0 M_s = 2$ Tesla, we have $\tau_y^s = 130$ kPa and $G^s = 340$ kPa. These are actually the order of magnitude which are obtained experimentally

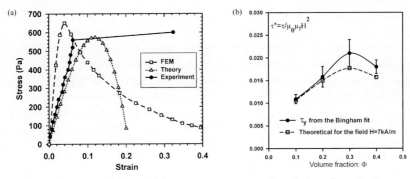

Fig. 7. Theories for yield stress (a) stress-strain curve for chains of steel spheres ○ experiment; □ finite elements; △ Eqs. (16)-(17) (b) normalized yield stress versus volume fraction for silica spheres (diameter 40-50μm) at low field domain (10-30 kA/m); solid curve: experiment; dashed curve: theory

both for saturation yield stress [24,15] and shear modulus [6]. Other models based on the same assumptions of a saturation zone give similar predictions [82,91].

Low permeability $\alpha < 5$

An other quite different situation occurs for low values of the ratio $\alpha = \mu_p/\mu_f$. A way to obtain this situation is to use a mixture of silica spheres in a ferrofluid. Then we have a suspension of magnetic holes ($\alpha < 1$) whose permeability for different volume fractions well follow the Maxwell-Garnett law [35]. In this case the dipolar approximation can be used and the interparticle forces being long ranged, the total magnetic energy is no longer sensitive to the change of the interparticle distance but rather to the deformation and rotation of the macroscopic aggregates in the field. If the magnetic moment of the anisotropic medium can be obtained as a function of the shape and orientation of the mesostructure then the energy and the magnetic stress can be derived since we have: $\tau = \frac{1}{2V} H_0 \frac{\partial m_z}{\partial \gamma}$ with m_z the component of the magnetic moment of the sample along the direction of the external field. The aggregates formed in the presence of the field are either cylinders or stripes of internal volume fraction Φ_a. The volume fraction $\varphi = \Phi/\Phi_a$ represents the part of the space occupied by the aggregates. If H is the field inside the suspension, then the predicted yield stress is [81,92]:

$$\frac{\tau^y}{\mu_0 \mu_f H^2} = \frac{3\sqrt{3}}{16} (\mu_s^*)^2 \frac{\varphi(1-\varphi)}{C_s + \mu_s^*(1-\varphi)} \tag{20}$$

with $C_s = 1$ if the particles are gathered into stripes and $C_s = 2$ if they form cylinders. The permeability $\mu_0 \mu_f$ is the one of the suspending fluid (here the ferrofluid) and $\mu_s^* = \mu_s/\mu_f - 1$.

A similar derivation was proposed independently for stripes, with a slight difference on the analytical result [93,81]. We have measured the dynamic yield stress by extrapolating at zero shear rate the stress versus shear rate curve, assuming a Bingham behavior $\tau = \tau_y + \eta \dot{\gamma}$. The comparison of the normalized experimental yield stress with the one predicted by Eq. (20) with $C_s = 2$ (cylindrical aggregates) is shown in Fig. 7b for different volume fractions and a field $H = 7$ kA/m. We see that Eq. (20) well captures the experimental behavior, especially taking into account that there are no free parameters (the internal volume fraction, Φ_a, has been set to 0.69 which is the volume fraction of a body cubic tetragonal structure). Besides the quantitative agreement it is worth noting that the yield stress presents a maximum with the volume fraction; this is because the difference in volume fraction between the aggregates and the effective medium is decreasing with the volume fraction and the shear stress is only a function of this difference. On the contrary for particles of high permeability, the stress is mainly sensitive to short range interparticle forces and increases linearly with Φ.

More recently this approach for a striped mesostructure was extended to cover both low and high values of α [94]. The permeability of the stripes is obtained from a formula valid for a cubic lattice of particles whatever α, and the internal permeability of the particles μ_p is approximated by a Frolich-Kennelly law in order to take into account the magnetic saturation.

Instead of simple chains arranged on a network (cubic or bct) some more complicated arrangements of chains have been studied either with dipolar approximation or with multipolar development which should be considered for high permeabilities [80,81,95,96]. The main point to remember is that the yield stress is not sensitive to the structure itself (single chains, double chains, planes or double planes give approximately the same values) but to the way it deforms. Besides the affine motion with equally spaced spheres we can consider a model with the formation of a single gap or with a single gap but starting at non zero strain. This last case where, due to extra particles, the structure can expand before breaking, gives a higher yield stress [32]. An other point which is still debated is relative to the position where the ruptures take place. In a chain model, due to the finite size, the force needed to separate two particles inside the chain is weaker at its extremity and the first rupture should take place close to the walls. For ER fluids with dipole images it is not so intuitive but still verified [97]. This is true if each extremity of the chain sticks to the wall, but if it slips and rotates the hydrodynamic tension being maximum at the center it will likely breaks at the center. Before breaking, the suspension can also slip on the walls at a lower stress which will depend on the roughness of the walls and also on the pressure exerted by the magnetic particles on the wall in the presence of the field. In this case a frictional shear stress (which is the normal stress times the friction coefficient) is observed [35]. Also it has been found that dense aggregates can stand high normal stresses obtained by moving a wall in the presence of the magnetic field. In this case very high frictional shear stresses following the law $\tau_y^e = \tau_y + kN_e$ where τ_y is the usual yield stress, N_e the external stress (up to $2MPa$) and k a friction coefficient close to 0.25 [98].

This presentation holds only for non Brownian particles ($\lambda \to \infty$). For finite values of λ the Brownian motion will play a role to weaken the structure in two different ways: by decreasing the length of the aggregates and by generating fluctuations of the gaps between the particles inside the chains, which will lower the rupture force. The first effect can be accounted for with the help of the theory leading to Eq. (8) or a similar one [99]. Then the only aggregates participating to the yield stress are those which have a length larger than the thickness of the cell [100].This model ignore the importance of the fluctuations of positions inside the chain which can considerably decrease the force needed for the rupture of a chain. It is possible to estimate the average distance between two particles inside a chain with the help of the Boltzmann distribution, but the breakdown of the percolated network is not sensitive to the average distance between the particles but rather to the largest fluctuation inside the chain. A Monte-Carlo calculation shows that the largest gap inside a chain of 200 particles at $\lambda = 300$ can be an order of magnitude greater than the average one and can explain the lower yield stress experimentally observed when λ is decreased [33]. On the other hand when a mechanical stress is imposed, the corresponding energy should be incorporated in the Boltzman distribution as is usually done in the standard theory of viscosity of simple liquids (the probability for a particle to jump from a potential well to an other is $e^{-(U_d \pm v\tau)/kT}$ with v the volume of a unit cell and τ the applied shear stress; the + sign being for the direction of the applied stress). This approach is used directly in [101] and incorporated in a chain model in [102]; no direct comparison with experimental results concerning the yield stress are done but rather concerning the Mason number dependence which is the subject of the next section.

4.3 Flow Regime

In steady shear flow the Bingham model: $\tau = \tau_d + \eta_0 \dot{\gamma}$ is widely used. If we call η_s the effective viscosity (defined by $\tau = \eta_s \dot{\gamma}$), then normalizing by the stress at zero field gives a relative viscosity: $\eta_r = \eta_s/\eta_0$ which decreases as $\tau_d/\eta_0\dot{\gamma}$ that is to say proportionally to Mn^{-1} where Mn is defined in Eq. (6). Actually experiments [107,35,76,83] as well as simulations [101,103] show that the Mason number does not allow to collapse all the results on the same curve. The reduced viscosity still follows a law $\eta_r = Mn^{-\nu}$ but with an exponent which can vary from 0.68 to 1 with a tendency to increase when λ is increased. In the limit ($\lambda \to \infty$) either a model based on ellipsoidal aggregates [75] or on chain of particles [102,35] predicts an exponent $\nu = 1$. These models are based on the idea that aggregates rotate in the shear flow and find their equilibrium angle when the hydrodynamic torque equilibrates the magnetic restoring torque. The rotation continues when the Mason number is increased until the radial hydrodynamic force at the center of the chain overcomes the attractive magnetic force. At this critical angle the aggregates break into two equal parts and the restoring magnetic torque - which is proportional to the volume of the aggregate - is divided by two. The hydrodynamic torque and the shear stress which are proportional to the cube of the length of the aggregate are divided by eight, so the two aggregate rotate backwards after their separation and will reform and break again at the same angle. Increasing further the shear rate, the equilibrium length will decrease until all the aggregates are destroyed. As pointed in [104] the two equations relative to the torque and to the radial force give a solution for the critical angle which does not depend on the size of the chain, but which increases if we consider multipolar forces rather than dipolar ones [35]. It is worth noting that this approach gives a Bingham law with a yield stress $\tau_d = C\mu_0\mu_f\beta^2 H^2$, where C is a constant of order one. The order of magnitude of this dynamic yield stress is the same as the one obtained in a dipolar approximation and taking into account multipolar interactions will increase the critical angle well above 45° which will still decrease the yield stress. The failure of the rotating chain model in predicting the dynamic yield stress for strong fluids could come from the interactions between aggregates which prevent them to reach the critical angle. These interactions are not directly introduced in the models.

Still it is the interaction between chains which leads to mesostructure formation (usually stripes of particles in the plane defined by the field and the velocity) at quite low strain ($\gamma = 0.15$) as observed experimentally in oscillating shear [67] or also in steady shear flow [106,73]. The average inclination of chains and their rupture is then a process mediated by the interactions between chains inside dense structures rather than by individual rotation and rupture.

Another model [107] based on an ellipsoidal shape of aggregates also uses a balance between the hydrodynamic and the magnetic torque to find the equilibrium angle θ, but the determination of the size of the aggregates is obtained by minimizing the total magnetic energy (which includes the surface energy). The result is a viscosity which, unlike the preceding model, should vary as: $\eta_r = 1 + CMn^{-2/3}$. This theory is restricted to low angles of inclination of the aggregates, or in other words to low Mason number. As the experiments show exponents which are between $2/3$ and 1 it can be tempting to conclude that the two models correspond to some limit (for instance $\nu = 2/3$ for low λ and $\nu = 1$ for $\lambda \to \infty$). The simulations made with Stokesian dynamics [101] do not comfort this point of view since, for $10 < \lambda < \infty$, the exponent is between 0.9 and 1 for low Mason numbers ($2 \cdot 10^{-5} < Mn < 10^{-3}$) instead of varying between $2/3$ and 1.

Introducing the parameter λ in the theory in order to model the effect of Brownian motion on the rheology is a quite difficult task. In [102] the chain model developed for $\lambda \to \infty$ is extended by introducing an activation energy which contribute to break the chains before they have reached their critical angle. A life time of the chains which depends on λ and Mn is introduced in this way and compared to $1/\dot{\gamma}$ in order to determine if the chains will break before reaching their equilibrium angle. This model predicts that the normalized field induced stress should scales as $(\lambda^{3/2} Mn)^{1/\lambda}$ but comparisons with experiments are lacking. A different model using the Eyring approach of viscosity is used in [101]. This model predicts a plateau at very low shear for the viscosity (which diverges as $e^{1.57\lambda}$). Above this low shear regime the apparent viscosity should only depend on the dimensionless shear rate: $(\dot{\gamma}a^2/D\lambda)e^{1.57\lambda}$. This prediction is well verified by numerical simulations in the range $2.5 < \lambda < 17$. For particles carrying a constant magnetic dipole, a low Mason number theory is proposed in [105] where the size distribution of chains as a function of λ is the one obtained at zero Mason number. Then the distribution function of orientations of the chains for each length is obtained from a Fokker-Planck equation balancing the fluxes due to hydrodynamic and magnetic torques and to Brownian relaxation. The hydrodynamic and magnetic stresses are then averaged over the distribution of lengths and over the distribution of orientations. Some results for a low value of λ-independent of field for constant dipoles - compare the increase of viscosity for field aligned on the velocity gradient and field aligned on the velocity, showing a maximum with the intensity of the field in the second case. Here too, no comparisons with experiments are done. Mixing thermal forces and shear forces to predict the equilibrium distribution of chains as a function of λ and Mn is difficult because the force associated with the flow is not the derivative of a potential energy. Nevertheless it is the case if we consider a chain of particles aligned on the direction of elongation in an elongational flow. Then it becomes possible to add the potential energy of the flow: $U_{fl} = -1/2\xi E_{ij} r_i r_j$ with ξ the Stokes coefficient, r, the separation vector between two particles and E_{ij} the velocity gradient tensor. The use of the thermodynamic theories is then possible and the evolution of the chain length as a function of λ and Mn is given in [108].

The knowledge of viscoelastic properties of MR fluids such as the storage and loss moduli, G', G'' is quite important for damping applications but few systematic studies are reported in the literature. As was first pointed in [109] the use of a dimensionless frequency which is equivalent to a Mason number in oscillating flow, and of a modulus normalized by the energy density of the field allows to collapse the data obtained at different fields on the same master curve. This is verified by experiments on MR fluids if λ is high enough [106], otherwise, as for steady flow, Brownian forces introduce an other time scale. The moduli G' and G'' do not depend markedly on frequencies whereas they strongly decrease with the strain [87]. The viscoelastic theories use the same approach as for steady flow: the aggregates are modeled by independent chains or ellipsoids. A kinetic chain model taking into account aggregation due to attractive dipole forces and rupture caused by hydrodynamic forces was proposed [110] by introducing a phenomenological equation for the evolution of the chain length. On the other hand the solution of the equation of motion of an ellipsoid in an harmonic shear flow - with a fixed length depending on λ - is an other possible approach [111]. A comparison of these models with experimental data on G', G'' is lacking.

Most of the models are based on the description of individual aggregates and ignore the effect of mesostructures on magnetorheology. This effect is obvious at low Mason number where it explains that the first stress-shear rate curve, obtained under steady shear flow after applying the field on an homogeneous suspension, is different from the second one (Fig. 8): the first flow curve is obtained starting from a mesostructure formed of cylinders, whereas the second and subsequent curves are obtained starting from a sheet like structure [106]. This sheet like structure gradually disappears when the Mason number is increased but at a Mason number around unity a new striped structure appears associated with a jump in viscosity [112]. This flow induced phase separation is due to the combination of the rotation of transient aggregates of particles in the

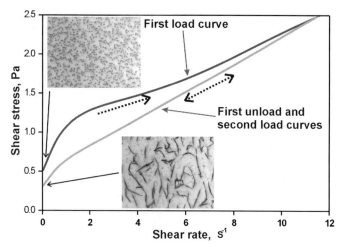

Fig. 8. Stress versus shear rate for suspension of magnetic polystyrene particles (d=0.5μm); applied field H=29 kA/m. The top-left photo corresponds to the structure before shearing and the bottom photo to the structure after shearing

shear (which begins above $Mn = 1$) and of an uniaxial applied magnetic field; this situation is actually reminiscent of the case of a rotating field discussed in the preceding section. More complicated mesostructures can be obtained by combining rotating field with axial field or crossed electric field and magnetic field or any other combination.

5 Conclusion

Careful experiments made with well characterized magnetic suspensions are still too scarce in order to check existing rheological models, especially for fluids with high saturation magnetization. It would be very useful for this purpose to synthesize well monodispersed magnetic particles in the range 0.1 to 1μm where it is possible to explore the whole range of λ without serious problems with sedimentation. Progresses in the long term stability of these fluids are more expected than a large increase in yield stress, except may be in specially confined geometries. Besides applications using viscosity control which are reviewed elsewhere [15,16] let us point out that the use of rotating fields or combined electric and magnetic fields to arrange colloidal particles in special structures, has also very promising applications in the domain of nanotechnologies.

References

1. W.M.Winslow, J. Appl. Phys. **20**, 1137 (1949)
2. J.Rabinow: AIEE Trans. **67**,1308 (1948)

3. J.D.Carlson, D.M.Catanzarite, K. A. St. Clair: "Commercial magneto-rheological devices". In: *Proceedings of the 5th International Conference on ER fluids, MR suspensions and associated technology, Sheffield, July 10-14, 1997,* ed. by W.A.Bullough (World Scientific 1996) pp. 20-28
4. A.Inoue, S.Maniwa, Y.Ide, H.Oda: Int. J. Mod. Phys. B **13**, 1966 (1999)
5. H.Orihara,M.Doi,Y.Ishibashi: Int. J. Mod. Phys. B **13**, 1949 (1999)
6. M.R.Jolly, J.D.Carlson, B.C.Munoz, T.A.Bullions: J. Intell. Mater. Sys. Struct. **7**, 613 (1996)
7. J.M.Ginder, M.E.Nichols, L.D.Elie, J.L.Tardiff: 'Magnetorheological elastomers; properties and applications'. In *Smart Structures and Materials 1999: Smart Materials Technologies, Newport Beach CA, 3-4 March 1999,* ed. by M.Wutttig, (SPIE Proceedings Vol. 3675, 1999) pp.131-138
8. G.Bossis, C.Abbo, S.Cutillas, S.Lacis, C Métayer: Int. J. Mod. Phys. **B, 15,** 564 (2001)
9. H.Block, J.P.Kelly: J. Phys. D: Appl. Phys. **21**, 1661 (1988)
10. A.P.Gast, C.F.Zukoski: Adv. Coll. Int. Sci. **30**, 153 (1989)
11. T.C.Jordan, M.T.Shaw: IEEE Trans. Elect. Insul. **24**, 849 (1989)
12. C.F.Zukoski: Ann. Rev. Mat. Sci. **23**, 45 (1993)
13. M.Parathasarty, D.J.Klingenberg: Material Science and Ingeniering **R17**, 57 (1996)
14. H.T.See: Korea-Australia Rheology Journal **11**, 169 (1999)
15. J.M.Ginder: MRS Bulletin **23**, 26 (August 1998)
16. P.J.Rankin, J.M.Ginder, D.J.Klingenberg: Curr. Opin. Colloidal Interface Sci. **3**, 373 (1998)
17. H.See: Applied rheology **11**, 70 (2001)
18. C.Kormann, H.M.Laun, H.J.Richter: Int. J. Mod. Phys. B **10**, 3167 (1996)
19. J.F.Brady, G.Bossis: Ann. Rev. Fluid. Mech. **10**, 111 (1988)
20. R.T.Bonnecaze, J.F.Brady: J.Rheol. **36**, 73 (1992)
21. D.M.Heyes, J.R.Melrose: Mol. Sim. **5**, 293 (1990)
22. V.I.Kordonsky, Z.P.Shulman, S.R.Gorodkin, S.A.Demchuk, I.V.Prokhorov, E.A.Zaltsgendler, B.M.Khusid, J. Magn. Magn. Mater. **85**, 114 (1990)
23. A.J.Margida, K.D.Weiss, J.D.Carlson: Int. J. Mod. Phys. B **10**, 3335 (1996)
24. P.P.Phulé, J.M.Ginder: Int. J. Mod. Phys. B, **13,** 2019 (1999)
25. J. David Carlson, "What makes a good MR fluid" In: *Proceedings of the - 8th International Conference on ER fluids, MR suspensions,Nice, July 9-13, 2001,* ed. by G.Bossis (World Scientific 2002) pp. 63-69
26. R.T.Foister US Patent 5,667,715, Sep 16,1997 «Magnetorheological fluids»
27. R.W.Chantrell, A.Bradbury, J.Popplewell, S.W.Charles: J. Appl. Phys. **53**, 2742 (1982)
28. R.E.Rosensweig: *Ferrohydrodynamics* (Cambridge University Press, New York 1985)
29. H.M.Laun, C.Kormann, N.Willenbacher: Rheol. Acta **35**, 417 (1996)
30. N.Rosenfeld, N.M.Wereley, R.Radakrishnan, T.S.Sudarshan: 'Behavior of magnetorheological fluids utilizing nanopowder iron'. In: *Proceedings of the 8th International Conference on ER fluids, MR suspensions and their applications, Nice, July 9-13, 2001,* ed. by G.Bossis (World Scientific, Singapore 2002) pp.452-458
31. B.E.Kashevskii, V.I.Kordonskii, I.V.Prokhorov: Magnetohydrodynamics **24**, 134 (1988)
32. G.Bossis, E.Lemaire: J.Rheol **35**, 1345 (1991)

33. E.Lemaire, A.Meunier, G.Bossis, J.Liu, D.Felt, P.Bashtovoi, N.Matoussevitch, "Influence of the particle size on the rheology of magnetorheological fluids".J.Rheol.39(1995)1011
34. B.J.De Gans, C.Blom, A.P.Philipse, J.Mellema: Phys. Rev E **60**, 4518 (1999)
35. O.Volkova, G.Bossis, M.Guyot, V.Bashtovoi, A.Reks: J. Rheology **44**,91 (2000)
36. W.I.Kordonsky, S.R.Gorodkin, E.V.Medvedeva: 'First experiment on magneto-electrorheological fluids'. In: *Electrorheological Fluids: Mechanisms Properties, Technology, and Applications, Proceedings of the 4th International Conference on ER Fluids, July 20 - 23, 1993 Feldkirch, Austria*, ed. by R.Tao, G.D.Roy (World Scientific, Singapore, 1994) pp.22-36
37. K.Koyama: Int. J. Mod. Phys. B **10**, 3067 (1996)
38. H.Takeda, K.Matsushita, Y.Masubuchi, J.I.Takimoti, K.Koyama: 'Mechanism of the synergistic effect in EMR fluids studied by direct observation'. In: *Proceedings of the 6th International Conference on ER fluids, MR suspensions and their applications, Yonezawa, July 22-25, 1997*, ed. by M.Nakano, K.Koyama (World Scientific 1998) pp. 571-577
39. F.Q.Jiang, Z.W.Wang, J.Y.Wu, L.W.Zhou, Z.Y.Ying: 'Magnetorheological materials and their application in shock absorbers'. In: *Proceedings of the 6th International Conference on ER fluids, MR suspensions and their applications, Yonezawa, July 22-25, 1997*, ed. by M.Nakano, K.Koyama (World Scientific 1998) pp. 494-501
40. J.M.Ginder, L.D.Elie, L.C.Davis: Magnetic fluid-based magnetorheological fluids, USA Patent 5,549,837 August, 27, 1996
41. L.Barsi, A.Büki, D. Szabo, M.Zrinyi: Progress in Colloid and Polymer Science **102**, 57 (1996)
42. A. Zrinyi, D. Szabo: Progress in Liquid Physics (II), 238 (2000)
43. T.Shiga, A.Okada, T.Kurauchi: J. Appl. Polym. Sci. **58**, 787 (1995)
44. T.Mou, G.A.Flores, J.Liu, J.Bibette, J.Richard: Int. J. Mod. Phys. B **8**, 2779 (1994)
45. M.Fermigier, A.P.Gast: J. Colloid and Interf. Sci. **154**, 522 (1992)
46. E.Lemaire, Y.Grasselli, G.Bossis: J. Phys II France 2, **359** (1992)
47. Y.Zhu, M.L.Ivey, P.Sheaffer, J.Pousset, J.Liu : Int. J. Mod. Phys. B **10**, 2973 (1996)
48. S.Cutillas, A.Meunier, E.Lemaire, G.Bossis: Int. J. Mod. Phys. B, **10,** 3093 (1996)
49. S.Cutillas, G.Bossis, E.Lemaire, A.Meunier, A.Cebers: Int. J. Mod. Phys., **13**, 1791 (1999)
50. M.Hagenbüchle, P Scheaffer, Y.Zhu, J.Liu: Int. J. Mod. Phys. B, **10,** 3057 (1996)
51. J.E.Martin, J.Odinek, T.C.Halsey: Phys. Rev. Lett. **69**, 1524 (1992)
52. A.O.Cebers: Magn.Gidrodinam. **2**, 36 (1974)
53. A.Y.Zubarev: Colloid Journal **61**, 338 (2001)
54. S.Fraden, A.J.Hurd, R.B.Meyer: Phys. Rev. Lett. **63**, 2373 (1989)
55. R.Jullien, R.Botet: *Aggregation and fractal aggregates* (World Scientific, Singapore 1987)
56. G.K.Batchelor: J. Fluid Mech. **44**, 419 (1970)
57. T.C.Halsey, W.Toor: J.Stat.Phys. **61,** 1257 (1990)
58. M.Mohebi, N.Jamaski, J.Liu: Phys. Rev. E **54**, 5407 (1996)
59. H.V.Ly, K.Ito, H.T.Banks, M.R.Jolly, F.Reitich: Int. J. Mod. Phys. B, **15**, 894 (2001)
60. R.Tao, J.M.Sun: Phys. Rev. Lett. **67**, 398 (1991)
61. T.J.Chen, R.N.Zitter, R.Tao: Phys. Rev. Lett. **68**, 2555 (1992)

62. A.van Blaaderen: J.Chem.Phys. **112**, 3851 (2000)
63. Y.Grasselli, G.Bossis, E.Lemaire: J. Phys II France **4**, 253 (1994)
64. J.Liu, E.M.Lawrence, A.Wu, M.L.Ivey, G.A.Flores, K.Javier, J.Bibette, J.Richard: Phys. Rev. Lett. **74**, 2828 (1995)
65. C.Flament, J.C.Bacri, A.Cebers, F.Elias, R.Perzynski: Europhys. Lett. **34**, 225 (1996)
66. W.B.Russel, D.A.Saville, W.R.Schowalter: *Colloidal dispersions* (Cambridge University Press, Cambridge 1991)
67. S.Cutillas, G.Bossis, A.Cebers: Phys. Rev. E **57**, 804 (1998)
68. W.R.Toor, T.C.Halsey: Phys. Rev. A **45**, 8617 (1992)
69. H.J.H.Clercx: J.Chem.Phys.**98**, 8284 (1993)
70. K.I.Morozov, A.F.Pshenichnikov, Y.L.Raikher, M.I.Shliomis: J. Magn. Magn. Mater. **65**, 269 (1987)
71. B.M.Berkovsky, V.I.Klikmanov, V.S.Filinov: J. Phys. C: Solid State Phys. **18**, L941 (1985)
72. T.C.Halsey, R.A.Anderson, J.E Martin: Int. J. Mod. Phys. B, **10**, 3019 (1996)
73. G.Bossis, P.Carletto, S.Cutillas, O.Volkova: Magnetohydrodynamics **35**, 371 (1999)
74. S. Melle, G.G.Fuller, M.A.Rubio: Phys. Rev. E **61**, 4111 (2000)
75. Z.P.Shulman, V.I.Kordonsky, E.A.Zaltsgendler, I.V.Prokhorov, B.M.Khusid, S.A.Demchuk: Int. J. Multiphase Flow **12**, 935 (1986)
76. B.J.de Gans, H.Hoekstra, J.Mellema: Faraday Discuss **112**, 209 (1999)
77. X.Tang, H.Conrad: J. Rheology **40**, 1167 (1996)
78. K.Walters: *Rheometry* (Chapman and Hall, London 1975)
79. E.Lemaire, G.Bossis, Y.Grasselli: Langmuir **8**, 2961 (1992)
80. G.L.Gulley, R.Tao: Phys. Rev. E **48**, 2744 (1993)
81. G.Bossis, E.Lemaire, O.Volkova, H.Clercx: J.Rheol. **41**, 687 (1997)
82. J.M.Ginder, L.C.Davis, L.D.Elie: Int. J. Mod. Phys B. **10**, 3293 (1996)
83. D.W.Felt, M.Hagenbuchle, J.Liu, J.Richard: J. Intell. Mater. Sys. Struct. **7**, 1167 (1996)
84. N.Felici, J.N.Foulc, P.Atten: Int. J. Mod. Phys. B, **8**, 2731 (1994)
85. S.Lacis, O.Volkova, G.Bossis: 'Application of FEM for magnetic force calculation between ferromagnetic spheres'. In: *Proceedings of the International Colloquium "Modelling of Material Processing", Riga, May 28-29, 1999,* ed. by A.Jakovics (University of Latvia, Riga 1999) pp. 222-227
86. P.J.Rankin, A.T.Horvath, D.J.Klingenberg: Rheol. Acta **38**, 471 (1999)
87. B.D.Chin, J.H.Park, M.H.Kwon, O.O.Park: Rheol. Acta **40**, 211 (2001)
88. Z.Y.Chen, X.Tang, G.C.Zhang, Y.Jin, W.Ni, Y.R.Zhu: 'A novel approach of preparing ultrafine magnetic metallic particles and the magnetorheology measurements for suspensions containing these particles'. In: *Proceedings of the 6th International Conference on ER fluids, MR suspensions and their applications, Yonezawa, July 22-25, 1997,* ed. by M.Nakano, K.Koyama (World Scientific, Singapore 1998) pp. 486-493
89. S.Lacis, E.Zavickis, G.Bossis: 'Magnetic interactions of chains formed by ferromagnetic spheres'. In: *Proceedings of the 8th International Conference on ER fluids, MR suspensions and their applications, Nice, July 9-13, 2001,* ed. by G.Bossis (World Scientific, Singapore 2002) pp.359-365
90. O.Volkova, G.Bossis, S.Lacis, M.Guyot 'Magnetorheology of a magnetic steel sphere suspension" *Proceedings of the 8th International Conference on ER fluids, MR suspensions and their applications, Nice, July 9-13, 2001,* ed. by G.Bossis (World Scientific, Singapore 2002) pp. 860-866

91. L.C.Davis: J. Appl. Phys. **85**, 3348 (1999)
92. E.Lemaire, G.Bossis, O.Volkova: Int. J. Mod. Phys. B, **10,** 3173 (1996)
93. R.E.Rosensweig: J. Rheol. **39**, 179 (1995)
94. X.Tang, H.Conrad: J. Phys. D: Appl. Phys. **33**, 3026 (2000)
95. X.Tang, Y.Chen, H.Conrad: J. Intell. Mater. Sys. Struct. 7, 517 (1996)
96. H.J.H.Clercx, G.Bossis: Phys. Rev. E **103**, 9426 (1995)
97. R.Tao, J.Zhang, Y.Shiroyanagi, X.Tang, X.Zhang: Int. J. Mod. Phys. B **15**, 918 (2001)
98. X.Tang, X.Zhang, R.Tao: "Enhance the yield shear stress of magnetorheological fluids" Proceedings of the 7th Int. Conf. on Electro-rheological Fluids and Magneto-rheological suspensions, Hawaï,Honolulu, July 19-23,1999 ed. R.Tao,World Scientific,2000 pp.3-10
99. M.A.Osipov, P.I.C.Teixeira, M.M.Telo de Gama: Phys. Rev. E **54**, 2597 (1996)
100. B.J.de Gans, N.J.Duin, D.van den Ende, J.Mellema: J. Chem. Phys. **113**, 2032 (2000)
101. Y.Baxter-Drayton, J.F.Brady: J. Rheol. **40**, 1027 (1996)
102. J.E.Martin: Int. J. Mod. Phys. B. **15**, 574 (2001)
103. J.R.Melrose: Mol.Phys. **76**, 635 (1992)
104. J.E.Martin, R.A.Anderson: J. Chem. Phys. **104**, 4814 (1996)
105. A.Yu.Zubarev, L.Yu.Iskakova: Phys. Rev E **61**, 5415 (2000)
106. O.Volkova, G.Bossis, E.Lemaire: 'Magnetorheology of model suspensions'. In: *Proceedings of the 6th International Conference on ER fluids, MR suspensions and their applications, Yonezawa, July 22-25, 1997*, ed. by M.Nakano, K.Koyama (World Scientific 1998) pp. 528-524
107. T.C.Halsey, J.E.Martin, D.Adolf: Phys. Rev. Lett. **68**, 1519 (1992)
108. [108] J.Pérez-Castillo, A.Pérez-Madrid, J.M.Rubi, G.Bossis: J. Chem. Phys. **113**, 6443 (2000)
109. D.J.Klingenberg: J. Rheol. **37**, 199 (1993)
110. J.E.Martin, J.Odinek: J.Rheol. **39**, 995 (1995)
111. A.Y.Zubarev: Colloid Journal **62**, 317 (2000)
112. O.Volkova, S.Cutillas, G.Bossis: Phys. Rev. Lett. **82**, 233 (1999)

Part IV

Applications

Targeted Tumor Therapy with "Magnetic Drug Targeting": Therapeutic Efficacy of Ferrofluid Bound Mitoxantrone

Ch. Alexiou[1], R. Schmid[1], R. Jurgons[1], Ch. Bergemann[2], W. Arnold[3], and F.G. Parak[4]

[1] Department of Otorhinolaryngology, Head and Neck Surgery, University Erlangen-Nürnberg, 91054 Erlangen, Germany
[2] Chemicell, 10823 Berlin, Germany
[3] Department of Otorhinolaryngology, Head and Neck Surgery, Klinikum rechts der Isar, Technical University of Munich, 81675 Munich, Germany
[4] Physics-Department E 17, Technical University of Munich, 85748 Garching, Germany

Abstract. The difference between success or failure of chemotherapy depends not only on the drug itself but also on how it is delivered to its target. Biocompatible ferrofluids (FF) are paramagnetic nanoparticles, that may be used as a delivery system for anticancer agents in locoregional tumor therapy, called "magnetic drug targeting". Bound to medical drugs, such magnetic nanoparticles can be enriched in a desired body compartment (tumor) using an external magnetic field, which is focused on the area of the tumor. Through this form of target directed drug application, one attempts to concentrate a pharmacological agent at its site of action in order to minimize unwanted side effects in the organism and to increase its locoregional effectiveness.

Tumor bearing rabbits (VX2 squamous cell carcinoma) in the area of the hind limb, were treated by a single intra-arterial injection (A. femoralis) of mitoxantrone bound ferrofluids (FF-MTX), while focusing an external magnetic field (1.7 Tesla) onto the tumor for 60 minutes. Complete tumor remissions could be achieved in these animals in a dose related manner (20% and 50% of the systemic dose of mitoxantrone), without any negative side effects, like e.g. leucocytopenia, alopecia or gastrointestinal disorders.

The strong and specific therapeutic efficacy in tumor treatment with mitoxantrone bound ferrofluids may indicate that this system could be used as a delivery system for anticancer agents, like radionuclids, cancer-specific antibodies, anti-angiogenetic factors, genes etc.

1 Introduction

Ferrofluids have been used in medicine since the 1960es for e.g. the magnetically controlled metallic thrombosis of intracranial aneurysms and magnetically guided selective embolization of the renal artery in case of a renal tumor. Permanent obliteration have been expected by producing an artificial thrombus in the target artery [1,2]. Iron catheters manipulated by an external magnetic field are in use since 1951. In literature the successful diagnosis of a congenital heart disease by the use of a magnetically guided catheter is described [3]. Another system using the magnetic guidance is the magnetic implant guidance system for stereotactic neurosurgery. A small magnetic NdFeB capsule within the brain has been moved from outside the body by six independently controlled coils. Certainly it could be used to deliver radioactivity, heat or chemotherapeutic drugs to a tumor in the brain [4]. Magnets have been also used to extract swallowed needles and nails or to remove iron objects from eyes or from other body parts [5]. Ferrofluids are also used as a contrast agent for MRI. Superparamagnetic iron oxide particles are a new class of MR contrast agents that have been shown to significantly increase the detectibility of hepatic and splenic tumors. No acute or subacute toxic effects were detected by histological or serologic studies in rats and beagle dogs who received a total of 3000 μmol Fe/kg, 150 times the dose proposed for MR imaging of the liver. These results indicate that ferrofluids are fully biocompatible potential contrast agent for MRI [6]. These particles are selectively taken up by cells of the mononuclear phagocytosing system (e.g. Kupffer's cells in the liver and macrophages in the spleen, lymph nodes, or bone marrow), depending on the particle design (e.g. coating or size). The value of these particles in the diagnostic evaluation of liver and spleen tumors has already been shown in animal experiments and clinical studies [7,8]. Ultrasound, CT, and MR imaging do not allow reliable differentiation between hyperplastic and tumorous lymph nodes or detection of small metastases in normal-sized lymph nodes. Superparamagnetic iron oxide particles are also used for MR lymphography for the detection of lymph node metastases. In tumor-bearing rabbits, different degrees of metastatic displacement of lymph nodes were discernible, and even small metastases (3 mm in diameter) could be visualized [9].

Magnetic properties of ferrofluids can be combined with targeting capability of affinity ligands for specific cell targeting and separation. Magnetic cell sorting is employed for both diagnostic and preparative applications in detection and isolation of various cell subpopulations including damaged and diseased cells. Attachment of monoclonal antibodies (immunomagnetic cell separation) and other specific ligands to magnetic particles enable the separation of different cell populations with relatively simple equipment, and the technology provides a relatively rapid and convenient method for processing large quantities of different cell subpopulations [10].

Ferrofluids are also an important subject in the development of an implantable artificial heart. A ferrofluidic actuator directly drives magnetic fluids simply by applying a magnetic field to the ferrofluids and does not require a bearing. Magnetic fluid in a U-shaped glass cylinder was placed in an air gap of a solenoid. When a magnetic flux of 0.32 T was applied to the interface of the ferrofluid and air, the ferrofluid was displaced and a pressure was obtained [11].

Ferrofluids coated with starch polymers can be used as biocompatible carriers in a new field of locoregional tumor therapy called "magnetic drug targeting" [12–14]. It is of great importance to target selective the antineoplastic agent to its tumor target as precisely as possible, to reduce the resulting systemic toxic side effects from generalized systemic distribution and to be able to use a much smaller dose, which could further lead to a reduction of toxicity.

In the present study, we confirm the concentration of ferrofluids in VX2 squamous cell carcinoma tissue of the rabbit using histological investigations and MR imaging. The biodistribution of these magnetic nanoparticles was studied by the use of ^{123}iodine-labeled particles. The therapeutic efficacy of "magnetic drug targeting" was studied using the rabbit VX2 squamous cell carcinoma model.

2 Materials and Methods

2.1 Ferrofluids

The ferrofluids used in the experiments were obtained from Chemicell (Berlin, Germany; German patent application no. 19624426.9) and consisted of a biocompatible colloidal dispersion formed by wet chemical methods from iron oxides and hydroxides to produce special multidomaine particles [15]. These ferrofluids are covered with hydrophilic starch polymers coupled with endstanding functional groups (e.g. phosphate) allowing ionic binding to many therapeutic drugs. The hydrodynamic diameter of the whole particle is about 100 nm Fig. 1 [12].

Organic as well as inorganic agents can ionically bound to the functional group, in this case phosphat. The chemotherapeutic agent mitoxantrone (Novantron®, Lederle, Wolfratshausen, Germany) is ionically bound to the endstanding groups of the starch coating (Fig. 2).

Mitoxantrone is a synthetic anthracendion that inhibits DNA and RNA synthesis by intercalating in DNA molecules, which causes strand breaks. It is an anticancer agent used in treatment of e.g. breast cancer, non-Hodgkin's lymphoma and various solid tumors (carcinoma and sarcoma)[16–18].

Fig. 1. Imaging of the starch coated magnetic nanoparticles by scanning electronic microscopy.

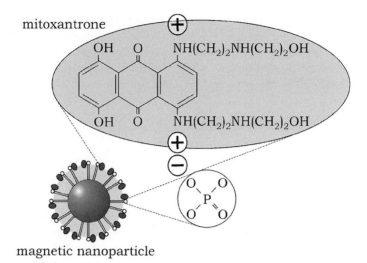

Fig. 2. Structural formula of mitoxantrone bound to magnetic nanoparticle [12].

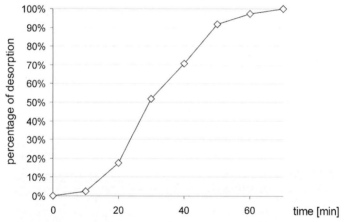

Fig. 3. Desorption of MTX measured by UV-visible-spectroscopy at wavelength of 648 nm, depending on time [12].

The body surface area and the dose of mitoxantrone were calculated according the recommendations in literature [19].

2.2 ^{123}I-labelled nanoparticles

Starch and iodine soluted in water result in the development of iodine/starch-complexes, which give an intensive blue colouring. Thereby the reduced iodine-molecules become deposited into the pockets of the starch-molecules (e.g. ß-amylase). Charge-shifting of electrons result in absorption of visible light, which causes the intensive blue colouring of the complex.

Ferrofluid (MAG-D 14/11) was mixed with ^{123}iodine. For the calculation of the activity, a definitive volume of the suspension was measured in the gamma-counter.

The experiments were performed in tumor-bearing rabbits with a catheter in the Arteria femoralis under the gamma-camera. The ferrofluid-^{123}iodine-suspension was injected into the tumor-supplying artery for 5 minutes, while the permanent-magnet was focused on the tumor-area. At the same time the signaldetection of ^{123}iodine started.

2.3 Magnetic Field

An electromagnet with a magnetic flux density of a maximum of 1.7 Tesla was used to produce an inhomogenous magnetic field. The magnetic flux density was focused onto the region of the tumor with a specially adapted pole shoe that was placed in contact with the surface of the tumor. On the tip of the pole shoe, the gradient has its maximum Fig. 4. The magnetic field was focused on the tumor during the ferrofluid injection of 10 minutes and for 60 min in total. The pole shoes of the magnet were specially modified in order to optimize flux density and gradient of the magnetic field (e.g. at a distance of 10 mm to the tip of the pole shoe the flux density was 1.0 Tesla).

A neodym-permanentmagnet with a magnetic flux density of 0,6 Tesla was used for the investigations with ^{123}Jod-labelled ferrofluids [12].

2.4 Surgical Intervention

The chemotherapy experiments were performed when the tumors had reached a volume of approximately 3500 mm^3.

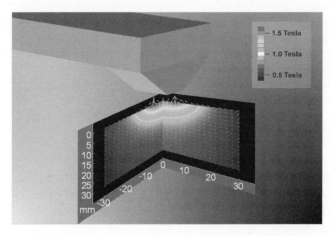

Fig. 4. Dependence of the magnetic flux density on the distance to pole shoe with the electromagnet [12].

For application of the chemotherapy, the femoral artery was cannulized and a catheter was placed approximately 2 cm distal to the inguinal furrow. The mitoxantrone bound ferrofluid was administered slowly over 10–15 minutes.

2.5 Experimental Protocols

The 26 animals were divided into six groups, depending on the type of treatment. Group 1 received an i.a. infusion of FF-MTX with the magnetic field at a dose equivalent to 20% and 50% of the systemic dose MTX (group 1a and group 1b, respectively). Group 2 received an i.a. infusion of MTX alone without the magnetic field at doses equivalent to 20%, 50%, 75% and 100% of the systemic dose. Group 3 received an i.a. infusion of FF alone with the magnetic field at equivalent doses compared with groups 1a and 1b. Group 4 received an i.v. infusion of FF-MTX with the magnetic field at doses equivalent to 20% and 50% of the systemic dose (group 4a and group 4b, respectively). Group 5 received an i.a. infusion of FF-MTX without the magnetic field at doses equivalent to 20% and 50% of the systemic dose, and group 6 was the control group without treatment.

After treatment, the tumor was measured every 3rd day by the same observer with a capiler ruler for a period of 3 months.

Blood samples were drawn by venipuncture every week. Immediately following intra-arterial infusion of 50% FF-MTX into the femoral artery, and following application of the magnetic field for a duration of 60 minutes, one animal was sacrificed and the tumor was fixed and stained with haematoxylin/eosin and Prussian Blue. Prussian blue, a specific stain to identify iron was used.

Following the 3 month observation period the remaining animals were sacrificed and the tumor, liver, kidneys, spleen, lungs, brain and inguinal lymph nodes were removed and examined histologically.

Enrichment of intra-arterially injected ferrofluids was investigated using conventional imaging techniques. In magnetic resonance imaging (MRI) the magnetic particles strongly reduce the transverse relaxation time (T2) and lower significantly the longitudinal relaxation time (T1). Therefore, magnetic nanoparticles are used as negative tissue-specific MRI contrast agents.

Six hours after ferrofluid application with an external magnetic field, a MRI was performed on four tumor bearing animals.

2.6 Statistical Analysis

The tumor volume was calculated using the formula for an elliptical mass ($1/6\pi$ $a^2 b$, a=width on the horizontal axis, b=length on the vertical axis). We considered change of volumes as percentages of tumor volumes (100%) found at day 0 (day of treatment). Statistical analysis for relative tumor volumes was performed using the one sample t-test (with a conservatively fixed value of 100% for the control group) and a t-test for two independent samples (Welch-test). For blood parameters (absolute values) we applied the t-test for two independent samples. The resulting two-sided p-values were considered significant at or below a value of 0.05. Between 0.01–0.05 the result was considered significant and highly significant if below 0.01.

3 Results

3.1 Tumor Volume

In the control group without treatment tumor volume increased and metastases appeared. The animals of group 1a, treated intra-arterially with 20% FF-MTX, had a complete tumor remission between the 15th and 36th day following treatment. The animals of group 1b (50% FF-MTX) had a decrease in tumor volume similar to group 1a. In group 2 (intra-arterial MTX alone, no magnetic field), lower dosages (20% and 50% of the systemic dose) did not result in tumor remission and enlarged, palpable inguinal lymph nodes were found after 48 days. At higher doses (75% and 100%), complete remission of tumor occurred.

The two group 3 animals (intra-arterial FF alone with the magnetic field, at 20% and 50%) demonstrated a progressive increase in tumor volume with palpable, enlarged inguinal lymph nodes (metastases).

The six animals of group 4 (intravenous injection via the ear vein of 20% and 50% FF-MTX with magnetic field) showed a slight tumor remission, but the reduction of volume was not statistically significant in comparison to the control group (not shown in figures).

The two animals of group 5 (intra-arterial FF-MTX 20% and 50%, without a magnetic field) showed a discontinuation of tumor growth and no evidence of metastases, but no remission of the tumor was seen (not shown in figures).

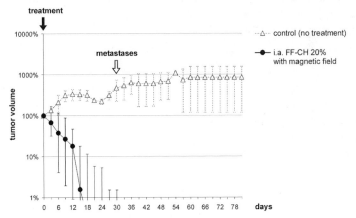

Fig. 5. Group 1a: Effect of i.a. application of FF-MTX [20% of the regular systemic dose (•)] on relative tumor volume after magnetic drug targeting compared with control group [group 6 (Δ), control(no treatment)]. Symbols, the median tumor volume; bars, the maximum and minimum; metastases, onset of metastases; treatment, the day of treatment (singular treatment)[12].

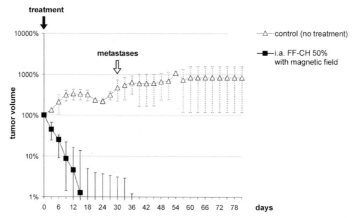

Fig. 6. Group 1b: Effect of i.a. application of ferrofluids bound to mitoxantrone (FF-MTX, 50% of the regular systemic dose),(●) on relative tumor volume after magnetic drug targeting compared with control group [group 6 (Δ), control (no treatment)]. Symbols, the median tumor volume; bars, the maximum and minimum values; metastases, onset of metastases; treatment, the day of treatment (singular treatment) [12].

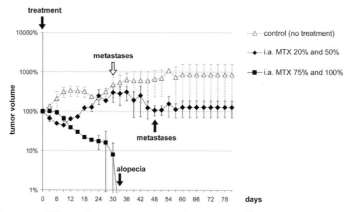

Fig. 7. Group 2: Effect of i.a. application of MTX - 20% and 50% (◊) and 75% and 100% (●) of the regular systemic dose - on relative tumor volume compared with control group [group 6 (Δ), control (no treatment)]. Symbols, the median tumor volume; bars, the maximum and minimum value; metastases, onset of metastases; alopecia, onset of alopecia; treatment, the day of treatment (singular treatment) [12].

At the time of treatment less then 5% of the animals showed a small necrotic fraction in the area of the tumor area [12].

3.2 Local and Systemic Effects

The general condition of the control group animals (limited to 2 animals for ethical reasons) worsened during the observation period and the animals developed

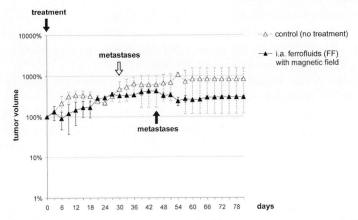

Fig. 8. Group 3: Effect of i.a. infusion of FF (50%) alone with magnetic field (▲). ∆, control (no treatment). Symbols, the median relative tumor volume; bars, the maximum and minimum values; metastases, onset of metastases; treatment, the day of treatment (singular treatment).

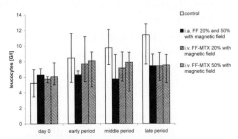

Fig. 9. Values of the WBC before treatment (day 0) and on days 3-15 (early period), 18-48 (middle period), and 51-81 (late period)after the respective treatment regimes. Columns, the median values; bars, the maximum and minimum values. The description of the bars, i.e. the respective group is mentioned beside the graphs [12].

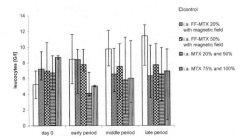

Fig. 10. Same as Fig. 12 for other groups mentioned beside the graphs.

pneumonia, which explains the increase of leucocytes as seen in Figs. 9 and 10 similar to description in the literature [20].

The animals of group 1 had no evident side effects such as alopecia, ulcers or muscular atrophy and their general condition (weight, food intake, excrement, urine, activity, fur condition) remained normal during the whole 3 month observation period compared to the physiological data of healthy animals (statement by Charles River, Sulzfeld, Germany). No significant changes in serum iron or leucocyte values were seen in this group (Fig. 9).

The urine of one animal in group 2 (50% MTX) showed blue-green discoloration and this animal developed mild alopecia in the region of the digits after 48 days. Both animals with low dose MTX (20% and 50%) had a decrease in leucocyte values, but this was not statistically significant (p=0.29). Both group 2 animals with high dose (75% and 100%) MTX had temporary blue-green urine discoloration, as well as a unilateral alopecia (palmar region of the digits to the knee joint) of the limb in which the tumor was implanted developing after 33 days. This hairless area developed cutaneous inflammation and ulceration, followed by mild alopecia of the fore limbs and head. The musculature of the treated limb became atrophic and the circumference was noticeably smaller (3 cm) at the end of the 3 month observation period. There was no marked difference in the severity of the side effects between the two animals, except for the fact that the animal with the higher MTX dose (100%) developed the changes several days sooner. Group 2 animals with 50%, 75% and 100% MTX steadily lost weight after an initial lag-phase and were underweight at the end of the observation period (mean value 1800 mg below the lower reference values according to the breeder's statement, Charles River, Sulzfeld, Germany). These animals became leucocytopenic ($\leq 2.95 \times 10^3/\mu l$) in the early phase (highly significant drop, p=0.004, Fig. 10), but recovered slightly in the middle and late periods [12].

None of the animals of groups 3 or 4 showed any significant changes in serum iron (not shown in figures) or leucocyte counts (group 3: Fig. 9; group 4a and b: Fig. 9) during the observation period when compared with initial values.

3.3 Histological Findings

Figure 11 up to 14 show cross sections of a VX2 tumor that was excised just after treatment with "magnetic drug targeting". Ferrofluids can be seen as blue pigment (Prussian Blue, Fig. 11, 12, 13) or as brown-black pigment (H.E., Fig. 14). Immediately after magnetic drug targeting, ferrofluids are evident in the vessels supplying the tumor and in macrophages. After the 3 month observation period, no viable tumor tissue was histologically evident in the animal group 1, with only fibrosis seen in the tumor implantation site. No metastases were found in the regional lymph nodes or in any other organs. Some FF particles were found in the spleen of the animals, but none were evident in the liver, lungs, brain or the implantation site and surrounding musculature and skin. No other macroscopic or histopathological changes were found in any of the investigated organs [12].

In group 2, the VX2 tumors of the 2 low dose animals were 8.644 mm^3 (50% MTX) and 2.497 mm^3 (20% MTX) in size with a large area of central necrosis and viable tumor at the periphery. The 2 animals with high dose MTX (75% and 100%) had no viable tumor at the implantation site. None of the other

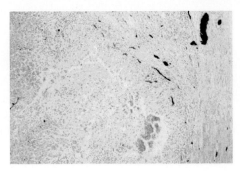

Fig. 11. Cross section of VX2 squamous cell carcinoma of the rabbit immediately after the magnetic drug targeting stained with Prussian Blue. Ferrofluids are visible in the vessels of the peripheral area of the tumor (original magnification x 100).

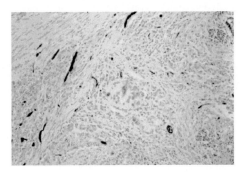

Fig. 12. Cross section of VX2 squamous cell carcinoma of the rabbit immediately after the magnetic drug targeting stained with Prussian Blue. Ferrofluids are visible in tumor supplying vessels in the connective tissue of the VX2 tumor (original magnification x 200).

investigated organs in the animals of group 2 (liver, kidneys, spleen, lungs, brain) had any pathological changes.

The tumors of both animals of group 3 measured 13.324 mm^3 and 17.649 mm^3, with a large area of central necrosis and viable tumor at the periphery. No FF particles were found within the tumor or in the surrounding musculature and skin. Some FF were found in the spleen. Metastases were found in the inguinal lymph nodes and liver of both animals. None of the other investigated organs (kidneys, spleen, lungs, brain) had any pathological changes [12].

3.4 MRI Imaging

The distribution and the enrichment of the intra-arterially injected ferrofluids were investigated in vivo by MRI after "magnetic drug targeting". In MRI the magnetic particles strongly reduce the transverse relaxation time (T2) and lower significantly the longitudinal relaxation time (T1). The MRI was made 6 h after treatment with "magnetic drug targeting" and the concentration of ferrofluids was seen by extinction of signal in the area of the tumor.

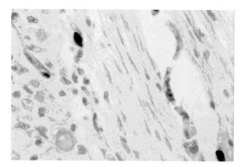

Fig. 13. Cross section of VX2 squamous cell carcinoma of the rabbit immediately after magnetic drug targeting stained with Prussian Blue. Proof of iron in macrophages and endothelium (original magnification x 1000).

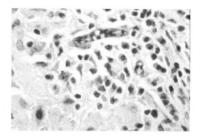

Fig. 14. Cross section of VX2 squamous cell carcinoma of the rabbit immediately after magnetic drug targeting stained with haematoxilin-eosin. Ferrofluids are visible intravascular and in a macrophage (Original magnification x 1000).

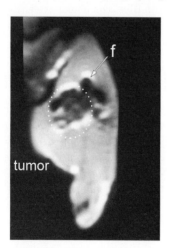
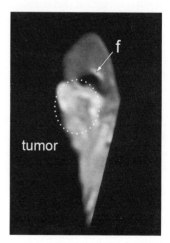

Fig. 15. (left) MRI of tumorous hind limb after i.a. application of ferrofluids; dotted circle: tumor region; f: head of the femur.
(right) MRI of tumorous hind limb after i.v. application of ferrofluids; dotted circle: tumor region; f: head of the femur [12].

Fig. 15 shows the left hind limb (implantation site) of two rabbits receiving intra-arterial or intravenous 50% FF-MTX, respectively, 6 hours after treatment. The tumor is situated at the medial portion of the hind limb (dotted circle), and the concentration of FF is seen by extinction of signal. Fig. 15(left) (intra-arterial FF-MTX) shows definite extinction of signal and Fig. 15(right) (intravenous FF-MTX) only a very discrete signal extinction. The area marked "f" is at the head of the femur and appears to be hypodense.

The tumor is situated at the medial portion of the hind limb (dotted circle), and the concentration of FF is seen by extinction of signal. Fig. 15(left) (intra-arterial FF-MTX) shows definite extinction of signal and Fig. 15(right) (intravenous FF-MTX) only a very discrete signal extinction. The area marked "f" is at the head of the femur and appears to be hypodense [12].

3.5 Biodistribution of ^{123}Iodine-Ferrofluids

In order to assess the distribution of the used nanoparticles under the influence of the magnetic field, ^{123}iodine was storaged in the starch-cover and was then injected into the tumor-supplying artery while the magnetic field was focused on the tumor for 60 minutes.

At the same time, the signal-detection of ^{123}iodine started. Around the tumor-area there was a "region of interest" (ROI) determined and the activity was measured over the time.

The injected whole-body-activity of the rabbit could not be measured with the gamma-camera, so the activity in the tumor-area was measured at different times and was related to the initial activity-maximum (relative activity in percent of the initial activity).

The scintigraphically detected signal after intra-arterial application has been shown to be at least twice as much higher in the magnetically focused region compared to the application without external magnetic field.

Enrichment of ^{123}I-labelled nanoparticles in the area of interest (VX2 tumor) after magnetic drug targeting:

The activity in the tumor region scintigraphically measured by gamma-camera and focused by the external magnetic field was after 30 minutes 38.2 %, after 40 minutes 31.8 %. and after 60 minutes 22.3 % (n=1). Without external magnetic field, the activity was 14.7 % after 30 minutes, 13.0 % after 40 minutes and 11.8 % after 60 minutes (n=1).

4 Discussion

Chemotherapy is a balancing act between efficacy and toxicity. Because of the relatively non-specific action of chemotherapeutic agents, there is almost always some toxicity to normal tissue. Therefore, it is of great importance to be able to selectively target the antineoplastic agent to its tumor target as precisely as possible, to reduce the resulting systemic toxic side effects from generalized

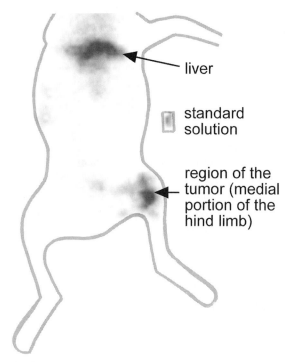

Fig. 16. Schematic drawing of a tumor-bearing rabbit under the γ-camera. Enrichment of ^{123}I-labelled nanoparticles in the area of interest (VX2 tumor) after "magnetic drug targeting". The image was taken 40 min after application, still showing stable concentration of ferrofluids within the tumor tissue [14].

systemic distribution and to be able to use a much smaller dose, which would further lead to a reduction of toxicity.

Regional chemotherapy via a regional artery administers a more concentrated dose of the active agent directly into the tumor [20]. The advantage of this approach is limited, however, by drain-off via the venous blood which limits exposure time and reduces the overall efficacy.

At present, i.a. delivery of chemotherapeutic agents is approved and well accepted for treatment of liver metastases [21] and has occasionally been used for other tumor types also (e.g., inoperable head and neck tumors); but it has often necessitated complicated, time-consuming operative procedures, including general anaesthesia [22].

"Magnetic drug targeting" means holding the chemotherapeutic agent due to the carrier (ferromagnetic particles) at the desired site of activity, thus increasing efficacy and diminishing systemic toxicity.

The goal of the present study was to concentrate ferrofluids coupled with chemotherapeutic agent in a desired target area using a magnetic field. The principle of "magnetic drug targeting" consists of two steps: The first step is the delivery of the mitoxantrone bound magnetic nanoparticles to the desired body compartment (tumor), the second is the release of the drug from its carrier [12–14]. An additional helpful factor is that microvascular permeability in neoplastic tissues is increased (8-fold compared with normal tissue) as is diffusion (33-fold) [23], which allows chemotherapeutic agents much easier to penetrate into tumor tissue. The metabolism of ferrofluid particles takes place in the liver and spleen in analogy to iron metabolism.

In this investigations, the authors found that this approach led to complete tumor remission with reduced doses of 20% and 50% MTX bound to FF (Figs. 5 and 6). The application was well tolerated by the animals and no signs of toxicity were detected. On the contrary, intra-arterial infusion of the same doses, 20% and 50%, of MTX alone (group 2, Fig. 7) resulted in no reduction of tumor volume and the animals developed metastases and suffered from chemotherapeutic side effects. Only when the dose of MTX alone was increased to 75% and 100%, a tumor remission was seen (Fig. 7), but this resulted in severe side effects (alopecia, ulcers, leucocytopenia), as seen in Fig. 10. Intravenous infusion of the FF-MTX complex was also ineffective since only a slight tumor remission that was not statistically significant resulted. The same result was observed after intra-arterially infused FF-MTX without an external magnetic field, since the tumor remained at the same size, without remission. Thus, the combination of intra-arterial infusion with a magnetic field was safe, effective, well tolerated by the animals and was very effective in treating the tumor although the dose of chemotherapeutic agent was markedly reduced [12–14].

At present, intra-arterial delivery of chemotherapeutic agents is approved and well accepted for treatment of liver metastases [22], and have occasionally also been used for other tumor types (e.g. inoperable head and neck tumors), but often necessitate complicated, time consuming operative procedures, including general anesthesia [22]. Experimentally, Swistel et al. [24] described encouraging results using intra-arterial chemotherapy for VX2 squamous cell carcinoma. They achieved complete tumor remission after intra-arterial application of adriamycin in 4 of 6 animals, whereas intravenous infusion of adriamycin caused severe toxicity and resulted in complete remission in only two cases.

A potential complication that could arise with the use of FF compounds is the fact that with larger particles embolization could occur, preventing a sufficient concentration of the chemotherapeutic agent from reaching the tumor. On the other hand, if the particles are too small, the external magnetic field might not provide sufficient attraction so that the particles are drawn into the tumor. The particles used in this study had a size of 100 nm (nanometer). No embolization was seen in the vascular system of the tumor. Our histological findings showing distribution of FF particles throughout the tumor strongly support the concept that large molecular weight substances such as chemotherapeutic agents or monoclonal antibodies can be effectively targeted to tumor tissue. In addition, the fact that the FF alone with a magnetic field failed to cause tumor remission (Fig. 8) indicates that therapeutic effect resulted from the action of the chemotherapeutic agent itself, rather than intratumoral embolization by the particles [12].

The electromagnet used for this study produced a magnetic flux density of a maximum of 1.7 Tesla which decreased depending on the distance to the pole shoe (Fig. 4). The magnetic gradient can be seen as a collection of vectors which point in the direction of increasing values as shown in Fig. 3 (yellow arrows). The arrow sizes correspond to the strength of the magnetic gradient. Both factors (direction and magnitude) reflect the inhomogenous character of the magnetic field, which is of key importance for "magnetic drug targeting". In previous studies it was suggested that a magnetic field strength of 8000 Gauss (0.8 Tesla) is sufficient to exceed linear blood flow in the intratumoral vasculature and allow 100% localization of magnetic carrier containing 20% magnetite [25]. In contrast, Goodwin and coworkers applied magnetic-targeted carriers (MTCs) intra-arterially in a swine model, focusing a magnetic field of only 250 to 1000 Gauss (0.025-0.1 Tesla, permanent neodymium magnet) to the desired compartments in the liver and lungs. The depth of this MTC targeting was 8 - 12 cm and the particle size 0.5 μm - 5 μm. With this model, MTCs with a predefined activity had a concentration of 67% in the liver and 50% in the lung localized by the magnet [26].

The magnetic field strength with a maximum of 1.7 Tesla used in the present investigation was the strongest ever applied for magnetic drug targeting. We achieved a high concentration of FF within the tumor after intra-arterial infusion of FF which was seen by histological (Figs. 11 to 14) and MRI methods (Fig. 15). The VX-2 squamous cell carcinoma was superficially exposed and had no migratory motion, as was the case with the liver and lung targets (breathing fluctuations) in the swine model of Goodwin and coworkers [26]. In addition these organs lie deeply in the body cavity (8-12 cm from the body surface), greatly complicating focusing a maximum of the magnetic flux density onto the tumor area. Two approaches to overcome this problem are possible,

1. the use of larger particles, as previously suggested by Bergemann et al. [27] or
2. using a stronger magnetic field.

The particles (FF-MTX) used in the present study were 100 nm in size and have shown good therapeutic results in smaller animals (mouse, rat) as well [28,29].

The strong magnetic field was very efficacious in combination with these particles, but further experiments (which we have already begun) should be performed using marked FF to clarify the optimal magnetic field strength and particle size [12].

A remarkable feature of using ionically bound pharmaceuticals is that the low molecular weight substances (molecular weight of mitoxantrone 517) is able to desorb from the carrier (FF) after a defined time span and can then pass through the vascular wall or interstitium into the tumor cells. This is important since once the FF-MTX complex has been directed to the tumor by the magnetic field, the drug must dissociate in order to act freely within the tumor. As shown in Fig. 2 MTX desorbs from the FF after 30 min (half life) and so 50% of the drug is free to act on the tumor after 30 min.

Dextran coated iron oxides have been shown to produce signal loss by MRI and have been used as a contrast medium for the detection of metastatic lymph nodes (negative contrast) [30]. We found total signal loss and therefore a very high concentration of FF by MRI after focusing by the magnetic field (Fig. 15 (left)) [12]. Recent studies have shown that intra-arterial application of radioactive-labeled, magnetic carriers with an external magnetic field resulted in retention of at least 50% in the target site [31]. In comparison, after intravenous injection, only very slight signal loss was seen, indicating a very low concentration (Fig. 15(right)). This underscores the advantage of intra-arterial versus intravenous infusion in magnetic drug targeting.

Biodistribution was studied by the use of ^{123}iodine-nanoparticles. The scintigraphically detected ^{123}iodine- signal after intra-arterial (artery leading the tumor, femoral artery) application has been shown to be significantly higher in the magnetically focused region compared to the application without external magnetic field (Fig. 16). 60 minutes after application there is still twice as much concentration of ^{123}iodine in the magnetically focused region compared to the application without the magnetic field [14].

Previous studies by Bacon et al. concerning FF with a particle size of 0.5-1.0 μm found no acute or chronic toxicity after the intravenous infusion of 250 mg iron/kg body weight in rats [32] and 1-3 mg iron/kg body weight in humans has been shown to be safe as well [33]. This agrees with our findings, since FF infusion was not associated with any signs of toxicity [12–14].

Magnetic microspheres loaded with the gamma-emitting radioisotope 90Y have also successfully been used as a form of radionuclide therapy. In one study, this compound was maneuvered within the body of a mouse to a subcutaneous lymphoma, resulting in eradication of the tumor [33]. Magnetic fluids have also been used for the so-called "magnetic fluid hyperthermia" that has been used to control the local growth of murine mammary carcinoma [35]. Additional modification of the magnetic particles, so that they could bind monoclonal antibodies, lectins, peptides or hormones could make delivery of these compounds more efficient and also highly specific. Targeted drug delivery using nanotechnology opens a new field in oncology science and improves basis science and clinical practice [36].

Acknowledgements

Supported by the Margarete Ammon Foundation, Munich, and grants from the Technical University of Munich, Germany.

References

1. Alksne JF, Fingerhut A, Rand R. Magnetically controlled metallic thrombosis of intracranial aneurysms. Surgery 1966; 60: 212-218.
2. Hilal SK, Michelsen WJ, Driller J, Leonard E. Magnetically guided devices for vascular exploration and treatment. Radiology 1974;113: 529-534.

3. Ram, W. and Meyer, H. Heart catheterization in a neonate by interacting magnetic fields. A new and simple method of catheterguidance. Catheterization and Cardiovascular Diagnosis, 22: 317-319, 1991.
4. McNeil, R. G., Ritter, R. C., Wang, B. et al. Functional design features and initial performance characteristics of a magnetic-implant guidance system for stereotactic neurosurgery. IEEE Transactions on Biomedical Engineering, 42: 793-801, 1995.
5. Luborsky, F.E. Recent advances in the removal of magnetic foreign bodies from the esophagus, stomach and duodenum with controllable permanent magnets. Am. J. Roentg. Rad. Ther. Nucl. Med., 1964.
6. Weissleder R, Hahn PF, Stark DD et al. MR imaging of splenic metastases: ferrite-enhanced detection in rats. AJR 1987; 149: 723-726.
7. Weissleder R, Stark DD, Engelstad BL et al. Superparamagnetic iron oxide: pharmacokinetics and toxicity. AJR 1989;152: 167-173.
8. Weissleder R, Elizondo G, Wittenberg J, Lee AS, Josephson L, Brady TJ. Ultrasmall superparamagnetic iron oxide. An intravenous contrast agent for assessing lymph nodes with MR imaging. Radiology 1990;175: 494.
9. Taupitz M, Wagner S, Hamm B, Dienemann D, Lawaczeck R, Wolf KJ. MR lymphography using iron oxide particles. Detection of lymph node metastases in the VX2 rabbit tumor model. Acta Radiol. 1993; 34: 10-15.
10. Hardingham JE, Kotasek D, Sage RE, Eaton MC, Pascoe VH. Detection of circulating tumor cells in colorectal cancer by immunobead-PCR is a sensitive prognostic marker for relapse of disease. Molec. Med. 195;1: 789-794.
11. Mitamura Y, Wada T, Keisuke S. A ferrofluidic actuator for an implantable artificial heart. Artif. Organs 1992; 16 (5): 490-495.
12. Alexiou, C., Arnold, W., Klein, R.J. et al. Locoregional cancer treatment with Magnetic Drug Targeting. Cancer Res 2000; 60: 6641-6648.
13. Alexiou, C., Arnold, W., Hulin, P. et al. Magnetic mitoxantrone nanoparticle detection by histology, X-ray and MRI after magnetic tumor targeting. JMMM 2001; 255: 187-193.
14. Alexiou, C., Schmidt, A., Klein, R.J. et al. Magnetic drug targeting: Biodistribution and dependency on magnetic field strength. JMMM in press.
15. Lübbe AS, Bergemann C, Huhnt W et al. Preclinical experiences with magnetic drug targeting: tolerance and efficacy. Cancer Res 1996; 56: 4694-4701.
16. Ho, A. D., Del Valle, F., Haas, R. et al. Sequential studies on the role of mitoxantrone, high-dose cytarabine, and recombinant human granulocyte-macrophage colony-stimulating factor in the treatment of refractory non-Hodgkin's lymphoma. Semin. Oncol., *17:* 14-18, 1990.
17. Hiddemann, W., Buchner, T., Heil, G. et al. Treatment of refractory acute lymphoblastic leukemia in adults with high dose cytosine arabinoside and mitoxantrone (HAM). Leukemia., *4*:637-640, 1990.
18. Freund, M., Wunsch-Zeddies, S., Schafers, M. et al. Prednimustine and mitoxantrone (PmM) in patients with low-grade malignant non-Hodgkin's lymphoma (NHL), chronic lymphocytic leukemia (CLL), and prolymphocytic leukemia (PLL). Ann. Hematol., *64:* 83-87, 1992.
19. Bistner SI, Ford RB, Raffe MR (eds.). Kirk and Bistner's Handbook of Veterinarian Procedures and Emergency Treatment, Ed 6, p. 907. Philadelphia: W.B. Saunders Co., 1995.
20. Stephens, F. O. Why use regional chemotherapy? Principles and pharmacokinetics. Reg Cancer Treat., *1:* 4-10, 1988.

21. Link, K.H., Kornmann, M., Formenti, A. et al. Regional chemotherapy of non-resectable liver metastases from colorectal cancer – literature and institutional review. Langenbecks Arch. Surg., *384:* 344-353, 1999.
22. v. Scheel, J.: Invasive procedures for antineoplastic chemotherapy. *In:* Naumann et al. (eds.): Head and Neck Surgery. Stuttgart, New York: Thieme, 1998.
23. Gerlowski, L.E. and Jain, R.K. Microvascular permeability of normal and neoplastic tissues. Microvascular Research *31:*288-305, 1986.
24. Swistel AJ, Bading JR and Raaf JH Intraarterial versus intravenous Adriamycin in the VX2 tumor system. Cancer (Phila.) 53: 1397-1404, 1984
25. Senyei, A., Widder, K., and Czerlinski, C. Magnetic guidance of drug carrying microspheres. J. Appl. Physiol., *49:* 3578-3583, 1978.
26. Goodwin, S., Peterson, C., Hoh, C., Bittner, C. Targeting and retention of magnetic targeted carriers. J. Magn. Magn. Mater., *194*: 132-139, 1999.
27. Bergemann, C., Müller-Schulte, D., Oster, J., à Brassard, L., Lübbe, A.S. Magnetic ionexchange nano- and microparticles for medical, biomedical and molecular biological applications. J. Magn. Magn. Mater.,*194*: 45-52, 1999.
28. Lübbe, A. S., Bergemann, Ch., Huhnt, W., Fricke, T., Riess, H., Brock, J. W., and Huhn, D. Preclinical experiences with magnetic drug targeting: tolerance and efficacy. Cancer Research, *56:* 4694-4701, 1996.
29. Lübbe, A. S., Bergemann, Ch., Riess, H. et al. Clinical experiences with magnetic drug targeting: a phase I study with 4'-epidoxorubicin in 14 patients with advanced solid tumors. Cancer research, *56:* 4686-4693, 1996.
30. Taupitz, M., Wagner, S., Hamm, B., Dienemann, D., Lawaczeck, R. and Wolf, K.J. MR Lymphography using iron oxide particles. Detection of lymph node metastases in the VX2 rabbit tumor model. Acta Radiologica, *34:*10-15, 1993.
31. Widder, K.J., Senyei, A.E. and Scarpelli, D.G. Magnetic microspheres: a model system for site specific drug deleivery in vivo. Proc. Soc. Exp. Biol. Med. *58:* 141-146, 1978.
32. Bacon, B.R., Park, D. D., Saini, S., Groman, E. V., Hahn, P. F., Compton, C. C., and Ferrucci, J. T. Ferrite particles: A new magnetic resonance imaging contrast agent. Lack of acute or chronic hepatoxicity after intravenous administration. J. Lab. Clin. Med., *110:* 164-171, 1987.
33. Rummeny, E., Weissleder, R., Stark, D. D., Elizondo, G. and Ferrucci, J. T. Magnetic resonance tomography of focal liver and spleen lesions. Experiences using ferrite, a new RES-specific MR contrast medium. Radiologe *28:* 380-386, 1988.
34. Häfeli, U.O., Pauer, G.J., Roberts, W.K., Humm, J.L. and Macklis, R.M. Magnetically targeted microspheres for intracavitary and intraspinal Y-90 radiotharpy. *In:*U. Häfeli and W. Schütt (eds.), Scientific and clinical applications of magnetic carriers, pp 501-516, New York and London: Plenum press, 1997.
35. Jordan, A., Scholz, R., Wust, P. et al. Effects of magnetic fluid hyperthermia (MFH) on C3H mammary carcinoma in vivo. Int. J. Hyperthermia, *13:* 587-605, 1997.
36. Partridge, M., Phillips, E., Francis, R., Li, S.R. Immunomagnetic seperation for enrichment and sensitive detection of disseminated tumour cells in patients with head and neck SCC. J. Pathol., *189:*368-377, 1999.

Magnetic Unit Systems

Historically numerous different magnetic unit systems are used. In this volume of *Lecture Notes in Physics* mainly the SI-system (chapters by Rosensweig, Blums, Zubarev, Thurm & Odenbach and Bossis) and the Gaussian system (chapters by Shliomis and Morozov & Shliomis) are employed.

To reduce the problems with the comparison of statements in the various chapters, the major units and their conversion are given in the table below.

Quantity	Symbol	Gaussian unit	Conversion factor	SI unit
magnetic induction	B	G	10^{-4}	T
magnetic field strength	H	Oe	$10^3/4\pi$	A/m
magnetization	M	emu/cm^3	10^3	A/m
	$4\pi M$	G	$10^3/4\pi$	A/m
mass magnetization	M	emu/g	1	Am^2/kg
magnetic moment	m	emu	10^{-3}	Am^2
volume susceptibility	χ	dim.less	4π	dim.less
permeability	μ	dim.less	$4\pi 10^{-7}$	H/m

An exception from these unit systems is made in the chapter by Müller and Liu where the Heaviside-Lorentz units, or the so-called *rational* units, are employed. In this system all four fields have the same dimension, square root of the energy density, ie $\sqrt{\mathrm{J/m^3}}$ in SI units and $\sqrt{\mathrm{erg/cm^3}}$ in Gaussian units, sensibly with $H = B$ and $D = E$ in vacuum. As a result, the factors of 4π, ϵ_0 and μ_0 vanish. To revert to SI units ($\hat{E}, \hat{H}...$), employ $\hat{H} = H/\sqrt{\mu_o}$, $\hat{B} = B\sqrt{\mu_o}$, $\hat{E} = E/\sqrt{\epsilon_o}$, $\hat{D} = D\sqrt{\epsilon_o}$, and $\hat{\varrho}_e = \varrho_e\sqrt{\epsilon_o}$, $\hat{j}_e = j_e\sqrt{\epsilon_o}$. To revert to the Gaussian system ($\tilde{E}, \tilde{H}...$), increase all four fields, and decrease all four sources, by the factor of $\sqrt{4\pi}$: $B = \tilde{B}/\sqrt{4\pi}$, (similarly for D, H, E); $\varrho_e = \sqrt{4\pi}\tilde{\varrho}_e$, (similarly for j, P, M)

Druck: Strauss Offsetdruck, Mörlenbach
Verarbeitung: Schäffer, Grünstadt

Lecture Notes in Physics

For information about Vols. 1–556
please contact your bookseller or Springer-Verlag

Vol. 557: J. A. Freund, T. Pöschel (Eds.), Stochastic Processes in Physics, Chemistry, and Biology. X, 330 pages. 2000.

Vol. 558: P. Breitenlohner, D. Maison (Eds.), Quantum Field Theory. Proceedings, 1998. VIII, 323 pages. 2000

Vol. 559: H.-P. Breuer, F. Petruccione (Eds.), Relativistic Quantum Measurement and Decoherence. Proceedings, 1999. X, 140 pages. 2000.

Vol. 560: S. Abe, Y. Okamoto (Eds.), Nonextensive Statistical Mechanics and Its Applications. IX, 272 pages. 2001.

Vol. 561: H. J. Carmichael, R. J. Glauber, M. O. Scully (Eds.), Directions in Quantum Optics. XVII, 369 pages. 2001.

Vol. 562: C. Lämmerzahl, C. W. F. Everitt, F. W. Hehl (Eds.), Gyros, Clocks, Interferometers...: Testing Relativistic Gravity in Space. XVII, 507 pages. 2001.

Vol. 563: F. C. Lázaro, M. J. Arévalo (Eds.), Binary Stars. Selected Topics on Observations and Physical Processes. 1999.IX, 327 pages. 2001.

Vol. 564: T. Pöschel, S. Luding (Eds.), Granular Gases. VIII, 457 pages. 2001.

Vol. 565: E. Beaurepaire, F. Scheurer, G. Krill, J.-P. Kappler (Eds.), Magnetism and Synchrotron Radiation. XIV, 388 pages. 2001.

Vol. 566: J. L. Lumley (Ed.), Fluid Mechanics and the Environment: Dynamical Approaches. VIII, 412 pages. 2001.

Vol. 567: D. Reguera, L. L. Bonilla, J. M. Rubí (Eds.), Coherent Structures in Complex Systems. IX, 465 pages. 2001.

Vol. 568: P. A. Vermeer, S. Diebels, W. Ehlers, H. J. Herrmann, S. Luding, E. Ramm (Eds.), Continuous and Discontinuous Modelling of Cohesive-Frictional Materials. XIV, 307 pages. 2001.

Vol. 569: M. Ziese, M. J. Thornton (Eds.), Spin Electronics. XVII, 493 pages. 2001.

Vol. 570: S. G. Karshenboim, F. S. Pavone, F. Bassani, M. Inguscio, T. W. Hänsch (Eds.), The Hydrogen Atom: Precision Physics of Simple Atomic Systems. XXIII, 293 pages. 2001.

Vol. 571: C. F. Barenghi, R. J. Donnelly, W. F. Vinen (Eds.), Quantized Vortex Dynamics and Superfluid Turbulence. XXII, 455 pages. 2001.

Vol. 572: H. Latal, W. Schweiger (Eds.), Methods of Quantization. XI, 224 pages. 2001.

Vol. 573: H. M. J. Boffin, D. Steeghs, J. Cuypers (Eds.), Astrotomography. XX, 434 pages. 2001.

Vol. 574: J. Bricmont, D. Dürr, M. C. Galavotti, G. Ghirardi, F. Petruccione, N. Zanghi (Eds.), Chance in Physics. XI, 288 pages. 2001.

Vol. 575: M. Orszag, J. C. Retamal (Eds.), Modern Challenges in Quantum Optics. XXIII, 405 pages. 2001.

Vol. 576: M. Lemoine, G. Sigl (Eds.), Physics and Astrophysics of Ultra-High-Energy Cosmic Rays. X, 327 pages. 2001.

Vol. 577: I. P. Williams, N. Thomas (Eds.), Solar and Extra-Solar Planetary Systems. XVIII, 255 pages. 2001.

Vol. 578: D. Blaschke, N. K. Glendenning, A. Sedrakian (Eds.), Physics of Neutron Star Interiors. XI, 509 pages. 2001.

Vol. 579: R. Haug, H. Schoeller (Eds.), Interacting Electrons in Nanostructures. X, 227 pages. 2001.

Vol. 580: K. Baberschke, M. Donath, W. Nolting (Eds.), Band-Ferromagnetism: Ground-State and Finite-Temperature Phenomena. IX, 394 pages. 2001.

Vol.581: J. M. Arias, M. Lozano (Eds.), An Advanced Course in Modern Nuclear Physics. XI, 346 pages. 2001.

Vol.582: N. J. Balmforth, A. Provenzale (Eds.), Geomorphological Fluid Mechanics. X, 579 pages. 2001.

Vol.583: W. Plessas, L. Mathelitsch (Eds.), Lectures on Quark Matter, XIII, 334 pages. 2002.

Vol.584: W. Köhler, S. Wiegand (Eds.), Thermal Nonequilibrium Phenomena in Fluid Mixtures. XVII, 470 pages. 2002.

Vol.585: M. Lässig, A. Valleriani (Eds.), Biological Evolution and Statistical Physics. XI, 337 pages. 2002.

Vol.586: Y. Auregan, A. Maurel, V. Pagneux, J.-F. Pinton (Eds.), Sound-Flow Interactions. XVI, 286 pages. 2002

Vol.587: D. Heiss (Ed.), Fundamentals of Quantum Information. Quantum Computation, Communication, Decoherence and All That. XIII, 265 pages. 2002.

Vol.588: Y. Watanabe, S. Heun, G. Salviati, N. Yamamoto (Eds.), Nanoscale Spectroscopy and Its Applications to Semiconductor Research. XV, 306 pages. 2002.

Vol.589: A. W. Guthmann, M. Georganopoulos, A. Marcowith, K. Manolakou (Eds.), Relativistic Flows in Astrophysics. XII, 241 pages. 2002

Vol.590: D. Benest, C. Froeschlé (Eds.), Singularities in Gravitational Systems. Applications to Chaotic Transport in the Solar System. XI, 215 pages. 2002

Vol.591: M. Beyer (Ed.), CP Violation in Particle, Nuclear and Astrophysics. XI, 334 pages. 2002

Vol.592: S. Cotsakis, L. Papantonopoulos (Eds.), Cosmological Crossroads. An Advanced Course in Mathematical, Physical and String Cosmology. XVI, 477 pages. 2002

Vol.593: D. Shi, B. Aktaş, L. Pust, F. Mikhailov (Eds.), Nanostructured Magnetic Materials and Their Applications. XII, 289 pages. 2002

Vol.594: S. Odenbach (Ed.), Ferrofluids. Magnetical Controllable Fluids and Their Applications. XI, 253 pages. 2002

Vol.595: C. Berthier, L. P. Lévy, G. Martinez (Eds.), High Magnetic Fields. Applications in Condensed Matter Physics and Spectroscopy. X, 493 pages. 2002

Vol.596: F. Scheck, H. Upmeier, W. Werner (Eds.), Noncommutative Geometry and the Standard Model of Elememtary Particle Physics. XII, 346 pages. 2002

Monographs
For information about Vols. 1–29
please contact your bookseller or Springer-Verlag

Vol. m 30: A. J. Greer, W. J. Kossler, Low Magnetic Fields in Anisotropic Superconductors. VII, 161 pages. 1995.

Vol. m 31 (Corr. Second Printing): P. Busch, M. Grabowski, P.J. Lahti, Operational Quantum Physics. XII, 230 pages. 1997.

Vol. m 32: L. de Broglie, Diverses questions de mécanique et de thermodynamique classiques et relativistes. XII, 198 pages. 1995.

Vol. m 33: R. Alkofer, H. Reinhardt, Chiral Quark Dynamics. VIII, 115 pages. 1995.

Vol. m 34: R. Jost, Das Märchen vom Elfenbeinernen Turm. VIII, 286 pages. 1995.

Vol. m 35: E. Elizalde, Ten Physical Applications of Spectral Zeta Functions. XIV, 224 pages. 1995.

Vol. m 36: G. Dunne, Self-Dual Chern-Simons Theories. X, 217 pages. 1995.

Vol. m 37: S. Childress, A.D. Gilbert, Stretch, Twist, Fold: The Fast Dynamo. XI, 406 pages. 1995.

Vol. m 38: J. González, M. A. Martín-Delgado, G. Sierra, A. H. Vozmediano, Quantum Electron Liquids and High-Tc Superconductivity. X, 299 pages. 1995.

Vol. m 39: L. Pittner, Algebraic Foundations of Non-Com-mutative Differential Geometry and Quantum Groups. XII, 469 pages. 1996.

Vol. m 40: H.-J. Borchers, Translation Group and Particle Representations in Quantum Field Theory. VII, 131 pages. 1996.

Vol. m 41: B. K. Chakrabarti, A. Dutta, P. Sen, Quantum Ising Phases and Transitions in Transverse Ising Models. X, 204 pages. 1996.

Vol. m 42: P. Bouwknegt, J. McCarthy, K. Pilch, The W3 Algebra. Modules, Semi-infinite Cohomology and BV Algebras. XI, 204 pages. 1996.

Vol. m 43: M. Schottenloher, A Mathematical Introduction to Conformal Field Theory. VIII, 142 pages. 1997.

Vol. m 44: A. Bach, Indistinguishable Classical Particles. VIII, 157 pages. 1997.

Vol. m 45: M. Ferrari, V. T. Granik, A. Imam, J. C. Nadeau (Eds.), Advances in Doublet Mechanics. XVI, 214 pages. 1997.

Vol. m 46: M. Camenzind, Les noyaux actifs de galaxies. XVIII, 218 pages. 1997.

Vol. m 47: L. M. Zubov, Nonlinear Theory of Dislocations and Disclinations in Elastic Body. VI, 205 pages. 1997.

Vol. m 48: P. Kopietz, Bosonization of Interacting Fermions in Arbitrary Dimensions. XII, 259 pages. 1997.

Vol. m 49: M. Zak, J. B. Zbilut, R. E. Meyers, From Instability to Intelligence. Complexity and Predictability in Nonlinear Dynamics. XIV, 552 pages. 1997.

Vol. m 50: J. Ambjørn, M. Carfora, A. Marzuoli, The Geometry of Dynamical Triangulations. VI, 197 pages. 1997.

Vol. m 51: G. Landi, An Introduction to Noncommutative Spaces and Their Geometries. XI, 200 pages. 1997.

Vol. m 52: M. Hénon, Generating Families in the Restricted Three-Body Problem. XI, 278 pages. 1997.

Vol. m 53: M. Gad-el-Hak, A. Pollard, J.-P. Bonnet (Eds.), Flow Control. Fundamentals and Practices. XII, 527 pages. 1998.

Vol. m 54: Y. Suzuki, K. Varga, Stochastic Variational Approach to Quantum-Mechanical Few-Body Problems. XIV, 324 pages. 1998.

Vol. m 55: F. Busse, S. C. Müller, Evolution of Spontaneous Structures in Dissipative Continuous Systems. X, 559 pages. 1998.

Vol. m 56: R. Haussmann, Self-consistent Quantum Field Theory and Bosonization for Strongly Correlated Electron Systems. VIII, 173 pages. 1999.

Vol. m 57: G. Cicogna, G. Gaeta, Symmetry and Perturbation Theory in Nonlinear Dynamics. XI, 208 pages. 1999.

Vol. m 58: J. Daillant, A. Gibaud (Eds.), X-Ray and Neutron Reflectivity: Principles and Applications. XVIII, 331 pages. 1999.

Vol. m 59: M. Kriele, Spacetime. Foundations of General Relativity and Differential Geometry. XV, 432 pages. 1999.

Vol. m 60: J. T. Londergan, J. P. Carini, D. P. Murdock, Binding and Scattering in Two-Dimensional Systems. Applications to Quantum Wires, Waveguides and Photonic Crystals. X, 222 pages. 1999.

Vol. m 61: V. Perlick, Ray Optics, Fermat's Principle, and Applications to General Relativity. X, 220 pages. 2000.

Vol. m 62: J. Berger, J. Rubinstein, Connectivity and Superconductivity. XI, 246 pages. 2000.

Vol. m 63: R. J. Szabo, Ray Optics, Equivariant Cohomology and Localization of Path Integrals. XII, 315 pages. 2000.

Vol. m 64: I. G. Avramidi, Heat Kernel and Quantum Gravity. X, 143 pages. 2000.

Vol. m 65: M. Hénon, Generating Families in the Restricted Three-Body Problem. Quantitative Study of Bifurcations. XII, 301 pages. 2001.

Vol. m 66: F. Calogero, Classical Many-Body Problems Amenable to Exact Treatments. XIX, 749 pages. 2001.

Vol. m 67: A. S. Holevo, Statistical Structure of Quantum Theory. IX, 159 pages. 2001.

Vol. m 68: N. Polonsky, Supersymmetry: Structure and Phenomena. Extensions of the Standard Model. XV, 169 pages. 2001.

Vol. m 69: W. Staude, Laser-Strophometry. High-Resolution Techniques for Velocity Gradient Measurements in Fluid Flows. XV, 178 pages. 2001.

Vol. m 70: P. T. Chruściel, J. Jezierski, J. Kijowski, Hamiltonian Field Theory in the Radiating Regime. VI, 172 pages. 2002.

Vol. m 71: S. Odenbach, Magnetoviscous Effects in Ferrofluids. X, 151 pages. 2002.

Vol. m 72: J. G. Muga, R. Sala Mayato, I. L. Egusquiza (Eds.), Time in Quantum Mechanics. XII, 419 pages. 2002.